# HERPESVIRUSES, THE IMMUNE SYSTEM, AND AIDS

# DEVELOPMENTS IN MEDICAL VIROLOGY

***Yechiel Becker, Series Editor***
***Julia Hadar, Managing Editor***



# DEVELOPMENTS IN MEDICAL VIROLOGY



Levine, P.H. (ed.) *Epstein-Barr Virus and Associated Diseases* (1985)
Becker, Y. (ed.) *Virus Infections and Diabetes Mellitus* (1987)
De Clercq, E. (ed.) *Clinical Use of Antiviral Drugs* (1988)
Revel, M. (ed.) *Clinical Aspects of Interferons* (1988)
Gilden, D.H. and Lipton, H.L. (eds.) *Clinical and Molecular Aspects of Neurotropic Virus Infection* (1989)
Aurelian, L. (ed.) *Herpesviruses, the Immune System and AIDS* (1990)

# DEVELOPMENTS IN MOLECULAR VIROLOGY

Becker, Y. (ed.) *Herpesvirus DNA* (1981)
Becker, Y. (ed.) *Replication of Viral and Cellular Genomes* (1983)
Becker, Y. (ed.) *Antiviral Drugs and Interferon: The Molecular Basis of Their Activity* (1983)
Kohn, A. and Fuchs, P. (eds.) *Mechanisms of Viral Pathogenesis from Gene to Pathogen* (1983)
Becker, Y. (ed.) *Recombinant DNA Research and Viruses. Cloning and Expression of Viral Genes* (1985)
Feitelson, M. *Molecular Components of Hepatitis B Virus* (1985)
Becker, Y. (ed.) *Viral Messenger RNA: Transcription, Processing, Splicing and Molecular Structure* (1985)
Doerfler, W. (ed.) *Adenovirus DNA: The Viral Genome and Its Expression* (1986)
Aloni, Y. (ed.) *Molecular Aspects of Papovaviruses* (1987)
Darai, G. (ed.) *Molecular Biology of Iridoviruses* (1990)

# DEVELOPMENTS IN VETERINARY VIROLOGY

Payne, L.N. (ed.) *Marek's Disease* (1985)
Burny, A. and Mammerickx, M. (eds.) *Enzootic Bovine Leukosis and Bovine Leukemia Virus* (1987)
Becker, Y. (ed.) *African Swine Fever* (1987)
De Boer, G.F. (ed.) *Avian Leukosis* (1987)
Liess, B. (ed.) *Classical Swine Fever and Related Viral Infections* (1987)
Darai,G. (ed.) *Virus Diseases in Laboratory and Captive Animals* (1988)
Campbell, J.B. and Charlton, K.M. (eds.) *Rabies* (1988)
Alexander, D.J. (ed.) *Newcastle Disease* (1988)
Wittmann, G. (ed.) *Herpesvirus Diseases of Cattle, Horses, and Pigs* (1989)
Petursson, G. and Hoff-Jorgensen, R. (eds.) *Maedi-Visna and Related Diseases* (1990)

# HERPESVIRUSES, THE IMMUNE SYSTEM, AND AIDS

edited by

Laure Aurelian, M.D.

The University of Maryland
Baltimore

Kluwer Academic Publishers
Boston/Dordrecht/London

**Distributors for North America:**
Kluwer Academic Publishers
101 Philip Drive
Assinippi Park
Norwell, Massachusetts 02061 USA

**Distributors for all other countries:**
Kluwer Academic Publishers Group
Distribution Centre
Post Office Box 322
3300 AH Dordrecht, THE NETHERLANDS

**Library of Congress Cataloging-in-Publication Data**

Herpesviruses, the immune system, and AIDS / edited by Laure Aurelian.
p. cm. — (Developments in medical virology : 6)
Includes bibliographical references.
Includes index.
ISBN 0-7923-0803-4
1. HIV (Viruses) 2. AIDS (Disease)—Pathogenesis. 3. Herpesvirus diseases. 4. Latent virus diseases—Immunological aspects. 5. Latent virus diseases—Pathogenesis. I. Aurelian, Laure. II. Series.
[DNLM: 1. Acquired Immunodeficiency Syndrome. 2. Herpesvirus Infections—immunology. 3. Immune System. WC 571 H5634]
QR414.6.H58H47 1990
616.97'92—dc20
DNLM/DLC
for Library of Congress 90-4795
CIP



Printed in the United States of America

In memory of my father, George Aurelian,
who taught me how to question
without forgetting how to dream.

## CONTENTS

## CONTRIBUTORS


Laure Aurelian, Ph.D.
Department of Pharmacology and Experimental Therapeutics
University of Maryland School of Medicine
Baltimore, Maryland 21201

P. D. Bednarik, Ph.D.
Oncology Center and Department of Molecular Biology
and Genetics
The Johns Hopkins University School of Medicine
Baltimore, Maryland 21205

W.H. Burns, M.D.
Oncology Center
The Johns Hopkins University
School of Medicine
Baltimore, Maryland 21205

Edouard M. Cantin, Ph.D.
Department of Neurology
City of Hope National Medical Center
Duarte, California

Ingrid Domke-Opitz, M.D.
Institut of Virus Research
German Cancer Research Center, INF 280
D6900 Heidelberg, FRG

David Hinton, M.D.
Department of Pathology
University of Southern California
School of Medicine
Los Angeles, California

Jeffrey Laurence, M.D.
Cornell University Medical College
Division of Hematology-Oncology
New York, N.Y. 10021

George K. Lewis, Ph.D.
Department of Microbiology and Immunology
University of Maryland School of Medicine
Baltimore, Maryland 21201

Stephen Martin, Ph.D.
Department of Microbiology and Nutrition
The Upjohn Company
Kalamazoo, Michigan 49001

Jose Menezes, D.V.M., Ph.D.
Laboratory of Immunovirology
Hospital Sainte-Justine and University of Montreal
Montreal, Canada

Paul S. Miller, Ph.D.
Department of Biochemistry
The Johns Hopkins University
School of Hygiene and Public Health
Baltimore, Maryland 21205

Harry Openshaw, M.D.
Department of Neurology
City of Hope National Medical Center
Duarte, California

Gary R. Pearson, Ph.D.
Department of Microbiology
Georgetown University Medical School
Washington, D.C.

P.M. Pitha, Ph.D.
Oncology Center and Department of Molecular Biology and Genetics
The Johns Hopkins University School of Medicine
Baltimore, Maryland 21205

Thomas C. Quinn, M.D.
Division of Infectious Diseases
The Johns Hopkins University School of Medicine
Baltimore, Maryland
Laboratory of Immunoregulation
National Institute of Allergy and Infectious Diseases
Bethesda, Maryland

Barry T. Rouse, D.V.M., Ph.D.
University of Tennessee
Department of Microbiology
College of Veterinary Medicine
Knoxville, Tennessee 37996

Marion Sarmiento, Ph.D.
Department of Veterinary Medicine and Surgery
The University of Texas
M.D. Anderson Cancer Center
Houston, Texas 77030

R.F. Siliciano, M.D., Ph.D.
Department of Medicine
The Johns Hopkins University School of Medicine
Baltimore, Maryland 21205

L.R. Stanberry, M.D.
Department of Pediatrics
University of Cincinnati
College of Medicine
Children's Hospital Research Foundation
Division of Infectious Diseases
Cincinnati, Ohio

Rainer Zawatzky, M.D.
Institut of Virus Research
German Cancer Research Center, INF 280
D6900 Heidelberg, FRG

*" I do not know what I may appear to the world, but to myself I seem to have been only like a boy playing on the sea-shore, and diverting myself in now and then finding a smoother pebble or a prettier shell than ordinary, whilst the great ocean of truth lay all undiscovered before me "*

Sir Isaac Newton

## PREFACE

During a recent visit to Baltimore, Professor Yechiel Becker pointed out that we are virtually ignorant of the immunopathogenesis of latent viral infections and urged me to develop a book that would address this question. This volume reflects my perception of the problem. Thus, seeking to reconcile the profound lymphopenia seen in AIDS with the absence of fulminant viremia and the unusually long latency period (average of 8-10 years), it has been postulated that infection with other viruses, most notably the herpesviruses, induces HIV expression thereby converting the latent HIV infection to a cytolytic one. Herpesviruses must meet three criteria in order to qualify for this role: (i) high prevalence of infection in the population at high risk of AIDS, (ii) direct or indirect contribution to immunosuppression and (iii) co-infection of clinically relevant cells and direct HIV activation. This interpretation assumes that control of herpesvirus infections will block HIV-induced immunopathogenesis.

The volume is organized with these thoughts in mind. First, Quinn summarizes epidemiologic and serologic findings on the association of these infections and presents evidence for herpesvirus participation in HIV acquisition and AIDS development. Next we consider virus-specific immunity and immunosuppression. Lewis reviews current concepts of antigen presentation and speculates on its potential role in the development of *protective* immunity. Molecular aspects of B and T lymphocyte responses to HIV, discussed by Siliciano, emphasize a novel pathway for depletion of $CD4^+$ T cells that involves cytolytic T lymphocytes (CTL) specific for the HIV envelope glycoprotein gp120. This concept of CTL involvement in immunopathology stands in direct contrast to that considered by Martin and Rouse which visualizes CTL as critical factors in the development of *protective* immunity.

Immune responses to CMV are considered by Burns who emphasises the contribution of CTL and natural killer cells to the control of infection. CMV infection of monocytes appears to be the critical factor in the immunosuppression that accompanies CMV infections. Pearson and Menezes review the immunobiology of EBV and provide a thoughtful analysis of EBV induced immunoregulatory events that might contribute to the profound immune deffect seen in AIDS patients. The final chapter on immunosuppression discusses mechanism(s) of suppressor T lymphocytes development and their role in the pathogenesis of HSV infection. It concludes with a summary of the potential levels of immune interaction between the herpesviruses and HIV.

Macrophages play a pivotal role in immunity both in terms of antigen presentation and as secretors of monokines that are involved in T cell activation, HIV expression and immunosuppression. They are also a potential reservoir for HIV. Domke-Opitz and Zawatsky present a scholarly review of the role of macrophages in HSV infection including monokine secretion, thereby identifying levels of potential interaction with HIV. Molecular aspects of HSV growth restriction by various populations of monocytes/macrophages is next presented by Sarmiento. Review of HIV interaction with macrophages and monokine-mediated HIV activation seems premature as this is a rapidly growing field. For additional information see Nakajima *et al.*, J. Immunol 142: 531, 1989; Wahl *et al.*, PNAS 86: 621, 1989; Clouse *et al.*, J. Immunol. 142:431, 143:470, 1989; Merrill *et al.*, J. Virol. 63: 4404, 1989; Schmittman *et al.*, Science 245:305, 1989.

The question of direct activation of the HIV promoter begins with a chapter by Openshaw and colleagues considering potential sites of *in vivo* interaction between HIV and the herpesviruses. It is followed by a comprehensive review of the trans-activation of HIV gene expression by Laurence that includes immunoregulatory monokines, cellular transcription factors and the newly discovered HHV-6 that may be most important if, as suggested by recent findings (Takahashi *et al.*, J. Virol. 63: 3161, 1989), it is $CD4^+$ T-cell tropic. Pitha and Bednarik conclude this section with an in-depth analysis of the molecular mechanism of HIV promoter activation by various herpesvirus genes, and the potential role of DNA methylation in HIV latency and trans-activation.

Prophylaxis and therapy have been exhaustively reviewed in the past. Therefore, I chose to concentrate on novel aspects that are at the forefront of present thinking. The potential prophylactic and therapeutic use of subunit vaccines is discussed by Stanberry and it includes the herpesviruses and HIV. The last chapter by Miller describes recent progress with a novel therapeutic approach in which modified *anti-sense* molecules are considered as selective antiviral agents.

This type of monograph is, perforce, the product of selectivity. The choice of topics and their restriction to certain aspects, were determined by previous coverage, available information and the attempt at unifying current knowdlege. Accordingly, the presented information is incomplete and I offer my apologies to those investigators whose work could not be covered here. Conceptual aspects are emphasized throughout the volume. Continued research in this area should provide answers to many of the questions that are raised. I hope that the volume will help the clinician to keep abreast of new developments that impact on our understanding of pathogenesis and may ultimately influence his clinical practice. It should also prove of interest to the virologist and immunologist interested in the role of immunity in the development and control of latent viral infections.

In closing I thank Professor Becker for initiating this monograph and Irene Gervis for her help in compiling the chapters. Extensive credit goes to the contributors who sought to overcome the barriers imposed by their respective disciplines in an attempt to bridge the gap between them.

# HERPESVIRUSES, THE IMMUNE SYSTEM, AND AIDS

# 1

## EPIDEMIOLOGIC AND SEROLOGIC EVIDENCE FOR A ROLE OF HERPESVIRUSES IN HIV INFECTION

THOMAS C. QUINN, M.D.

**Divison of Infectious Diseases, The Johns Hopkins University School of Medicine, Baltimore, Maryland; and Senior Investigator, Laboratory of Immunoregulation, National Institute of Allergy and Infectious Diseases, Bethesda, Maryland.**

### ABSTRACT

Since the recognition of AIDS, herpesviruses have played a significant role in the transmission, clinical presentation, and pathogenesis of HIV infection. Herpesviruses, including cytomegalovirus (CMV), herpes simplex virus (HSV), Epstein-Barr virus (EBV) and varicella-zoster virus (VZV) are among the most common opportunistic infections observed in AIDS patients. The immunosuppressive effects of herpesviruses and their recurrent nature have led many investigators to study the immunomodulating effects of these viruses and their direct effect on HIV infection. *In vitro* studies have demonstrated that latent HIV infection can be activated during HSV and CMV infections resulting in increased viral replication. Seroepidemiologic studies have demonstrated a strong association with recent seroconversion to HSV-2 and the subsequent acquisition of HIV infection, suggesting that genital ulcerations due to HSV may enhance the transmissibility of HIV infection during sexual intercourse. This association between genital HSV-2 infection and HIV infection was found to be independent of age, sex, risk group, or the number of sexual partners. Additional studies have also suggested that herpesvirus infections, particularly HSV-2 and VZV are poor prognostic factors and may accelerate the immunosuppression associated with HIV infection leading to the development of clinical AIDS. These studies therefore suggest that accurate diagnosis and therapy of infections with herpesviruses may reduce the efficiency of HIV transmission and possibly slow the progression of HIV infection to AIDS.

## INTRODUCTION

Shortly after the recognition of the acquired immunodeficiency syndrome (AIDS) in 1981 (1-3), it was hypothesized that herpesviruses may play an important role in the etiology or pathogenesis of this newly identified syndrome (4,5). This theory was based on the fact that infection with herpesviruses was a prominent clinical feature in nearly all patients with AIDS (3-5). Chronic mucocutaneous herpes simplex virus (HSV) infections were one of the first opportunistic infections described in patients with AIDS (3), and both cytomegalovirus (CMV) and HSV infections were extremely common in individuals identified to be at highest risk for acquiring AIDS, such as homosexual men, intravenous drug users and hemophiliacs (4-8). CMV and Epstein-Barr virus (EBV) were also prominent infections which were suspected as possible etiologic agents of the prolonged fever, wasting, and lymphadenopathy that often precedes AIDS, frequently referred to as the chronic lymphadenopathy syndrome (9,10). Subsequent clinical studies have indeed demonstrated that infections with HSV, CMV, EBV, and even varicella-zoster virus (VZV) are frequent opportunistic infections which occur among AIDS patients (11-14). Several of the opportunistic infections caused by herpesviruses include encephalitis, chorioretinitis, hairy leukoplakia, esophagitis, enteritis, colitis, Burkitt's lymphoma, primary CNS lymphoma, zoster, and there has even been speculation about the role of CMV in the pathogenesis of Kaposi's sarcoma (15,16). Furthermore, the herpesviruses, particularly CMV and EBV have been known to be strongly associated with immunosuppression, particularly of cell-mediated immune functions, which further supported the hypothesis that herpesviruses may contribute to the immune defects that characterize AIDS.

## THE ROLE OF HERPEVIRUSES IN AIDS

During the early years of the epidemic, an intensive search for an etiologic agent and for a better understanding of pathogenesis of AIDS was undertaken. In 1982-83, the Centers for Disease Control (CDC) conducted a case-control study in New York City, San Francisco, Los Angeles and Atlanta to better characterize the syndrome and to identify possible risk factors for immunosuppression, malignancy and infection in these patients (17,18). For this study a case was defined as a living homosexual man between ages 15 and 60 years, who had biopsy proven Kaposi's sarcoma, *Pneumocystis carinii* pneumonia or both. Persons with known predisposing risk factors for immunodeficiency such as malignancies other than Kaposi's sarcoma or prior treatment with immunosuppressive drugs were excluded. For each patient, they identified four controls, matched for age, sex, sexual orientation, race, and metropolitan area of residence. These were classified as a homosexual friend who was not a sexual partner, randomly chosen homosexual patients at a sexually transmitted disease clinic, and homosexual patients chosen randomly from the files of several local practicing physicians. In this study, Jaffe *et al.* (17) found that the most important variable associated with AIDS was the number of male sex partners per year. Compared with controls, "cases" were also more likely to have been exposed to feces during sex, to have had syphilis and non-B hepatitis, to have been treated for enteric parasites, and to have used various ilicit substances. As part of this study, biologic specimens were also obtained from patients and controls for laboratory examinations to determine the character, relative frequency, and extent of immunosuppression, and the relative frequency of past or current infection with several microorganisms known to have had infected patients with AIDS or suspected of being associated with either immunosuppression or malignancy (18).

IgG antibody to EBV capsid antigen was the only herpesvirus antibody titer found to be significantly higher for patients than controls, grouped separately or combined (Table 1). However, over 90% of all patients and controls had detectable antibody to CMV, HSV-1, HSV-2, and EBV. The EBV viral capsid antigen titer difference was statistically significant when values and patients were compared with those in clinic controls (odds ratio = 1.66; 95% C.I. = 1.18-2.32), private practice controls (odds ratio = 1.88 (95% C.I. = 1.19-2.97), friend controls (unmatched analysis only, $p < 0.05$), and with all of the control groups combined (OR = 1.73; 95% C.I. 1.29-2.33). CMV complement fixation titers were also significantly higher for cases and controls, but only when compared with all controls combined (OR = 1.61; 95% C.I. = 1.09-2.39).

Titers of antibodies to CMV, measured by indirect hemagglutination, and HSV-1 and HSV-2, by complement fixation and indirect hemagglutination, of cases did not differ significantly from those of the control groups independently or combined. In addition, cultures of patients urine and throat swab specimens yielded CMV more frequently than those of combined controls, 13 (25%) of 52 patients having an isolation made from 1 or more sites versus 9 (7%) of 124 combined controls (OR = 3.29; 95% C.I. = 1.13-9.62).

**Table 1. Antibody Prevalence Rates to Herpes viruses in AIDS Patients and Controls ***

| | | Controls | | |
|---|---|---|---|---|
| Agent | AIDS Cases | Friend | STD Clinic | Private Practice |
| CMV | 50/50 | 19/19 | 60/61 | 36/37 |
| HSV-I | 50/50 | 18/19 | 58/61 | 36/37 |
| HSV-II | 49/50 | 18/19 | 55/61 | 36/37 |
| EBV-capsid antigen-IgG | | | | |
| ≥ 10 | 49/49 | 19/19 | 60/60 | 37/37 |
| ≥ 80 | 11/49 | 1/19 | 3/60 | 2/37** |
| early antigen | | | | |
| ≥ 10 | 11/31 | 4/17 | 11/44 | 3/7 |
| VZV ≥ 2 | 28/37 | 17/17 | 53/58 | 34/35** |

* Adapted from Rogers et al (18)

** $p < 0.05$

Rogers *et al.* (18) concluded that the increased titers of antibody to EBV capsid antigen and CMV complement fixing antibodies, and increased isolation of CMV in cases, were probably a direct result of immunosuppression since herpesviruses are known to cause latent infections and to become reactivated during immunosuppression. This is especially true in persons with defective cell-mediated immunity. In this particular study, determination of whether production of virus particles preceded or followed the onset of immunosuppression in the study patients was not possible because the patients were identified and tested after the onset of illness. The authors suggested instead that the serologic results probably reflected a distinctive life style as an important risk factor for the subsequent acquisition and development of AIDS.

The second major study to examine the role of herpesvirus infections in AIDS was by Quinnan *et al.* (5), who studied 34 patients with AIDS, 9 patients with generalized lymphadenopathy, and 13 asymptomatic homosexual men. Herpesvirus infections were common in all groups, but CMV and EBV infections were much more common than VZV infections (Table 2). In patients with AIDS, CMV was present in 33 of 34 patients and EBV was present in 30 of 32 patients. By comparison, active infections were diagnosed by culture for HSV and VZV in less than 25% of patients with AIDS. HSV and VZV infections occurred only in patients with opportunistic infections, not in patients with Kaposi's sarcoma without opportunistic infections, patients with lymphadenopathy, or the "healthy" homosexual men. In patients with the chronic lymphadenopathy syndrome, CMV was present in 7 of 9 patients and EBV in 9 of 9. In asymptomatic homosexual men CMV was identified in 5 of 13 patients and EBV in 6 of 8.

**Table 2. Frequency of Isolation of Herpesviruses in Homosexual Men With AIDS, Chronic Lymphadenopathy or Without Evidence of Disease.** *

| Study Group | CMV | HSV | VZV | EBV |
|---|---|---|---|---|
| AIDS | | | | |
| Kaposi's Sarcoma | 9/10 | 0/10 | 0/10 | 10/10 |
| Opportunistic | 24/24 | 8/24 | 4/24 | 20/22 |
| Chronic Lymphadenopathy | 7/9 | 0/9 | 0/9 | 9/9 |
| Asymptomatic | 5/13 | 0/13 | 0/13 | 6/8 |

* Adapted from Quinnan et al (5)

In summary, CMV infections were frequently related to disease and death and the authors concluded that the prevalence of these herpesviruses in AIDS and chronic lymphadenopathy may be the result of an important interaction between these viruses and the cause of AIDS. However, a major weakness with both of the above studies is that many of the "controls" may have been infected with an etiologic agent of AIDS which would make interpretation of the results more difficult. Furthermore, prospective studies are required to determine whether these herpesviruses are important in the pathogenesis of AIDS or are secondary opportunistic manifestations of a suppressed cell mediated immune system.

**THE HUMAN IMMUNODEFICIENCY VIRUS**

In 1983-84 intensive virologic investigations lead by Montagnier and Gallo resulted in the identification of the human immunodeficiency virus (HIV) as the etiologic agent of AIDS (19,20). As sensitive serologic tests for detection of antibody to HIV were developed, it became clear that 100% of AIDS patients and a high proportion of individuals in high risk populations were infected with HIV (21-22). Repeated isolation and serologic detection of HIV in patients with AIDS or with risk factors for AIDS, and the selective cytopathic activity of HIV for T-helper lymphocytes *in vitro*, argued convincingly for these early HIV isolates as the causative agent of AIDS (23,24). Extensive epidemiologic studies worldwide demonstrated that HIV was transmitted sexually, including homosexual, bisexual and heterosexual activities; by parenteral transmission such as among IV drug users, infected blood transfusions and factor 8 and factor 9 concentrates; and by perinatal transmission, either *in utero*, at the time of delivery, or possibly postnatally (25,26).

Both *in vivo* and *in vitro* studies demonstrated that the CD4 positive lymphocyte (T-helper lymphocyte) is the primary target for HIV. The specific tropism for CD4 positive cells is based on the affinity of the viral gp120 envelope protein for the CD4 molecule on the target cell (27). Indeed a variety of cells besides T-helper/inducer lymphocytes are now known to express the CD4 molecule and to be infected by HIV. These include monocytes, follicular dendritic cells, Langerhans cells and possibly other cell types within the central nervous system. These cells generally express smaller amounts of CD4 on their surface and the corresponding cytopathic effects typically found with CD4 positive lymphocytes are not generally observed. Within the central nervous system of infected individuals, it appears that the predominant cell type infected with HIV is the monocyte/macrophage (28).

Once infected with HIV, individuals are believed to be infected and infectious for life in the absense of any effective treatment. Properties inherent to the retroviral life cycle contribute to this persistence. Following attachment and infection of the CD4 positive cell, DNA replicated from the HIV genome is integrated into host cell chromosome, and expression of the viral genome is down-regulated as it is in other lentivirus infections, such as in visna virus infection of sheep. This is supported by the observation that very few CD4 positive lymphocytes express viral message at any one point in time and that the general course of HIV disease is characterized by a protracted progressive T4 lymphocyte loss. Natural history studies of HIV demonstrated that clinical AIDS represents only one spectrum of this viral infection, and that the mean incubation period from time of exposure to the time of development of AIDS is approximately 8 years for homosexual men and for adults receiving infected blood transfusions (29). Prior to the development of AIDS, most HIV infected individuals remain asymptomatic or develop symptoms such as fatigue and chronic lymphadenopathy. During this period, HIV remains latently integrated in host genome primarily within CD4 positive lymphocytes, but also within other immune cells such as macrophages.

**BIOLOGIC INTERACTIONS BETWEEN HIV AND HERPESVIRUSES**

*In vitro*, infection of human peripheral blood lymphocytes and CD4 positive cell lines by HIV can result in an explosive lytic infection or in transient cytopathology followed by an out-growth of virus containing cells that either express very low levels of viral message or remain entirely latent until activated (30). Activating signals include treatment with halogenated pyrimidines *in vitro*, but can also include concurrent infection by other viruses such as the herpesviruses (EBV, CMV, HSV, VZV), HTLV-I, hepatitis B-virus, as well as activating signals inherent to the immune response (31-33). For the HIV infected individual, the wide range of antigenic stimuli to which one is frequently exposed would be expected to serve as physiologically relevant cellular activators of viral expression. Thus, one possibility in which herpesviruses could enhance the pathogenesis of HIV infection in AIDS is by serving as an activation signal for latently infected CD4 positive cells, resulting in enhanced viral replication, cytolysis and immunosuppression. With developing immunosuppression, the reactivation of latent herpesvirus infections such as HSV or CMV, might serve to further activate other latently infected cells, thus accelerating the disease process.

While *in vivo* evidence for accelerated pathogenesis of HIV due to herpesviruses is limited, *in vitro* studies have shown that the HIV long terminal repeat (LTR) promoter can be directly activated during infection with HSV-1. Gendelman *et al.* (31) first demonstrated in cotransfection experiments that molecular clones containing several DNA viruses including HSV-1 and VZV increased chloramphenicol acetyltransferase (CAT) expression directed by HIV-LTR. Transactivation of HIV-LTR was specifically stimulated by HIV *tat* gene product, and this effect was maximally increased when *tat* plus HSV-1 DNA were transfected into cells with HIV-LTRcat plasmid. In subsequent experiments, Mosca *et al.* (32) and Skolnik *et al.* (34) demonstrated that HSV-1, HSV-2 and CMV could reactivate and stimulate HIV-LTR and that this effect was independent of the *tat* protein. Similarly, Skolnik *et al.* (34) also showed that HIV enhanced productive CMV infection suggesting that biological interactions between the herpesviruses and HIV are bidirectional. However, in these transfection experiments, it is not entirely clear whether HIV is specifically or nonspecifically activated during HSV infection, and studies involving cells latently infected with HIV itself will be necessary in order to confirm these provocative studies (35).

## HERPESVIRUSES AS CO-FACTORS OF HIV INFECTION

Another level at which herpesviruses may serve as cofactors in HIV infection is by enhancing either the susceptibility or infectiousness of HIV infection. In the case of HSV infection, small genital ulcerations may allow direct mucosal access for HIV through small breaks in the skin during sexual intercourse with an infected individual. Alternatively, the immune response to a mucosal HSV lesion, which is primarily lymphocytic, may contain infected CD4 positive lymphocytes, thus enhancing the HIV presence within these ulcerations and leading to enhanced viral transmission to an uninfected individual. Multiple studies in Africa, Europe and the United States have demonstrated that the history or presence of genital ulcer disease is a major risk factor for HIV, independent of other risk behavior such as number of sexual partners, prostitution or age (25,26).

In a study of prostitutes in Nairobi, Kenya, Kreiss *et al.* (36), reported the presence of genital ulcers in 50% of HIV seropositive women and in 27% of HIV seronegative, and logistic regression analysis demonstrated a 3.3-fold relative risk of HIV infection in women with genital ulcers. In contrast, sexual activity (number of sexual acts per year) was unrelated to HIV infection. Several subsequent studies in Nairobi also demonstrated strong associations of genital ulcerative disease in heterosexual men with HIV seropositivity and with HIV seroconversion (37,38). In Zimbabwe, a past history of genital herpes in the male index case was a risk factor for serologic concordance among 75 married couples (39). These data suggest that genital ulcerative disease may be an independent risk factor for HIV infection among East African heterosexuals. While the presence of genital ulcers may be invariably linked to behavioral factors, it is also possible that the association is causal. Genital ulcers may alter an individual susceptibility to HIV by providing a portal of entry for the virus. Alternatively, individuals with HIV infection are more susceptible to genital ulcers or have more severe ulcers and are more likely to seek treatment at STD clinics where these studies are often performed. Thus, only prospective studies will fully be able to address these issues. However, a number of recent case-control studies in homosexual men and heterosexual men and women have supported the hypothesis that genital ulceration due to HSV plays an important role in the transmission of HIV and these are reviewed below (40-49).

In a study by Holmberg *et al.* (41), a select group of homosexual men were selected who had recently seroconverted to HIV and who had at least two serum samples drawn before their HIV seroconversion (one from the year before HIV seroconversion). They also enrolled HIV seronegative control men who also had serum samples drawn in the same years. HIV seroconverters frequently had antibodies to CMV (95%), EBV (89%), HSV-1 (56%) and HSV-2 (47%) prior to seroconversion. However, similar percentages of initial antibody positivity were found among those who remained HIV seronegative: CMV (86%), EBV (98%), HSV-1 (51%) and HSV-2 (39%). Antibody titers to CMV, EBV, and HSV-1 did not significantly rise or differ between the men who seroconverted to HIV as compared to those who remained seronegative. However, they found a strong association between HSV-2 positivity and acquisition of HIV. They found that 32 (68%) of 47 HIV seroconverters had antibodies to HSV-2, whereas only 26 (46%) of 57 men who remained HIV seronegative had HSV-2 antibody (Fisher's exact test, $p < 0.02$, 1 tailed) (Table 3). Of the 62 HIV seroconverters with at least 1 serum sample for analysis, the last tested serum sample was HSV-2 positive in 42 (68%); while only 27 (44%) of 61 men who remained HIV seronegative had HSV-2 antibodies (Fisher's exact test, $p < 0.01$, 1 tailed).

**Table 3. Serologic Results Of HSV Infections In Association With HIV Seroconversion. ***

| | Serum Specimen | | |
|---|---|---|---|
| Virus | 1 | 2 | 3 |
| HSV - 2 Antibody | | | |
| HIV Seroconversion (n=47) | 47% | 58% | 68%** |
| HIV Seronegative (n=57) | 39% | 39% | 46% |
| HSV - 1 Antibody | | | |
| HIV Seroconversion (n=42) | 56% | 64% | 62% |
| HIV Seronegative (n=51) | 51% | 57% | 59% |

* Adapted from Holmberg et al (41)

** $P < 0.02$

Seroconversion to HSV-2 between any two specimens was found in 11 (42%) of the 26 HIV seroconverters, but in only 5 (14%) of 35 men who remained HIV seronegative (Fisher's exact test, $p < 0.02$, 1 tailed) (Table 3). This association was independent of age and number of life-time partners, but men who seroconverted to HIV were more likely to have had receptive anal intercourse than men who remained HIV seronegative (36.1% vs. 8.5%, $p < 0.001$). Log linear analysis to control for these putative confounding variables and for years of serum samples still showed a significant association between HSV-2 seropositivity and HIV seroconversion ($p = 0.02$). Interestingly, there was no substantial difference between HIV infected and uninfected men and their history of symptomatic HSV infection or other anal/genital infections. Five (16%) of 32 HSV-2 and HIV seropositive men gave a history of prior anal/genital herpes, whereas three (12%) of 26 HSV-2 positive, but HIV seronegative men gave such a history. In summary, the authors demonstrated HSV-2 seroconversion in most men before they acquired HIV antibodies, suggesting that genital herpes may predispose some persons to HIV infection. In clinically symptomatic persons, HSV-2 infection causes visible blistering, erosion, and bleeding of penile, rectal and vaginal mucosa. Such breaks in the integrity of the skin might allow easier access for HIV to the blood stream of persons being infected by HIV.

In a subsequent study by Stamm *et al.* (42), the association of HSV-2 infection and syphilis, the two most common causes of genital ulcerations in the U.S., with HIV infection was studied in 200 men previously enrolled in a study of acute proctitis between 1983 and 1986 and in 111 asymptomatic homosexual men undergoing HIV serologic testing and counseling. In this study, a history of reported genital, oral, or anal rectal herpes was correlated with HIV infection. Serologic evidence of previous HSV-2 infection as determined by Western blot for HSV-2 glycoprotein (gG-2) (43,44) was demonstrated in 66% of 100 HIV seropositive men compared to 30 (35%) of 85 HIV seronegative subjects ($p < 0.0001$). In contrast, there was no association between HIV and antibody to HSV-1. Following logistic regression analysis, the association of HIV seropositivity with a history of HSV infection and with serologic evidence of HSV-2 infection persisted after controlling for age, number of recent and life-time sexual partners and years of sexual activity (Table 4). In an evaluation of the association between the time of acquisition of HSV and HIV, they demonstrated that HIV antibody was present in 69% of those who had HSV-2 antibodies at study entry compared to 43% of those who had only HSV-1 antibodies and 44% of those who presented with first episode anal/genital HSV at study entry ($p < 0.05$ when the group with previous HSV-2 infection was compared with all other groups). These differences remain significant after controlling for age, years of sexual activity and number of sexual partners.

**Table 4. Association of HSV Infection With HIV Infection Using Logistic Regression to Adjust for Indexes of Sexual Activity. ***

| Factor | Odds Ratio | P valve |
|---|---|---|
| Proctitis - Enteritis Patients | | |
| History of syphilis | 3.0 | .009 |
| History of gential HSV | 2.1 | .07 |
| Any history of HSV | 2.3 | .01 |
| HSV - 1 antibody | 1.6 | .18 |
| HSV - 2 antibody | 3.3 | .0003 |
| Syphilis antibody | 8.4 | .03 |
| Asymptomatic Homosexual Men | | |
| HSV - 1 antibody | 1.4 | .44 |
| HSV - 2 antibody | 8.5 | <0.0001 |

* Adapted from Stamm et al (42)

Among asymptomatic homosexual men, serologic evidence for HSV-2 infection was present in 36 (73%) of 49 HIV seropositive men and in 17 (32%) of 60 HIV seronegative men ($p < 0.0001$). Antibody to HSV-1 was not statistically different in HIV infected vs. HIV uninfected men. Logistic regression analysis showed that both the association between HSV-2 antibody and HIV seropositivity persisted after controlling for the number of sexual partners in the preceding 2 and 12 months. Thus, as in the study by Holmberg *et al.* (41)

discussed above, Stamm and colleagues (42) provided further serologic evidence that HSV-2 infection, particularly anal/rectal HSV infection usually precedes HIV infection and that the relative risk of HIV infection appears to increase after acquisition of HSV-2.

In a subsequent study to assess whether herpesviruses serve as cofactors in the development of AIDS and other HIV associated diseases among HIV infected individuals, Holmberg and colleagues (45) reclassified HIV infected individuals in their study according to clinical classification. The study was limited to 34 pairs of men who had several specimens drawn before the development of HIV related diseases. They found no significant differences in the rises in antibody titers to CMV, EBV or seroconversion to HSV-1 or HSV-2 between the men with HIV associated diseases and those without. Unfortunately, this was a purely serologic study which is difficult to interpret since many of the individuals in the latter clinical stages of HIV infection such as AIDS, have profound immunosuppression which may directly affect the level of antibody titer to these viruses.

In contrast to the above study, patients with VZV infection appear to have an increased risk of developing AIDS during HIV infection. For example, in one study in Lusaka, Zambia (46), 212 patients with herpes zoster and HIV infection were followed prospectively. Sixty-two (29.7%) of the 212 herpes zoster patients progressed to AIDS which was significantly higher than the 9.5% rate observed in 200 patients with progressive generalized lymphadenopathy, or the 10.1% of the 168 patients with other features of AIDS related conditions. Age, sex, and number of sexual partners, past STD's, etc., were not significantly different among the cohorts. While these data are highly suggestive that herpes zoster during HIV infection is an indicator of a poor prognosis and rapid progression to AIDS, it is not clear whether the appearance of herpes zoster is reflective of a more immunosuppressed state of HIV infection or whether the clinical appearance of herpes zoster results in activation of CD4 positive lymphocytes and further HIV viral replication, thus accelerating the natural history of HIV.

In another study, Kaslow *et al.* (47), also sought to determine whether HSV or VZV infection accelerated the immunodeficiency observed in HIV seropositive men. They compared HIV seropositive, AIDS free homosexual men with clinical HSV or VZV histories during a six month follow-up period compared to HIV seropositive men without HSV or VZV infection. For those patients with HSV infection, the mean CD4 count declined 36 in the 6 months before, 32 in the 6 months around, and 26 in the 6 months after HSV infection. For VZV, the mean decline in CD4 positive cells was 41, 27, and 39 before, around and after the attack of zoster, respectively. Thus, the overall loss of CD4 positive cells during the period of viral infection was slightly greater in those with HSV infection (94, $p = 0.03$) and in those with VZV infection (107, $p = 0.1$) than in the uninfected men (CD4 positive cell count is 59). Therefore in this slightly better controlled study, additional evidence was provided that HIV induced immunodeficiency progressed slightly faster in men with clinically expressed HSV and VZV infection.

While the above studies demonstrated a strong association between prior HSV-2 infection and HIV acquisition in homosexual men, it was of interest to determine whether the same association could be demonstrated for heterosexuals with HIV infection. In a retrospective seroepidmeiologic study performed among African heterosexual men living in Belgium or Africa, Hermans *et al.* (40) demonstrated that 82% of HIV infected patients had HSV-2 antibodies, whereas only 38% of HIV seronegative persons had HSV-2 antibodies (p , 0.001). However, these results were not controlled for sexual activity. In another study among 1000 pregnant women in Haiti, Boulos *et al.* (48) found a 9.2% seroprevalence for HSV-2. In univariate analysis, antibodies to HSV-2 were significantly more common in HIV-1 infected patients (88.4%) and HTLV-I infected patients (81.8%) than in HIV and HTLV-I uninfected patients (53.4%) ($p < 0.0001$). These data further confirm the association between HSV-2 and HIV but also suggest that genital herpes or the sexual activities leading to infection with genital herpes may also play a role in HTLV-I infection.

In 1987, Quinn *et al.*, screened 4,028 patients attending two sexually transmitted disease clinics in Baltimore for the prevalence and associated risk factors for HIV infection (49). Two hundred and nine (5.2%) of these patients were HIV seropositive. HIV seropositivity rates were higher among men (6.3%) than women (3.0%) ($p < 0.001$), and among blacks (5.0%) than whites (1.2%) ($p < 0.02$). Among male patients, HIV infection was associated with homosexuality and IV drug use, sex with an IV drug user, a history of syphilis, genital herpes and genital warts, as well as serologic evidence for syphilis. However, following logistic regression analysis, history of genital herpes no longer remained significant. Due to the fact that historical evidence for prior genital herpes is frequently unreliable, they undertook a serologic case-control study to determine the prevalence of HSV-2 infection among patients with and without HIV infection attending the STD clinics. For this analysis, the patient population was comprised of 182 HIV seropositive cases and 367 age, race and gender matched controls. Although a history of genital herpes was relatively rare among these patients with only 13 (2.3%) reporting a history of genital herpes, patients with a history of genital herpes were significantly more likely to be HIV infected than patients without this history (85% in patients with HIV infection vs. 56% in patients without HIV, $p = 0.05$). Serologic evidence of prior infection with HSV-1 and HSV-2 was quite common in both HIV infected and uninfected patients. HSV-1 antibody prevalence was similar among HIV infected and uninfected patients even when men were further stratified by sexual preference. Antibodies to HSV-2 were also more common in HIV infected patients than patients without HIV infection, although these differences were only significant for heterosexual men (Table 5). HSV-2 antibodies were present in 81% of HIV infected homosexual/bisexual men vs. 55% of those who were not infected with HIV. Among heterosexual men, HSV-2 antibody was present in 61.7% of HIV infected men vs. 46.4% of HIV uninfected men ($p = 0.013$). Similarly, in women HSV-2 antibodies were present in 77.4% of HIV infected women vs. 58.8% of HIV uninfected women ($p = 0.09$).

## REFERENCES

1. Gottlieb, M.S., Schroff, R., Schanker, H.M., Weisman, J.D., Fan, P.T., Wolf, R.A. and Saxon, A. N. Engl. J. Med. 305:1425-1431, 1981.
2. Masur, kH., Michelis, M.A., Greene, J.B., Onorato, I. Stouwe, R.A., Holzman, R.S., Wormser, G. Brettman, L., Lange, M., Murray, H.W. and Cunningham-Rundles, S. N. Engl. J. Med. 305:1431-1438, 1981.
3. Seigal, F.P., Lopez, C., Hammer, G.S., Brown, A.E., Kornfeld, S.J., Gold, J., Hassett, J., Hirschman, S.Z., Cunningham-Rundles, C., Adelesberg, B.R., Parham, D.M., Siegal, M., Cunningham-Rundles, S. and Armstrong, D. N. Engl. J. Med. 305:1439-1445, 1981.
4. Mintz, L., Drew, W.L., Miner, R.C. and Braff, E.H. Ann. Intern. Med. 99:326-329, 1983.
5. Quinnan, G.V. Jr., Masur, H., Rook, A.H., Armstrong, G., Frederick, W.R., Epstein, J. Manischewitz, J.F., Macher, A.M., Jakson, L. and Ames, J. JAMA 252:72-77, 1984.
6. Drew, W.L., Mintz, L., Miner, R.C., Sands, M. and Ketter, B. J. Infect. Dis. 143:188-192, 1981.
7. Drew, W.L., Conant, M.A., Miner, R.C., Huang, E.S., Ziegler, J.L., Grounwater, J.R., Gullett, J.H., Volberding, P., Abrams, D.I. and Mintz, L. Lancet 2:125-127, 1982.
8. Lange, M., Klein, E.B., Kornfield, H., Cooper, L.Z. and Grieco, M.H. JAMA 252:1908-1910, 1984.
9. Center for Disease Control Task Force on Kaposi's Sarcoma and Opportunistic Infections. N. Eng. J. Med. 306:248-252, 1982.
10. Centers for Disease Control. MMWR 31:249-251, 1982.
11. Macher, A.M., Reichert, C.M., Straus, S.E., Longo, D.L., Parrillo, J., Lane, H.C., Fauci, A.S., Rook, A.H., Manischewitz, J.F. and Quinnan, G.V. Jr. New. Engl. J. Med. 309:1454, 1983.
12. Welch, K., Finkbeiner, W. Alpers, C.E., Blumenfeld, W., Davis, R.L., Smuckler, E.A. and Beckstead, J.H. JAMA 252:1152-1159, 1984.
13. Fauci, A.S., Macher, A.M., Longo, D.L., Lane, H.C., Rook, A.H., Masur, H. and Gelmann, E.P. Ann. Intern. Med. 100:92-106, 1984.
14. Jacobson, M.A. and Mills, J. Ann. Intern. Med. 108:585-594, 1988.
15. Giraldo, G., Beth, E. and Huang, E.S. Int. J. Cancer 26:24-29, 1980.
16. Mamor, M. Friedman-Kein, A.E., Zolla-Pazner, S., Stahl, R.E., Rubinstein, P., Laubenstein, L., William, D.C., Klein, R.J. and Spigland, I. 100:809-815, 1984.
17. Jaffe, H.W., Choi, K., Thomas, P.A., Haverkos, H.W., Auerback, D.M., Guinan, M.E., Rogers, M.F., Spira, T.J., Darrow, W.W., Kramer, M.A., Friedman, S.M., Monrow, J.M., Friedman-Kien, A.E., Laubenstein, L.J., Marmor, M., Safai, B., Dritz, S.K., Crispi, S.J., Fannin, S.L., Orkwis, J.P., Kelter, A., Rushing, W.R., Thacker, S.B. and Curran, J.W. Ann Intern. Med. 99:145-151, 1983.
18. Rogers, M.F., Morens, D.M., Stewart, J.A., Kaminski, R.M., Spira, T.J., Feorino, P.M., Larsen, S.A., Francis, D.P., Wilson, M. and Kaufman, L. Ann. Intern. Med. 99:151-158, 1983.
19. Barre-Sinoussi, F., Chermann, J.C., Rey, F., Nugeyre, M.T., Chamaret,

S., Gruest, J., Dauguet, C., Axler-Blin, C., Vezinet-Brun, F., Rouzinoux, C., Rozenbaum, W. and Montagnier, L. Science 220:868-871, 1986.

20. Popovic, M., Sarngadharan, M.G., Read, E. and Gallo, R.C. Science 224:497-500, 1984.
21. Schupback, J., Popovic, M. Gilden, R.V., Gonda, M.A., Sarngadharan, M.G. and Gallo, R.C. Science 224:503-505, 1984.
22. Sarngadharan, M.G., Popovic, M. Bruch, L., Schupback, J. and Gallo, R.C. Science 224:506-508, 1984.
23. Gallo, R.C., Salahuddin, S.Z., Popovic, M., Shearer, G.M., Kaplan, M., Haynes, B.F., Palker, T.J. Refield, R., Oleske, J. and Safai, B. Science 224:500-503, 1984.
24. Zagury, D., Bernard, J., Leonard, R., Cheynier, R., Feldman, M., Sarin, P.S. and Gallo, R.C. Science 231:850-853, 1986.
25. Curran, J.W., Jaffe, H.W., Hardy, A.M., Morgan, W.M., Selik, R.M. and Dondero, T.J. Science 239:610-616, 1988.
26. Piot, P., Plummer, F.A.,Mhalu, F.S., Lamboray, J.L., Chin, J.L. and Mann, J.M. Science 239:573-579, 1988.
27. Klatzmann, D., Barre-Sinoussi, F., Nugeyre, M.T., Danquet, C., Vilmer, E., Griscelli, C., Brun-Veziret, F., Rouzioux, C. and Gluckman, J.C. Science 225:59-63,1984.
28. Koenig, S., Gendelman, H.E., Orenstein, J.M., Dal-Canto, M.C., Pezeshkpour, G.H., Yungbluth, M., Janotta, F., Aksamit, A., Martin, M.A. and Fauci, A.S. Science 233:1089-1093, 1986.
29. Lui, K.J., Darrow, W.W. and Rutherfod, G.W. 3d. Science 240:1333-1335, 1988.
30. Folks, T., Powell, D.M., Lightfoote, M.M., Benn, S., Martin, M.A. and Fauci, A.S. Science 231:600-602, 1986.
31. Gendelman, H.E., Phelps, W., Feigenbaum, L., Ostrove, J.M., Adachi, A., Howley, P.M., Khoury G., Ginsberg, H.S. and Martin, M.A. Proc. Natl. Acad. Sci. U.S.A. 83:9759-9763, 1986.
32. Mosca, J.D., Bednarik, D.P., Raj, N.B., Rosen, C.A., Sodroski, J.G., Haseltine, W.A. and Pitha, P.M. Nature 325:67-70, 1985.
33. Folks, T.M., Justement, J., Kinter, A., Dinarello, C.A. and Fauci, A.S. Science 238:800-802, 1987.
34. Skolnik, P.R., Kosloff, B.R. and Hirsch, M.S. J. Infect. Dis. 157:508-514, 1988.
35. Latchman, D.S. Nature 325:487, 1985.
36. Kreiss, J.K., Koech, D., Plummer, F.A., Holmes, K.K., Lightfoote, M., Piot, P., Ronald, A.R., Nkinya-Achola, J.O., DCosta, L.J. and Robert, P. N. Engl. J. Med. 314:414-418, 1986.
37. Greenblat, R.M., Lukehart, S.A., Plummer, F.A., Quinn, T.C., Critchlow, C.W., Ashley, R.L., DCosta, L.J. and Nkinya-Achola, J.O. AIDS 2:47-50, 1988.
38. Plummer, F.A., Simonsen, J.N. and Ngugi, E.N. Abstract M.8.4. in Third International Conference on AIDS, Washington, D.C., 1987.
39. Katzenstein, D.A., Lutif, A., Bassett, M.T. and Emmanuel, J.C. Abstract M8.3 in third International Conference on AIDS, Washington, D.C., 1987.
40. Hermans, P. Lee, F.K. and Poncin, M. Abstract THP 61 in Third Conference on AIDS, Washington, D.C. June 1-3, 1987.

41. Holmberg, S.D., Stewart, J.A., Gerber, A.R., Byers, R.H., Lee, F.K., OMalley, P.M. and Nahmias, A.J. JAMA 259:1048-1050, 1988.
42. Stamm, W.E., Handsfield, H.H., Rompalo, A.M., Ashley, R.L., Roberts, P.L. and Corey, L. JAMA 260:1429-1433, 1988.
43. Lee, F.K., Coleman, R.M., Pereira, L., Bailey, P.D., Tatsuno, M. and Nahmias, A. J. Clin Microbiol. 22:641-644, 1985.
44. Ashley, R.L., Militoni, J., Lee, F., Nahmias, A. and Corey, L. J. Clin. Microbiol. 26:662-667, 1988.
45. Holmberg, S.D., Gerber, A.R., Stewart, J.A., Lee, F.K., OMalley, P.M. and Nahmias, A.J. Lancet 2:746-747, 1988.
46. Hira, S., Tembo, G., Wadhawan, D., Kamanga, J., Macvacva, R. and Perine, P. Abstract WAP 54 in Fifth International Conference on AIDS, Montreal, Canada, 1989.
47. Kaslow, R.A., VonRaden, M. and Deloria, M. Abstract Thap 100 in Fifth International Conference on AIDS. Montreal, Canada, 1989.
48. Boulos, R., Halsey, N., Ruff, A., Adiren, M., Quinn, T.C. and Nahmias, A. Abstract WAP 7 in Fifth International Conference on AIDS, Montreal, Canada, 1989.
49. Quinn, T.C., Glasser, D. and Cannon, R.O. N. Engl. J. Med. 318:197-203, 1988.
50. Nahmias, A., Corey, L., Lee, F., Aumech, N., Cannon, R. and Holber, S. Abstract THP 119 in Fifth International Conference on AIDS, Montreal, Canada, 1989.

# 2

## THE CELL BIOLOGY OF ANTIGEN PROCESSING: A REVIEW

G. K. Lewis

**Department of Microbiology and Immunology, University of Maryland School of Medicine, Baltimore, Maryland, 21201**

### ABSTRACT

In the last 13 years the major mechanisms whereby the immune system generates an extremely large number of different antibodies and T cell antigen receptors were identified. By contrast, it is only in the last 4 years that any detailed understanding of a third aspect of antigen recognition has emerged; namely, how an antigen presenting cell creates a specific ligand for recognition by the T cell antigen receptor. Fortunately, the methods of molecular and cell biology, combined with current immunological paradigms, offer a unique opportunity to attack the mechanisms of antigen presentation. The following review will attempt to place antigen processing in the context of our current knowledge of the expression of the MHC molecules that serve as "receptors" for immunogenic peptides on the antigen presenting cell (APC) side of an immunologic cell couple.

### INTRODUCTION

Cellular immunology is currently at a cross-roads where its issues and paradigms are fast becoming those of mainstream cell biology. For this reason, it is my goal to tie the phenomena commonly referred to as "antigen processing" and its parent discipline, antigen presentation, to the broader issues that confront all students of cell biology. In so doing, I will begin by briefly sketching the historical development of the antigen presentation paradigm, turning next to current observations and interpretations of processing and presentation experiments. At the end of each section I will end by posing what I perceive to be the next questions. Throughout the review, I will draw from experiments on synthetic antigens, peptides, proteins, and viruses. However, to make the review as general as possible I have not attempted an analysis of each different processing system-- in return, I hope that clarity is the product of selectivity and issue my apologies in advance to those investigators whose work could not be included.

**BACKGROUND TO ANTIGEN PROCESSING**

A vague picture of a role for antigen processing in adaptive immunity has been a part of immunology since the time of Metchnikoff who studied the localization and persistence of tetanus toxin (1). The picture was brought into sharper focus in the middle of this century by three kinds of studies.

First, using increasingly better protein labels, ranging from the initial studies with colored azo-proteins (2), to fluorescent proteins (3), to radioactive proteins (4), antigen localization studies provided increasingly better information about the types of cells that take up antigen versus the types of cells that produce antibody. Early on, after it was recognized that macrophages are favored sites of antigen deposit (2), it was mistakenly thought that they produced antibody (see 5 for a review, 6). Indeed, the seminal observations of Fagraeus (7) implicating plasma cells as antibody producing cells were cooly received. Later studies clearly established the lymphoid origins of antibody (reviewed in 8), but this false start regarding the macrophage is an example of the major problem with antigen localization studies; namely, in knowing whether or not the antigen that is traced in the experiment is immunologically relevant.

The second type of study shed light on what form immunologically relevant antigen might assume. These studies, done initially be Gell and Benacerraf (9), and confirmed by several other groups (10,11), showed that native and denatured protein antigens cross-react in delayed type hypersensitivity reactions whereas they do not in immediate hypersensitivity reactions. These observations were the first to show that the cells mediating delayed type hypersensitivty (ie. T cells) could recognize a protein antigen in a non-native conformation. Subsequent studies by several groups (12,13) showed that macrophages take up native protein antigens and effectively denature them for recognition by T cells.

The third type of study was done using cytotoxic T cells (CTL). After the first demonstration of alloreactive CTL (14), several investigators used the methodology to study virus (15,16) and hapten (TNP)-modified self-specific CTL (17). In general, there are few confirmed reports of exogenous protein antigens forming effective ligands on target cells lysed by CTL (18), suggesting that the ligand recognized on the target cell by a CTL is formed only by direct modification of self-proteins (in the case of TNP-modified self) or more usually by the endogenous synthesis of an allo- or viral protein. This issue will be discussed in more depth below, but there is now evidence that for an exogenous protein to sensitize for class I restricted lysis, it must be delivered to the cytoplasm of the target cell (19,20,21). In the next section it will be argued that differences in the processing of endogenous antigens vs. exogenous antigens pertain more to the MHC restriction element used than to the functional phenotype (ie. help vs. cytotoxicity).

These observations illustrate a recurrent theme of this chapter. The theme is that the general rules of protein uptake, internalization, synthesis, sorting, degradation, and transport back to the cell surface that are understood by all cells underlie the specific rules of immunogenicity that are uniquely understood by cells of the immune system.

## EXOGENOUS VS ENDOGENOUS PROCESSING PATHWAYS

A key distinction will be made at the outset between the processing and presentation of exogenously acquired antigens versus the processing and presentation of endogenously acquired antigens. The observations supporting this distinction are that soluble protein antigens (ie. exogenously acquired antigens) are only presented via class II MHC molecules (see 22 and 26 for a hypothesis and review). A comparable pathway is less obvious for Class I molecules. There is one recent report of Class I restricted cells specific for soluble protein (18), but most of the evidence argues that such proteins must be delivered to the cytoplasm by either direct delivery through osmotic shock (19) or viral receptors (20) for processing and presentation to occur. It is clear, however, that soluble exogenous peptides can be presented by either Class I (23,24,25) or Class II (22,26) pathways. Thus, any differences between Class I and Class II pathways, real or apparent, must be explained by the fates of these two restriction elements in the antigen presenting cell.

Accordingly, the analysis of current literature will be developed along two lines; the processing and presentation of exogenous proteins and the processing and presentation of endogenous proteins. However, before taking up these topics, it is appropriate to consider the ultimate product of antigen processing --- the ligand recognized by the T cell antigen receptor.

## THE LIGAND RECOGNIZED BY THE T CELL ANTIGEN RECEPTOR

It is now known that the final product of antigen processing is the creation of the ligand recognized by the the T cell antigen receptor. Once it became possible to study the interactions of highly radioactive antigens with putative antigen receptors on T and B cells using autoradiography, it was rapidly found that B cells bind naked antigen with avidities that approximate those expected for antigen-antibody reactions (27,28), and that the interactions are inhibitable with anti-immunoglobulin reagents. Parallel studies with T cells were more controversial.

Several groups reported success using highly radioactive synthetic poly-amino acid antigens such as (T,G)A-L (29,30), but a number of other investigators failed in this regard. Subsequently, others succeeded in getting T cells that mediate suppression to bind antigen coated plates (31,32,33), leading to the idea that suppressor cells are distinct lineages from conventional MHC restricted T cells that only bind to processed antigen presented by antigen presenting cells. The phenotype of suppressor cells remains controversial, however, several groups have now provided convincing evidence that some Class I (34,35,36) and some Class II (36,37,38,39) MHC restricted T cells can bind antigen directly.

Studies by Siliciano, *et. al.* (34,35,36) and by Rao, *et. al.* (37,38,39) have shown antigen binding by T cell clones using either fluoresceinated polymers or azobenzearsonate-conjugated proteins , respectively. In both studies, evidence was presented that binding was mediated by the T cell antigen receptor. In the studies of Siliciano, *et. al.* (36), the antigen receptor was retarded on fluorescein affinity columns while in the studies of Rao, *et. al.* (39) receptor specificity was transferred to other T cells by a retroviral construct carrying the Va3 gene. Furthermore, recent studies by the groups of Janeway *et. al.* (40), and Kappler, *et. al.* (41,42) have uncovered a unique set of antigens bound by MHC restricted T cells called "super-antigens".

These antigens include members of the staphylococcal enterotoxin family that are known to polyclonally activate T cells. They are thought to bind to particular Vb segments and to monomorphic structures on Class II MHC molecules, cementing a collaborative bond that mimics a conventional T cell receptor-peptide-MHC complex. Thus, it appears that certain antigens, derived either synthetically or from common pathogens, can bind directly to a restricted set of T cell antigen receptors. The studies of Siliciano, *et. al.* (34,35) show clearly that the epitope density and size of the antigen dramatically affect binding. It will be interesting to understand the structural variables at play in such instances. Furthermore, in the case of such antigens, it will be interesting to learn the degree to which they activate T cells both in the presence and absence of antigen presenting cells.

However interesting direct antigen binding by T cells might be, it stands in stark contrast to the more usual finding that T cells recognize antigen on the surface of an antigen presenting cell and in the context of MHC. Almost from the start these findings led to the idea that there must be some antigenic specificity in the MHC molecules themselves (43). Although the idea was initially hard to accept because of the limited number of allelic forms of MHC molecules in any single individual, it now appears to be correct.

Once it became clear that Class II MHC molecules are the Ir gene products (44), several lines of evidence increasingly pointed toward their specificity. First, antigenic competition studies at the APC level (45,46,47) were widely interpreted as evidence for competition for MHC molecules, although no direct chemical data were presented for the competition and competition was not observed for all T cell clones (47). Second, studies in the cytochrome-c system wherein the rank order of stimulation by different cytochromes-c for a single T cell hybrid changed when changing the Class II alleles on the APC, strongly argued for specificity at the Class II level (48, 49,50). These results were extended in the same system using a novel transition-point assay (51) that allowed dissociation of APC from T cell contributions to a dose response curve. Third, Babbitt *et. al*. (52,53) and Buus, *et. al*. (54) provided direct evidence for the specific and competitive binding of antigenic peptides to Class II MHC molecules. Similar observations were recently reported for Class I molecules (55). Thus, it is now relatively certain that the ligand created via antigen processing by the antigen presenting cell is a complex between an antigenic peptide and a MHC molecule. This picture is supported by the recent publication of the HLA-A2 crystal structure (56,57) in which a putative binding site is clearly delineated in the structure. Mutagenesis studies support the idea that the HLA-A2 cleft is indeed a binding site for peptides (summarized in 57).

One puzzling aspect of the ligand is the kinetics of its formation *in vitro*. In the initial studies by Babbitt, *et. al*. (52) the affinity of the peptide-Class II interaction as measured by equilibrium dialysis was approximately 10mM. Kinetic measurements by Buus, *et. al*. (58) revealed a similar affinity and illustrated a peculiar aspect of the reaction. Both the forward (kf) and reverse (kr) rate constants are extremely slow, in the range of $1\ M^{-1}\ sec^{-1}$ for kf and $10^{-5}\ sec^{-1}$ for kr. Thus, the apparent physical chemistry of the ligand is unusual in that once a peptide has bound to MHC, it is slow to let go. This raises a problem of just how much free MHC exists on the surface of an APC and how a presumed excess of non-immunogenic "self-peptides" is kept from swamping the MHC molecules required for the immunogenic "foreign-peptides". The reality of the problem is suggested by the studies of Babbitt *et. al*. (53) who showed that a non-immunogenic murine lysozyme peptide analog of the immunogenic hen egg-white lysozyme peptide could competitively bind to Class II. While the paradox remains unresolved, there are several possible solutions.

First, the "self-peptide"-MHC complexes may not form *in vivo*. Second, the APC may have some mechanism for selectively ridding itself of "self-peptide"-MHC complexes. Third, a predominance "self-peptide" - MHC complexes may co-exist with a minority of "foreign-peptide"-MHC complexes. In this instance, the T cell antigen receptor system will have to be sensitive enough to recognize and respond to a low concentration of the relevant peptides. At the moment, the third possibility is the most tenable, but to verify it, the affinity of the T cell antigen receptor must be characterized.

Currently, there is scant information on the affinity of the T cell antigen receptor for peptide-MHC complexes. The most informative studies done so far used an allospecific anti-H-2Kb hybridoma and a hybrid H-2Kb/Q10b molecule to estimate Kd of approximately $10^{-7}$ M/L by inhibition of activation (59). This Kd approximates that estimated by Siliciano *et. al*. (34) in the fluorescein system and suggests that the T cell receptor affinity approximates that of antibody. In the studies using soluble class I molecules, no evidence was obtained for direct binding to T cells (59,60) while it was shown that the molecules had to be in a multivalent form to induce T cell activation (60), clearly pointing to receptor aggregation as a key step in T cell activation by antigen. Similar conclusions were reached earlier for anti-receptor antibodies (61,62).

## PROCESSING AND PRESENTATION OF CLASS I RESTRICTED ANTIGENS

Processing and presentation of Class I restricted antigens has been studied almost entirely with cytotoxic T cells. For this reason, this discussion will center on the processing and presentation events leading to a sensitized target cell for Class I restricted CTL. Little can be said of the processing and presentation requirements for the induction of CTL precursors or of the requirements for Class I restricted cells that have other functional phenotypes [ie. helper cells (63,64)].

### Synthesis and Expression of Class I Molecules

Class I molecules are expressed on the surfaces of essentially all nucleated somatic cells (65). As a rule, cell surface Class I is associated with

$\beta$2-microglobulin ($\beta$2-m) (66) which has long been thought to be a requirement for Class I expression. Early studies (66) using the $\beta$2-m deficient lymphoblastoid cell line Daudi showed that its Class I expression could be rescued by fusion with a $\beta$2-m competent cell line, suggesting that $\beta$2-m association is a necessary step in the surface expression of Class I. The simplest interpretation of these observations was that $\beta$2-m associates with Class I in the ER and that this step is required for the transport from the ER to the cis-Golgi. Two alternative hypotheses have also been advanced for the role of $\beta$2-m in Class I expression. First, studies using a cell free translation system (67) suggested that $\beta$2-m makes Class I more resistant to proteolysis, thereby increasing its intracellular stability and surface expression. Second, both monoclonal antibodies (68,69) and anti-minor MHC antigen T cell clones (70) selectively detect $\beta$2-m associated Class I molecules, suggesting that $\beta$2-m plays a necessary role in the folding of Class I molecules. Paradoxically, there are reports of Class I expression in $\beta$2-m negative cell lines (71,72) showing that $\beta$2-m expression is not absolutely required for Class I expression and therefore arguing against all 3 of the above hypotheses.

Very recently, studies by Hansen *et. al.* (73) have shown that two forms of Class I can co-exist in $\beta$2-m competent cells. Using a monoclonal antibody specific for Ld they showed that one form of Class I can exist in both $\beta$2-m associated or $\beta$2-m free states while the second form is unable to associate with $\beta$2-m. Both forms of Class I had undergone oligosaccharide processing in the Golgi and were expressed on the cell surface, although the $\beta$2-m free form was expressed at relatively low levels as compared to the $\beta$2-m associated form. These data clearly show that $\beta$2-m association is not absolutely necessary for surface expression and point toward the existence of multiple forms of Class I on a single cell surface. Interestingly, pulse-chase studies showed that the two forms of Class I are independently determined at the level of the ER. Based on these observations, Hansen *et. al.* (73) proposed that $\beta$2-m acts early in the synthesis of Class I molecules where it selectively induces a conformational change in the outer two domains of Class I. The $\beta$2-m associated form is selectively transported to the cell membrane and the $\beta$2-m free form is apparently trafficked elsewhere. Earlier data reported by Allen *et. al.* (74) showed that Dd molecules expressed in $\beta$2-m$^-$ cell lines are not recognized by Dd restricted T cells, arguing that only the $\beta$2-m associated forms have the ability to bind nominal antigen. The studies of Hansen *et. al.* (73) showed that Ld can exist in either $\beta$2-m associated or free forms once it has been "conformed" by $\beta$2-m, but it is not known whether or not both forms can equally participate in Class I restricted recognition.

In summary, several lines of evidence show that while Class I expression can be dissociated from $\beta$2-m expression, association of Class I with $\beta$2-m in the ER imparts a conformation that is "remembered" by Class I during intracellular maturation and selectively allows the surface expression of this form of Class I on the cell surface. Furthermore, the $\beta$2-m conformed state of Class I might be the immunologically competent form that is recognized by both allospecific and virus specific T cells (73,74). If the function of $\beta$2-m is to constrain Class I conformation to its immunologically relevant form early in biosynthesis, it raises the possibility that Class I molecules may become able to bind antigenic peptides while still in the ER. This possibility is made even more intriguing by the recent results of Lippincott-Schwartz *et. al.* (75) who have found a new pathway for degradation of newly synthesized proteins. This pathway is in a pre-Golgi compartment, possibly in the ER itself, and rapidly degrades newly synthesized proteins that are unable to form mature complexes. It is possible that the early folding of Class I into its immunologically relevant form, induced by $\beta$2-m, allows the immune system to survey this early step in protein synthesis for foreign epitopes. A recent and exciting report adds an additional dimension to this idea.

Quite recently, Townsend *et. al.* (76) reported studies using a mutant cell line deficient in Class I expression which showed that Class I expression may be directly affected by peptide binding. In these studies, the deficient cells could be induced to express Class I by incubation with peptides leading to peptide-MHC complexes that are recognized by CTL. The data strongly support the idea that the antigenic peptide itself is required for the proper folding, assembly, and transport of Class I molecules to the cell surface. This result further supports the idea that Class I molecules allow the immune system to survey an early step in protein synthesis and degradation for the presence of foreign sequences.

Internalization of Class I Molecules

Because Class I molecules are expressed on all nucleated cells, their cell surface expression is likely to be heterogeneous with respect to half-lives and modes of internalization. That this is indeed the case is shown in studies on the "spontaneous" internalization of Class I molecules. Early studies by Machy *et. al.* (77,78) and later studies by Tse and Pernis (79) and by Tse, *et. al.* (80) revealed the rapid and *de novo* internalization of Class I molecules in T cells or T cell lymphomas but not in other cell types including B cells. Macrophages may be an exception to the rule in that Dasgupta *et. al.* (81) recently showed the internalization of Class I-anti-Class I complexes via coated pits in cells of the monocyte/macrophage lineage. In a recent study by Machy *et. al.* (82) using T cells, approximately 40% of the surface Class I molecules are internalized in endocytic type vesicles. Other studies suggest that these vesicles may selectively include other surface molecules such as the transferrin receptor (79,80). The vesicles are rapidly internalized into an acidic environment as judged by the quenching of vesicular fluorescence and its reversal in chloroquine treated cells (80). The vesicles then become progressively more neutral in pH and are trafficked to the cell membrane with a $t_{1/2}$ of approximately 20 minutes (79,80).

Currently, the reason why some cells spontaneously internalize their Class I molecules, whereas others do not, is not known. In the case of T cells, where the phenomenon is best documented, it is possible that spontaneous internalization is related to the problem of how a Class I restricted T cell deals with its own Class I molecules that may bind the relevant antigen. It is possible to create Class I restricted CTL targets by "soaking" cells in antigenic peptides. *In vivo*, if peptides can sensitize target cells, then it is possible that a Class I restricted T cell may be sensitized by a peptide to which it will respond. How, then, does a Class I restricted T cell antigen receptor handle Class I-peptide complexes that it can recognize when the two sets of molecules are on the same cell surface? Does the CTL kill itself? Apparently, the answer to the latter question is no because class I restricted CTL clones make poor targets themselves (83). The mechanism for this refractoriness appears to be their resistance to perforins. It further underscores the need for directionality in recognition, whereby a CTL is not "short circuited" by recognition of Class I-peptide on its own surface.

Types of Antigens that Induce Class I Restricted Responses

Early on, it was found that typical exogenous protein antigens, so conveniently studied by "helper" assays, are poor subjects for CTL assays. As they are commonly done, the two assays largely reflect Class I restricted responses vs. Class II restricted responses, so it has become common dogma that soluble protein antigens do not induce Class I restricted responses. There is a single published exception to this dogma (18), but it stands as a useful rule for distinguishing the types of antigen that elicit Class I vs. Class II restricted responses. What, then, can be said of the types of antigens that induce Class I restricted responses?

First, and foremost, is the necessity of protein antigens to be delivered to the appropriate cytoplasmic compartment. It is currently unclear what this compartment is, but it now seems clear that it is not the same compartment to which soluble protein antigens that elicit Class II restricted responses are delivered. In early studies on the creation of Class I restricted T cell targets, it was found that the antigen had to be covalently coupled to the target cells [as for TNP-modified self (17)] or had to be synthesized by the target cell itself (19,20,21).

In the case of viral antigens, inhibitors of protein synthesis inhibit the formation of effective targets by virus infection (84,85), leading to the idea that biosynthesis is a necessary step. It was subsequently found that target cells could be created by killed virus (86,87) provided that the virus harbors a fusion protein that is either functional at neutral pH (eg. Sendai virus) or that is activated by exposure of the virus to acidic pH (influenza virus). In the case of influenza virus an additional step of inactivating the receptor destroying enzyme (eg. the neuraminidase) is required for inactivated virus to sensitize target cells (87). Thus, the general picture holds that only those antigens that are able to reach the correct cytoplasmic compartment can create effective targets for Class I restricted T cells. It is also important to note that soluble proteins such as ovalbumin that function well in Class II restricted responses will also function well in Class I restricted responses when they are synthesized by the target cell as shown recently in transfection studies (21). Although the creation of effective ligands from proteins for Class I restricted responses requires cytoplasmic delivery of the protein, the ligands can also be created using soluble peptides (23,24,25,26). In this regard, it was recently shown that approximately 0.3% of purified HLA-A2 molecules could bind antigenic peptides derived from influenza virus (55). Recently, Yewdell and Bennink (87) showed that the drug Brefeldin A inhibits the presentation of influenza virus proteins but not of influenza virus peptides. Brefeldin A is most likely inhibiting processing of the proteins because it specifically disrupts the cis-medial Golgi causing the return of the associated proteins to the endoplasmic reticulum. It is noteworthy that exocytosis of any ER resident Class I molecules would be inhibited.

One of the most interesting aspects of Class I restricted recognition (indeed, of Class II restricted recognition as well) is that it allows internal proteins of cells and viruses to be recognized by T cells. Although it is currently thought that the T cell receptor only recognizes antigen-MHC complexes that are on the cell surface, the complexes need not come from the outside of the cell or even the outside of a virion. One of the first examples is the finding that Class I restricted CTL can recognize influenza virus nucleoprotein which is found only inside the virion (reviewed in 88). Because Class I restricted recognition requires the appropriate "targeting" of antigenic proteins to an as yet uncharacterized cytoplasmic locus, it is possible that the replication cycle of some viruses has evolved to circumvent this pathway of recognition. For example, there is little confirmed evidence for Class I restricted responses to certain virus classes including picornaviruses and the gD-1 protein of herpes. Thus, the ability to sample different subcellular compartments by Class I and Class II restricted pathways likely affords a broader base for host protection.

**PROCESSING AND PRESENTATION OF CLASS II RESTRICTED ANTIGENS**

The processing and presentation of Class II restricted antigens has been studied using several different readouts of T cell activation including cytotoxicity, proliferation, and lymphokine production. Both untransformed T cell clones and malignant T cell hybrids have been used for the studies cited below. To date the class of MHC restriction appears to be a better predictor of the processing pattern than the functional readout used to assess antigen recognition. For this reason, I will freely interchange conclusions obtained with the various systems used to study Class II restricted responses.

Synthesis and Expression of Class II Molecules

A fundamental distinction between Class I and Class II MHC molecules is their tissue distribution. In contrast to Class I molecules, Class II molecules are expressed constitutively on relatively few cell types including B cells (89), monocytes (90), dendritic cells (91), and, in humans, T cells (92). On the other hand, Class II expression is inducible in a variety of cell types (see 93), especially consequent to pathological states. Its constitutive expression has been most intensively studied in lymphoblastoid B cells.

Two features of Class II expression distinguishing it from Class I are immediately apparent. First, Class II expression requires the assembly of a heterodimer comprised of appropriate $\alpha$ and $\beta$ chains (94). What do I mean by appropriate? Early studies of Class II expression in F1 hybrids demonstrated mixed heterodimers encoded by a single locus comprised of an $\alpha$ chain from one parent and a $\beta$ chain from the other parent (95,96). $\alpha$ and $\beta$ chains from different loci associate very poorly or not at all (however, see 97 for a discussion of exceptions). Later studies, within $\alpha$ locus, showed allelic preferences for certain $\alpha$-$\beta$ pairs (98, 99). The allelic preference was mapped to the polymorphic N-terminal domain of the $\beta$ chain for I-A (99), thus the region most probably involved in antigen recognition by Class II molecules also specifies an additional general property of these molecules, the haplotypic preference of chain association. The evolutionary significance of this interesting paradox has been discussed in detail by Germain *et. al*. (100).

Second, early studies by Jones (94) showed that Class II molecules are associated with a third molecule called the invariant chain I. Later studies showed that I can be expressed as a proteoglycan with chondroitin sulfate side chains (101,102). The requirement for invariant chain synthesis in class II antigen presentation is a matter of debate (103,104,105). However, the recent studies of Stockinger *et. al.* (105) argue strongly that invariant chain greatly increases the efficiency of antigen presentation. It should be pointed out that Class II molecules can assemble and be transported to the cell surface in the absence of detectable invariant chain synthesis (103). The most tantalizing clue as to the role of invariant chain comes from its subcellular distribution.

Invariant chain is associated with Class II only inside the cell and not when Class II is on the cell surface (94). It has been suggested that invariant chain prevents Class II from binding peptides until Class II reaches the correct compartment (104), most probably a post-Golgi compartment where the exocytic and endocytic pathways meet (105). This is discussed in more detail in the following paragraph.

Intracellular Fate of Class II Molecules

What is the intracellular fate of Class II molecules in cells permissive for Class II expression? This question can be framed differently for each cell type that expresses Class II, so it should be pointed out that most of what is known about Class II expression comes from studies on lymphoblastoid B cells and fibroblast transfectants.

Class II molecules appear to associate with invariant chain early after synthesis in the endoplasmic reticulum (106) and are transported to the trans-Golgi (107) compartment in vesicles that can be stained with both anti-Class II and anti-invariant chain antibodies (108). In comparison to Class I molecules, Class II molecules require longer times [1 to 4 hours for Class II (106,108)] to arrive at the cell surface once synthesized, and only 30 minutes of this time are spent traversing the Golgi (106). Pulse-chase and carbohydrate trimming studies by the Cresswell group (107) argue that the Class II-invariant chain complex is held at a site at the trans-Golgi or at a post-Golgi but pre-cell surface, for at least 2 hours prior to its dissociation into Class II plus invariant chain and the surface expression of Class II.

Elegant studies by Cresswell (109) strongly argue that this pre-cell surface compartment intersects with the endocytic pathway. In these studies, a neuraminidase-transferrin complex was used to mark the endocytic pathway and it was shown that neuraminidase "clipped" sialic acid from both Class II and its associated invariant chain. This study raises again the provocative idea that the role of invariant chain might be to "constrain" Class II in a non-antigen binding conformation until it reaches the correct compartment which intersects the endocytic pathway, thus allowing Class II to selectively sample this pathway.

This hypothesis raises the strong prediction that the binding sites of much of the Class II arriving at the cell surface may already be occupied by endocytically derived peptides. It should be possible to demonstrate this association by pulsing Class II positive cells with labeled peptides that are known to bind to Class II in solution. This observation could also account for the low percentage of solubilized Class II that will bind an antigenic peptide in solution (54). It should be pointed out, however, that peptides may not require processing [although they may still be processed (110)] and are presented by fixed antigen presenting cells (111,112). Assuming that Class II transport to the cell surface no longer occurs in such cells, a portion of the surface borne Class II molecules must be able to bind the antigen. For a viable cell this may mean that both the endocytic and extracelluar compartments are sampled by Class II molecules. Whatever the ultimate conclusions are with respect to function, the above experiments point to fundamental differences between Class I and Class II molecules with respect to intracellular transport and expression.

## What Types of Antigens Induce Class II Restricted Responses?

In contrast to Class I restricted responses, Class II restricted responses are readily demonstrated for soluble protein antigens derived extracellularly. The general picture is that such antigens must be taken up by the cell either by fluid phase pinocytosis (113), or receptor mediated transport (114,115),(see 116, 117, and 118 for immunoglobulin as the receptor), and degraded to a form that will associate with Class II. As mentioned above, some determinants do not require processing including, peptides (119), some proteolytically clipped proteins (120), and even some determinants on very large proteins such as fibrinogen (121). It is probable that the association with Class II is the rate limiting step in just how far a protein needs to be degraded before it can be presented.

Since many T cell determinants are buried inside of proteins and require processing to liberate them so that they can interact with MHC molecules, differential processing can have dramatic effects on immunogenicity. When multiple T cell epitopes are potentially present, those that are most easily liberated by processing or that are most readily accessible to MHC after limited processing may constitute dominant determinants. There are several examples of processing affecting immunogenicity and dominance for both Class I (122) and Class II restricted responses (123,124,125,126). The control of clonal dominance and immunogenicity of individual T cell determinants in multideterminant antigens is likely to provide the key insights necessary for the rational design of vaccines. Indeed, the recent studies of Wachsman *et. al.* (127, 128) who showed differential protection against herpes infections using vaccinia constructs that express the herpes simplex virus gD-1 gene product under the control of different promoters can be interpreted in such a way.

## SPECULATIONS ON THE ROLE OF ANTIGEN PROCESSING IN IMMUNOGENICITY

The above discussion focuses on what is currently known about the cell biology of antigen processing. However, it remains to reconcile this knowledge with the phenomena of protective immune responses against infectious organisms. Of necessity, any such response will be comprised of two elements. First, an element to discriminate self from non-self is required, and, second, an effector class to rid the host of the parasite is also required (see 129 for a more complete, prior derivation of this concept). Failure to generate either the self/non-self discriminator or the appropriate effector class results in successful colonization of the host by the parasite. The elements that discriminate self from non-self are well known and include antibodies, T cell antigen receptors, and MHC molecules. Likewise, most of the effector classes are equally well known and a partial list includes mononuclear phagocytes, cells of the granulocytic series, and the proteins of the complement system. What is less known is how the two elements adaptively interact to produce protective immunity. In this section, I will briefly speculate that antigen processing can play a key role in the adaptive integration of self/non-self discrimination with the selection of effector class. I will also speculate that such integration exerts selective pressure on the invasive and replicative strategies of parasites. Antigen presentation can affect specificity-effector class integration via two related mechanisms. First, different APCs may process antigen differently yielding different peptide-MHC complexes from the same immunogen. Furthermore, different APCs almost certainly address different effector classes creating an association between specificity and effector class selection. Second, any invasive or replicative strategy on the part of the parasite that interferes with MHC expression or function can result in evasion of immunity by the loss of response in the effector classes dependent on that particular APC.

Mechanism 1: Differential Processing Affects the Association Between Specificity and Effector Class

A key paradox emerging from the above hypothesis is that response specificity can apparently select the effector class. For example, in an experimental model of leishmaniasis it was found that an immunogenic fraction 1 soluble extract of *Leishmania major* induced a T cell response that is non-protective (130). By contrast, fraction 9 of the same extract induced a T cell response that is protective. Later studies (131), showed that fraction 1 selectively induces Th2 T cells whereas fraction 1 selectively induces Th1 T cells and the latter confer protection upon adoptive transfer. Thus, the response specificity (for fraction 1 or fraction 9) selects effector class (Th2 or Th1, respectively), and only one effector class (Th1 cells) leads to protection. The major question emerging from such results is how does this selection occur? One possibility is that Th1 and Th2 cells draw differentially from the antigen receptor pool; another is that the APC determines the outcome. At this point, for the first possibility to be correct, different $CD4^+$ T cell subsets would have to draw differentially from the V gene pool. This is made unlikely in that there is emerging evidence that Th1 and Th2 cells may derive from a common precursor meaning that this differentiation would also involve some type of "V-gene" sorting mechanism in the periphery. Such a mechanism seems unlikely. The second possibility is that the effector class is determined by the cell biology of the APC and that the two antigenic fractions are presented by different APCs. Thus, any discussion of specific protective responses must focus on the precise nature of the antigen presenting cell. The apparent connection between specificity and effector class would have to come from the differential processing of the immunogens by the different APCs.

The differential selection of effector class by antigen processing might also affect the regulation of an immune response. For example, an antigenic determinant that selectively turns on suppressor cells offers a way for the parasite to evade a protective response. Indeed, the cells that get turned on in a protective response might have dual functions; one to protect, the other to dampen non-protective responses. Examples exist of both possibilities. For the former, effective immunity against HSV-1 in mouse and guinea pig models requires that suppression be avoided (132,133). For the latter, consider the *Leishmania major* model discussed above. Th1 cells mediate protection, but there is also evidence that Th1 cells can suppress the activity of Th2 cells via interferon (IFN)-$\gamma$ which inhibits the stimulation of Th2 cells by interleukin-4 (IL-4) (134). Thus, one effector class may regulate the activity of another. This hypothesis can also explain the old phenomenon of the inverse relationship between delayed type hypersensitivity [mediated by Th1 cells in the mouse (135)] and antibody production [in the mouse Th2 cells provide the broadest spectrum of lymphokines involved in help (136)]. It will be of great interest to extend these sorts of studies into models of HIV and herpesviruses.

Mechanism 2: Interference with MHC Expression Selectively Affects Effector Classes

This is actually a variant of the first mechanism where some effector classes are rendered inaccessible due to the products of antigen processing. The recent report of Townsend *et. al.* (76) suggest that the antigenic peptide plays an active role in the proper folding of Class I molecules for assembly and transport to the cell surface. This surprising result has profound implications for the hypothesis that antigen presentation affects the integration of specificity and effector class selection. For example, if an immunogenic peptide is required for the correct folding of Class I and the transport of the complex to the cell surface, might not there be a class of non-immunogenic peptides that inhibit the correct folding of Class I? Liberation of such peptides from a protein via processing might render an APC population ineffective for presenting not only the antigen in question but other antigens as well. Indeed, there are several examples where infection of APCs by herpesviruses decreases MHC expression (137,138), and while such decrease can be explained by trans-activating negative control elements in the viral genome, the possibility that it could also be due to the creation of "inhibitor" peptides via processing should also be entertained. Indeed, the need for a pathogen to evade one or another effector class should provide strong selective pressure over evolutionary time for fixing such sequences in the parasite genome.

**SUMMARY**

In the preceding paragraphs, I have attempted an overview of the history and general features of antigen processing. Historically, the paradigm of antigen processing arose from studies attempting to localize and identify antibody secreting cells by tracking antigen and from studies on the specificity of immunologic reactions. These studies led to the reluctant conclusion that in addition to lymphocytes, which have specific receptors for antigen, antigen presenting cells must also specifically recognize antigen. Thus, the "real" role of MHC molecules was elucidated-- they act as antigen receptors, forming the peptide-MHC complexes that are recognized by the T cell antigen receptor.

Thus, the biochemistry and cell biology of antigen processing became the biochemistry and cell biology of MHC expression. Indeed, this last statement most accurately reflects the current state of processing and presentation studies. Processing and presentation studies are currently at an exciting stage where the necessary paradigms and tools are in hand to achieve a real and deep understanding of how these phenomena underlie immunogenicity. I hope that the above review is reflective of this position and stimulates the reader to consider the importance of antigen processing when reading the chapters to follow.

## REFERENCES


1. Metchnikoff, E. Ann. Inst. Pasteur. 11:801-899, 1897.
2. Sabin, F.R. J. Exp. Med. 70:67-83, 1939.
3. Coons, A.H., Synder, J.C., Cheever, F.S. and Murray, E.S. J. Exp. Med. 91:31-38, 1950.
4. Fine, J., and Seligman, A. M. J. Clin. Invest. 22:285-303, 1943.
5. McMaster, P.D. Sites of Antibody Formation. In: **The nature and significance of the antibody response** (Ed. A.M. Pappenheimer), Columbia University Press. New York. 1953, pp.13-24.
6. Adler, F.L., Fishman, M. and Dray, S. J. Immunol. 97:554-558, 1966.
7. Fagraeus, A. J. Immunol. 58:1-13, 1948.
8. Nossal, G.J.V., and Ada, G. **Antigens, Lymphoid Cells and the Immune Response.** Academic Press, New York, 1971.
9. Gell, P.G.H. and Benacerraf, B. Immunology 2:64-70, 1959.
10. Ishizaka, K., Kishimoto, T., Delespesse, G. and King, T.P. J. Immunol. 113:70-74, 1974.
11. Schirmacher, V. and Wigzell, H. J. Exp. Med. 113:1635-1642, 1974.
12. Unanue, E.R. and Askonas, B.A. J. Exp. Med. 127:925-932, 1968.
13. Frei, P.C., Benacerraf, B. and Thorbecke, G.J. Proc. Natl. Acad. Sci. USA 53:20-24, 1965.
14. Cerottini, J.C., Nordin, A.A. and Brunner, K.T. Nature 227:72-74, 1970.
15. Marker, O. and Volkert, M. J. Exp. Med. 137:1511-1520, 1973.
16. Zinkernagel, R.M. and Doherty, P.C. J. Exp. Med 138:1266-1269, 1973.
17. Shearer, G. M. Eur. J. Immunol. 4:527-531, 1974.
18. Staerz, U. D., Karasuyama, H. and Garner, A.B. Nature 329:449-451, 1987.
19. Moore, M.W., Carbone, F.R. and Bevan, M.J. Cell 54:777-785, 1988.
20. Townsend, A. , Bastin, J. , Bodmer, H. , Brownlee, G. , Davey, J., Gotch, F.,Gould, K. , Jones, I., McMichael, A. J. and Rothbard, J. Phil. Trans. of the Royal Society of London Series B: Biological Sciences 323:527-533, 1989.
21. Germain, R.N. Nature 322:687-689, 1986.
22. Germain, R.N. Cell 54:441-444, 1988.

23. Townsend, A., Rothbard, J., Gotch, F.M., Bahadur, G., Wraith, D. and McMichael, A.J. Cell 44:959-958, 1986. .
24. Carbone F.R. , Moore M.W. , Sheil, J.M. and Bevan, M.J. J. Exp. Med. 167:1767-1779, 1988.
25. Allen, P.M., Matsueda, G.R., Evans, R.J., Dunbar, J. B., Marshall, G.R. and Unanue, E.R. Nature 327:713:717-720, 1987.
26. Guillet, J.G., Lai, M.Z., Briner, T.J., Buus, S., Sette, A., Grey H.M., Smith, J.A. and Gefter, M.L. Science 235:865-870, 1987.
27. Davie, J. M. and Paul, W E. J. Exp. Med. 124:495-504, 1972.
28. Davie, J.M. and Paul, W.E. J. Exp. Med. 135:660-667, 1972 .
29. Roelants, G , Forni, L. and Pernis B. J. Exp. Med. 137:1060-1066, 1973.
30. Hammerling, G. and McDevitt, H. J. Immunol. 112:1726-1730, 1974.
31. Taniguchi, M. and Miller, J. F.A.P. J. Exp. Med. 146:1450-1454, 1977.
32. Lewis, G.K. and Goodman, J.W. J. Exp. Med. 148:915-921, 1978.
33. Bach, B. Benacerraf, B. and Greene, M. I. J. Immunol. 121:1460-1453,1978.
34. Siliciano, R.F., Keegan, A., Dintzis, R. Z., Dintzis, H.M. and Shin, H. S. J. Immunol. 135:906-914, 1985.
35. Siliciano, R.F., Colello, R. M., Keegan, A., Dintzis, R. Z., Dintzis, H. M. and Shin, H.S. J. Exp. Med. 162:768-773, 1985.
36. Siliciano, R. F., Hemesath, T. J., Pratt, J.C., Dintzis, R.Z., Dintzis, H.M., Acuto, O., Shin, H.S. and Reinherz, E.L. Cell 47:161-171, 1986.
37. Rao, A., Faas, S.J. and Cantor, H. Cell 36:889-895,1984.
38. Rao, A., Ko, W.W.-P., Faas, S.J. and Cantor, H. Cell 36:879-888, 1984.
39. Tan, K.-N., Datlof, B.M., Gilmore, J.A., Kronman, A.C., Lee, J.H., Maxam, A.M. and Rao, A. Cell 54:247-261, 1988.
40. Janeway, C.A. Jr. , Yagi, J., Conrad, P.J. , Katz, M.E. , Jones, B. and Vroegop, S. Immunol. Rev. 107:61-88, 1989.
41. White, J., Herman, A., Pullen, A.M., Kubo, R., Kappler, J.W. and Marrack, P. Cell 56:27-35, 1989.
42. Pullen, A.M., Kappler, J.W. and Marrack, P. Immunol. Rev. 107:125-139, 1989.
43. McDevitt, H.O. and Benacerraf, B. Adv. Immunol. 11:31-54, 1968.
44. Benacerraf, B. and McDevitt, H.O. Science 175:273-276, 1972.
45. Werdelin, O. J. Immunol. 129:1883-1888, 1982.
46. Rock, K.L., and Benacerraf, B. J. Exp. Med. 157:1618-1623, 1983.
47. Godfrey, W. L., Lewis, G.K. and Goodman, J.W. Mol. Immunol. 21:969-974, 1984.
48. Hansburg, D., Heber-Katz, E., Fairwell, T. and Appella, E. J. Exp. Med. 158:25-32, 1983.
49. Heber-Katz, E., Hansburg, D. and Schwartz, R.H. J. Mol. Cell. Immunol. 1:3-8, 1983.
50. Schwartz, R.H. Adv. Immunol. 3:237-244, 1985.
51. Ashwell, J.D. and Schwartz, R. H. Nature 320:176-179, 1986.
52. Babbitt, B.P., Allen, P.M., Matsueda, G., Haber, E. and Unanue, E. R. Nature 317:359-360, 1985.
53. Babbitt, B.P., Matsueda, G.R., Haber, E., Unanue, E.R. and Allen, P.M. PNAS 83:4509-4513, 1986.
54. Buus, S., Sette, A., Colon, S.M., Miles, C. and Grey, H.M. Science

235:1353-1358, 1987.
55. Chen, B.P. and Parham, P. Nature 337:743-745, 1989.
56. Bjorkman, P.J. , Saper, M.A. , Samraoui, B. , Bennett, W.S. , Strominger, J.L. and Wiley, D.C. Nature 329:506-512, 1987.
57. Bjorkman, P.J. , Saper, M.A. , Samraoui, B. , Bennett, W.S. , Strominger, J.L. and Wiley, D. C. Nature 329:512-518, 1987.
58. Buus, S., Sette, A. and Grey, H.M. Immunol. Rev. 98:115-114, 1987.
59. Schneck, J. , Maloy, W.L. , Coligan, J.E. and Margulies, D.H. Cell 56:47-55, 1989.
60. McCluskey, J ., Boyd, L.F. , Highet, P.F. , Inman, J. and Margulies, D.H. J. Immunol. 141:1451-1455, 1988.
61. Meuer, S.C., Hodgdon, J.C., Hussey, R.E., Protetis, J P, Schlossman, S. and Reinherz, E.L. J. Exp. Med. 158:988-997, 1983. .
62. Kaye, J. A, Porcelli, J.S., Tite, J., Jones, B. and Janeway, C.A. J. Exp. Med. 158:836-856, 1983.
63. Rosenberg, A.S. , Mizuochi, T. and Singer, A. J. Exp. Med. 168:33-45, 1988.
64. Mizuochi, T. and Singer, A. J. Immunol. 140:1421-1424, 1988.
65. Snell, G, and Stimpfling, J.H., In: **Biology of the Laboratory Mouse** (Ed. N.M. Green), McGraw-Hill, New York, 1966, p.p. 457.
66. Krangle, M.D., Orr, H.T. and Strominger, J.L. Cell 18:979-991, 1978.
67. Ploegh, H. L., Cannon, L.E. and Strominger, J.L. Proc. Natl. Acad. Sci. USA 76:2273-2276, 1979.
68. Kahn-Perles, B., Boyer, C., Arnold, B., Sanderson, A.R., Ferrier, P. and Lemonnier, F.A. J. Immunol. 127:916-921, 1987.
69. Kievits, F. and Ivanyi, P. Human Immunol. 20:115-121, 1987.
70. Rammensee, H.G., Robinson, P.J., Crisanti, A. and Bevan, M.J. Nature 319:502-504, 1986.
71. Martinko, J.M., Anderson, S.J. and Potter, T.A. Immunogenetics 23:64-69, 1986.
72. Potter, T.A., Boyer, C., Schmitt-Verhulst, A.M., Golstein, P., and Rajan, T.V. J. Exp. Med. 160:317-321, 1984.
73. Hansen, T. H., Myers, N. B. and Lee. D. J. Immunol. 140:3522-3527, 1988.
74. Allen, H., Fraser, J., Flyer, D., Calvin, S. and Flavell, R. Proc. Natl. Acad. Sci. USA 83:7447-7451, 1986.
75. Lippincott-Schwartz, J., Bonifacino, J.S., Yuan, L.C. and Klausner, R. D. Cell 54:209-220, 1988.
76. Townsend A , Ohlen C , Bastin J , Ljunggren HG , Foster, L. and Karre, K. Nature 340:443-448, 1989.
77. Machy, P., Barbet, J. and Leserman, L. Proc. Natl. Acad. Sci. USA 79:4184-4152, 1982.
78. Machy, P., Truneh, A., Gennaro, D. and Hoffstein, S. Nature 328:724-726, 1987.
79. Tse, D.B., Cantor, C.R., McDowell, J. and Pernis, B. J. Mol. Cell. Immunol. 2:315-329, 1986.
80. Tse, D.B. and Pernis, B. J. Exp. Med. 159:193-207, 1984.
81. Dasgupta , J.D., Watkins, S., Slayter, H. and Yunis, E.J. J. Immunol. 141:2577-2580, 1988.

82. Machy, P. and Truneh, A. Mol. Immunol. 26:687-696, 1989.
83. Blakely, A., Groman, K., Ostergaard, H., Torbett, B., Svoboda, K. and Clark W.R. J. Exp. Med. 166:1070-1077, 1987.
84. Kosznowski, V., Gething, M.J. and Waterfield, M. Nature 267:160-162, 1977.
85. Braciale, T.J. and Yap, K. L. J. Exp. Med. 147:1236-1244, 1978.
86. Hosaka, Y., Sasao, F. and Ohara, R. Vaccine 2:245-249, 1985.
87. Yewdell, J. W., and Bennink, J. R. Science 244:1072-1074, 1989.
88. Townsend, A. and Bodmer, H. Ann. Rev. Immunol. 7:601-624, 1989.
89. Kearney, J.F., Cooper, M.D., Klein J., Abney, E.R., Parkhouse, R.M.E. and Lawton, A.R. J. Exp. Med. 146:297-304, 1977.
90. Beller, D.I., Kiely, J.M. and Unanue, E.R. J. Immunol. 124:1426-1431, 1980.
91. Klareskog, L., Tjernlund, U.M., Forsum, U. and Peterson, P.A. Nature 26:248-250, 1977.
92. Daar, A.S., Fuggle, S.V., Fagre, J.W., Ting, A., and Morris, P.J. Transplantation 38:293-299, 1984.
93. Londei, M., Bottazzo, G.F. and Feldmann, M. Science 228:85-89, 1985.
94. Jones,P. P., Murphy, D.B. and McDevitt, H.O. J. Exp. Med. 148:925,934 1978.
95. Fathman, C.G. and Infante, P.D. Immunogenetics 8:577-580, 1979.
96. Fathman, C.G. and Kimoto, M. Immunol. Rev. 54:57-80, 1981.
97. Lotteau, V., Teyton, L., Burroughs, D. and Charron, D. Nature 329:339-341, 1987.
98. Germain, R.N. and Quill, H. Nature 320:72-75, 1986.
99. Germain, R.N., Bentley, D.M. and Quill, H. Cell 43:233-242, 1985.
100. Germain, R.N., Braunstein, N.S., Lechler, R.I., Miller, J., Ronchese, F., and Sant, A.J. In: **Processing and presentation of antigens** (Eds. B. Pernis, S. C. Silverstein and H. Vogel),Academic Press, New York, 1988, pp 99-122.
101. Sant, A.J., Schwartz, B.D.,and Cullen, S.E. J. Exp. Med. 158:1979-1992, 1983.
102. Sant, A.J., Cullen, S.E. and Schwartz, B.D. J. Immunol. 135:416-422, 1985.
103. Miller, J. and Germain, R.N. J. Exp. Med. 164:1478-1489, 1986.
104. Sekaly, R.P., Jacobson, S., Richert, J.R., Tonnelle, C., McFarland, H.F. and Long, E.O. Proc. Natl. Acad. Sci. USA 85:1209-1212, 1988.
105. Stockinger, B, Pessara, U., Lin, R.H., Habicht, J., Grez, M. and Koch, N. Cell 56:683-689, 1989.
106. Machamer, C.E. and Cresswell, P. J. Immunol. 129:2564-2569, 1982.
107. Cresswell, P. , and Blum, J. In: **Processing and presentation of antigens** (Eds. B. Pernis, S. C. Silverstein and H. Vogel), Academic Press, New York, 1988, pp. 43-51.
108. Kelner, D.K. and Cresswell, P. J. Immunol. 137:2632-2639, 1986.
109. Cressswell, P. Proc. Natl. Acad. Sci. USA 82:8188-8192, 1985.
110. Fox, B.S., Carbone, F.R., Germain, R.N., Paterson, Y. and Schwartz, R.H. Nature 331:538-540, 1988.
111. Shimonkevitz, R., Kappler, J., Marrack, P. and Grey H. M. J. Exp. Med. 158:303-310, 1983.

112. Shimonkevitz, R., Colon, S., Kappler, J., Marrack, P. and Grey H. M. J. Immunol. 133:2067-2071, 1984.
113. Swanson, J.A., and Silverstein, S.C. In: **Processing and presentation of antigens** (Eds. B. Pernis, S. C. Silverstein and H. Vogel), Academic Press, New York, 1988, pp 15-27.
114. Wileman, T., Harding, C.L. and Stahl, P. Biochem. J. 232:1-14, 1985.
115. Stahl, P. and Schwartz, A.L. J. Clin. Invest. 77:657-662, 1986.
116. Chestnut, R.W. and Grey, H.M. J. Immunol. 126:1075-1079, 1981.
117. Rock, K.L., Benacerraf, B. and Abbas, A.K. J. Exp. Med. 160:1102-1113, 1984.
118. Lanzavecchia, A. Nature 314:537-539, 1985.
119. Streicher, H., Berkower, I., Busch, M., Gurd, F.R.N. and Berzofsky, J.A. Proc. Natl. Acad. Sci. USA 81:6381-6385, 1984.
120. Berkower, I., Buckenmeyer, G.K. and Berzofsky, J.A. J. Immunol. 136:2498-2503, 1986.
121. Lee, P., Matsueda, G.R. and Allen, P.M. J. Immunol. 140:1063-1068, 1988.
122. Gotch, F., McMichael, A.J. and Rothbard, J. J. Exp. Med. 168:2045-2057, 1988.
123. Berkower, I., Kawamura, H., Matis, L.A. and Berzofsky, J.A. J. Immunol. 135:2628-2634, 1985.
124. Brett, S.J., Cease, K.B. and Berzofsky, J.A. J. Exp. Med. 168:357-373, 1988.
125. Gammon, G., Shastrin, N., Cogswell, J., Wilbur, S., Sadegh-Nasseri, S., Kzrych, U., Miller, A. and Sercarz, E. Immunol. Rev. 98:53-73, 1987.
126. Shastri, N., Miller, A. and Sercarz, E. J. Immunol. 136:371-376, 1986.
127. Wachsman, M. , Aurelian, L. , Smith, C.C. , Perkus, M.E. and Paoletti, E. J. Infect. Dis. 159:625-634, 1989.
128. Wachsman, M., Aurelian, L. , Hunter,. JC. , Perkus, M. E. and Paoletti, E. Bioscience Reports 8:323-34,1988.
129. Langman, R. E. **The Immune System**. Academic Press, NY, 1989.
130. Scott, P., Pearce, E, Natovitz, P and Sher, A. J. Immunol. 139:3118-3122, 1987.
131. Scott, P., Natovitz,P, Pearce, E., Sher, A., and Coffman, R.L. In: **Vaccines '88** (Eds. R. A. Lerner, H. Ginsberg, R.M. Chanock, and F. Brown), Cold Spring Harbor Press, 1989, 295-299.
132. Aurelian, L. , Yasumoto, S. and Smith, C.C. J. Virol. 62:2520-2524, 1988.
133. Yasumoto, S., Hayashi, Y. , Aurelian, L. J. Immunol. 139:2788-2793, 1987.
134. Gajewski, T. F. and Fitch, F. W. J. Immunol. 140:4245-4250. 1988.
135. Mossmann, T. R. Cherwinski, H. Bond, M. W., Giedlin, M. A. and Coffman, R. L. J. Immunol. 126:2348-2354, 1986.
136. Cherwinski, H. and Mossmann, T.R. J. Immunol. 138:3688-3692, 1987.
137. Jennings, S.R., Rice, P.I. Kloszewski, E.D., Anderson, R. W., Thompson, D.L. and Tevethia, S.S. J. Virol. 56:757-766, 1985.
138. Masucci, M.G., Torsteinsdottir, S., Colombarni, J., Brautbar, C., Klein, E. and Klein G. Proc. Natl. Acad. Sci. USA 84:4567-4571, 1987.

# 3

## MOLECULAR ASPECTS OF HUMAN B AND T LYMPHOCYTE RESPONSES TO HIV

R. F. SILICIANO

**Department of Medicine, Johns Hopkins University School of Medicine Baltimore MD 21205**

### ABSTRACT

The human immunodeficiency virus type 1 (HIV-1) induces readily detectable virus-specific B and T lymphocyte responses in humans. These responses may control the infection for various periods of time but ultimately fail to prevent the development of AIDS. This chapter reviews of the human immune response to HIV-1, focusing on those immune mechanisms that may exert a beneficial anti-viral effect and those that may contribute to the pathophysiology of HIV-1 infection by promoting the depletion of $CD4^+$ T lymphocytes. Particular emphasis is placed on a novel mechanism for the depletion of $CD4^+$ T cells which involves cytolytic T lymphocytes (CTL) specific for the HIV-1 envelope glycoprotein gp120.

### INTRODUCTION

Infection with the human immunodeficiency virus type 1 (HIV-1) (1,2) initiates a very complex series of host-virus interactions. The ultimate consequence of these interactions is that infected individuals become profoundly immunodeficient. The acquired immunodeficiency disease resulting from HIV-1 infection is characterized by numerous immunologic abnormalities, the most prominent of which are severe quantitative and qualitative defects in the $CD4^+$ T cell compartment (3,4). Immunodeficiency develops despite the presence of readily detectable B and T lymphocyte responses to HIV-1. In fact, in many ways, HIV-1 is a highly immunogenic virus. Virtually all infected individuals develop antibody responses to several of the protein products of the HIV-1 genome, notably the envelope glycoproteins (5,6). Even more striking is the finding that most infected individuals also have very high levels of virus-specific cytolytic T lymphocyte (CTL) activity (7,8). The frequency of HIV-specific CTL in the peripheral blood of seropositive individuals is so high that cytolytic activity can be detected in freshly isolated peripheral blood mononuclear cell preparations without any *in vitro* restimulation (7).

The presence of a vigorous anti-viral immune response, at least early in the course of HIV-1 infection, raises a number of important questions: To what extent do these responses control HIV-1 infection, and why are they ultimately unsuccessful? How do the multitude of immunologic abnormalities in infected individuals influence anti-HIV-1 responses, and how do anti-viral responses change with disease progression? Finally, to what extent do immune responses to HIV-1 contribute to the pathogenesis of immunodeficiency in AIDS patients? This chapter will consider the nature of B and T cell responses to HIV-1 in infected individuals in the context of the issues cited above. For a comprehensive discussion of the immune response to HIV-1, the reader is referred to several excellent reviews (3,4,9,10).

## ANTIBODY RESPONSES TO HIV-1 IN INFECTED INDIVIDUALS

### General Aspects.

Almost all HIV-1 infected individuals develop readily detectable circulating antibodies to one or more of the proteins derived from the *gag*, *pol*, and *env* genes of HIV-1 (5,6,11,12). The presence of these antibodies provides the basis for current diagnostic tests for HIV-1 infection (reviewed in 12). Antibodies to virtually all of the protein products of the HIV-1 genome have been detected in serum samples from infected individuals, but a given infected individual does not necessarily have antibodies to all of these proteins (13). For example, the current CDC guidelines for the interpretation of diagnostic Western blots recommend that the test be considered positive if bands for p24 or gp41 are present either alone or in combination with other characteristic bands.

There is considerable interest in the issue of the length of time between initial exposure to HIV-1 and the development of a readily detectable antibody response to HIV-1 (seroconversion). The interval can be as short as a few weeks (12). However, because of the propensity of HIV-1 to establish latent infection, this interval is in some cases prolonged. HIV-1 can be detected by culture methods in some individuals at risk for AIDS as long as 35 months prior to seroconversion (14). This prolonged interval may indicate that some critical threshold level of HIV-1 replication must be reached before seroconversion takes place, and that in some individuals this threshold is not reached for prolonged periods.

Seroconversion is sometimes associated with a mononucleosis-like illness (15). In such patients, IgM antibodies to HIV-1 appear about 5 days after the onset of the mononucleosis-like illness and decline to low levels after about 3 months. IgG antibodies appear about 11 days after onset and persist.

The central pathophysiologic feature of HIV-1 infection, the loss of $CD4^+$ T cells, first becomes detectable about the time of seroconversion in some infected individuals. The loss of $CD4^+$ T cells occurs over a period of several years during which time the infected individual remains seropositive for HIV-1 infection but is usually asymptomatic. The degree of loss of $CD4^+$ T cells is an excellent predictor of the time to onset of frank AIDS (16,17). There are several potential mechanisms for the loss of $CD4^+$ T cells including direct viral cytopathic effects (2), syncytium formation (18,19), and cytolytic mechanisms that are dependent on host B and T cell responses to HIV-1. These cytolytic mechanisms are discussed in detail below. Significantly, the number of $CD4^+$ T cells in the peripheral blood does not begin to decline before seroconversion, even in individuals who have been virus-positive (by culture) for prolonged periods (14). Thus, the development of a readily measurable immune response to HIV-1 coincides with the onset of the process of $CD4^+$ cell depletion. A question of critical importance is whether both events are independent consequences of HIV-1 infection or whether the immune response to HIV-1 contributes to $CD4^+$ cell depletion.

Some changes in the patterns of antibody responses are seen over the course of HIV-1 infection. In rare cases, formerly seropositive individuals have reverted to seronegativity (20). In some infected individuals, antibodies to the *gag*-related p24 protein become undetectable although antibodies to the *env* proteins persist (5,11,13). This pattern is often associated with the presence of detectable p24 antigen in the circulation, indicative of high levels of HIV-1 replication. The appearance of p24 antigen in the circulation is a good predictor of progression to AIDS (17).

Neutralizing antibodies.

Given that HIV-1 infection generally induces a vigorous humoral immune response, it is important to considerable whether this response exerts a beneficial anti-viral effect. This issue has been addressed in *in vitro* studies of the neutralization of HIV-1 by sera from infected individuals. Neutralization is typically measured by incubating a stock HIV-1 preparation with serum from an infected donor and then measuring residual infectivity of the treated virus preparation. Infectivity is quantitated by culturing susceptible cell lines in the presence of the treated virus preparation, with subsequent assay of syncytium formation, reverse transcriptase activity, or viral p24 antigen production in the infected culture. Another assay for neutralizing activity uses as a readout the inhibition of syncytium formation following co-cultivation of $CD4^+$ T cells and HIV-1 infected cells.

Several early studies demonstrated the presence of neutralizing antibodies in the sera of AIDS patients and individuals at risk for AIDS (21,22). In general, levels of neutralizing antibodies were low even when high levels of antibodies to HIV-1 envelope glycoproteins were present (22). Robert-Guroff *et al.* found that 60% of AIDS patients had neutralizing antibodies with titers ranging from 1/10-1/540 (21). Interestingly, levels of neutralizing antibodies were higher in patients with ARC and higher still in healthy seropositive individuals, a finding which has been confirmed in some (23) but not all (24) subsequent studies. For example, in a recent study of a large number of HIV-1 infected individuals, neutralizing activity was detected in 92% with mean titers of around 1/64 (24). Significantly, in this study, neutralizing titers correlated poorly with clinical course or with ELISA assays for antibodies reactive with disrupted HIV-1 virions. Longitudinal studies suggest that neutralizing antibodies may first appear several months after seroconversion and after the process of $CD4^+$ cell depletion has begun (23,25). Once induced, neutralizing antibodies may slow the rate of $CD4^+$ cell depletion (25).

Although antibodies to various HIV-1 proteins are present in infected individuals, it is expected that antibodies to the envelope glycoproteins would be of major importance in neutralizing viral infectivity *in vivo* since these antibodies should react with intact HIV-1 virions. In principle, neutralization might result from steric blocking of the interaction of HIV-1 with its cellular receptor CD4. It is also possible that neutralizing antibodies might block subsequent post-binding events involved in the fusion of the viral envelope with the plasma membrane. Some progress has been made recently in identifying epitopes recognized by neutralizing antibodies, but the mechanisms by which antibodies to these epitopes cause virus neutralization are for the most part unknown.

Epitopes recognized by neutralizing antibodies have been best defined in studies in which such antibodies are induced by immunization of experimental animals. Not surprisingly, the gp120 envelope protein has been the target antigen used to induce neutralizing antibodies in most of these studies. Immunization of various experimental animals with purified viral or recombinant gp120 has been shown to elicit neutralizing antibodies (23,26-31). In addition, the neutralizing activity present in sera of HIV-1 infected humans can be partially reduced by passing the sera over a gp120 affinity column, suggesting that the gp120 envelope protein is a target of at least some of the neutralizing antibodies induced during HIV-1 infection in humans (29; P. Berman, personal communication).

An intriguing difference between neutralizing antibodies found in the sera of HIV-1 infected humans and neutralizing antibodies induced by immunization of experimental animals with gp120 relates to the strain specificity of neutralization. Most strains of HIV-1 can be neutralized by sera from most HIV-1-infected humans (23,24,26). In contrast, anti-gp120 antisera raised in experimental animals are generally restricted to the immunizing strain in neutralizing activity (26,29,31). Similarly, murine monoclonal antibodies to gp120 which recognize an epitope contained within residues 279-472[1] have high-titer neutralizing activity which is type-specific (32). The ability of sera from HIV-1-infected humans to neutralize a wider range of HIV-1 strains than experimentally-induced antibodies may be a reflection of the fact that at least a portion of the neutralizing antibodies in such sera are directed at epitopes in the other envelope subunit, gp41, which shows much less sequence heterogeneity than gp120. Another explanation is that there are many distinct but related molecular clones of HIV-1 in each infected individual (33), so that the immunizing epitopes are more variable in natural infection.

Analysis of neutralization is complicated by the fact that some HIV-1 strains, including the standard IIIb laboratory strains, are more resistant to neutralization than most other strains (26). In addition, even subtle changes in gp120 structure can affect neutralization by such sera from infected individuals. In a recent study of the neutralization of distinct but related clones of HIV-1 derived from the IIIb isolate, sera from infected humans and gp120-immunized goats showed marked heterogeneity in their capacity to neutralize these closely related variants, suggesting that even a single amino acid change in gp120 can affect recognition by neutralizing antibodies (34).

A consistent feature of the neutralizing antibody response is that the level of neutralizing antibody is generally much lower than the level of antibodies reactive with the HIV-1 envelope proteins as determined by ELISA or Western blot. This suggests that much of the antibody response to the envelope proteins is directed at epitopes that are not directly involved in critical functions of these proteins. To understand the molecular nature of the neutralizing antibody responses, it is necessary to define the epitopes recognized by neutralizing antibodies. Several laboratories have recently identified a B cell epitope within gp120 which elicits antibodies that neutralize viral infectivity and/or inhibit virus-induced fusion. Rushe *et al.* (35) have shown that immunization of goats with recombinant gp160 elicits fusion-inhibiting antibodies. All of the fusion-inhibiting activity in the antisera can be absorbed

---

1) Amino acid numbering begins with the first Met in the *env* open reading frame. Unless otherwise stated, numbering refers to the BH10 strain of HIV-1.

by a synthetic peptide (RP135) representing residues 307-330 of gp120. This epitope is contained within a hypervariable region of gp120, and not surprisingly, fusion-inhibiting antibodies are isolate-restricted in this activity. Interestingly, fusion-inhibiting antibodies in the serum of an HIV-1 infected chimp also reacted with this peptide. Similar results were obtained by Parker and colleagues (36) who were able to elicit isolate-specific neutralizing antibodies in goats using as an immunogen the gp120 peptide 303-321 coupled to carrier proteins. Antibodies to this peptide were detected in 21% of sera from seropositive humans. Goudsmit and colleagues (37) have shown that HIV-infected chimps develope type-specific antibodies to an epitope in gp120 including residues 307-321. They suggest that this epitope, which appears to be associated with neutralization, is part of a loop between Cys residues at positions 296 and 331. This loop has a conserved $\beta$-turn flanked on either side by variable residues.

Recently, Ho and colleagues have identified another gp120 epitope recognized by neutralizing antibodies (38). Rabbit antisera to the gp120 peptide 247-267, which is in a conserved region of the molecule, neutralized three different HIV-1 isolates. Little antibody reactive with this peptide was found in sera of HIV-1 seropositive individuals, suggesting that this epitope is not immunogenic in the context of natural infection in humans. Interestingly, these antisera did not block binding of HIV-1 to $CD4^+$ cells. These results are consistent with the observations of Willey and colleagues (39) who found that *in vitro* mutagenesis of residues in the region 261-263 abolished the infectivity of the virus without altering *env* biosynthesis, binding to CD4, or ability to induce syncytia. It is possible that this neutralizing epitope is involved in some critical post-attachment event, possibly the interaction of gp120 with a second (as yet unidentified) cell surface receptor.

Kennedy and colleagues have identified another B cell epitope in gp120 which is recognized by neutralizing antibodies (40). A peptide representing the C-terminus of gp120 and the N-terminus of gp41 (gp160 residues 496-525) was used to generate rabbit antisera and a mouse monoclonal antibody. These antibodies reacted with gp120 and recognized a gp120 epitope present on the surface of infected cells. The rabbit antisera had weak neutralizing activity. Reactivity to this peptide was found in about a third of HIV-1-infected individuals.

Ho and colleagues have used a systematic approach to look for other linear neutralizing epitopes in the HIV-1 envelope glycoproteins. They synthesized a large series of peptides spanning much of the gp160 sequence (23). These peptides were used to immunize mice, and the resulting antisera were tested for neutralizing activity. Four peptides were found to induce low titer neutralizing antisera: 298-314, 458-484, 616-632, and 728-751. Whether any of the neutralizing antibodies in sera from infected humans were directed at these epitopes was not determined. Two of these epitopes (616-632 and 728-751) are located within the transmembrane envelope glycoprotein gp41. Other studies have also shown that certain antibodies to gp41 can neutralize HIV-1 *in vitro* (41,42). Kennedy and colleagues produced a rabbit antiserum to a conserved hydrophilic peptide from gp41 representing amino acids 735-752. This antiserum had low titer (1/32) neutralizing activity (41). Mouse monoclonal antibodies generated against the same peptide neutralized several different strains of HIV-1 *in vitro* (42). Sera from HIV-1 seropositive individuals contain antibodies reactive with this peptide. Recently, recombinant poliovirus vectors containing this sequence inserted into a surface accessible region of the poliovirus capsid protein have been generated (43). The recombinant poliovirus can absorb HIV-1 neutralizing antibody from HIV-seropositive sera, and, when inoculated intradermally into rabbits, can induce low titer (<80) neutralizing antibodies against divergent HIV-1 strains. It should be pointed out that the putative membrane-spanning hydrophobic domain of gp41 is contained within residues 683-705. The gp41 epitope described above (735-752) is C-terminal to this putative transmembrane domain but may nevertheless have an external location with respect to the viral envelope since the gp41 polypeptide chain may loop back outside the membrane in this region. Interestingly, some monoclonal antibodies to the p17 core protein are neutralizing (44). How antibodies to an internal virion component neutralize is unclear.

Another region of gp41 involved in the neutralizing antibody response has recently been identified by an immunoselection approach. *In vitro* culture of HIV-1 in the presence of neutralizing antiserum results in the outgrowth of clones resistant to neutralization by the antiserum used for the selection (45). The critical amino acid substitution responsible for resistance was shown to be an Ala to Thr replacement at position 582. This residue is located N-terminal to the transmembrane domain of gp41 and is therefore presumed to be have an exterior location. In addition, Oldstone and colleagues have identified an adjacent region of gp41 (residues 598-609 of LAV gp160) as a major B cell epitope (46). Sera from virtually all HIV-1-infected individuals contain antibodies reactive with a synthetic peptide representing this region of gp41. High titer rabbit antisera to this peptide are weakly neutralizing.

It is of interest to compare the location of epitopes recognized by neutralizing antibodies with those involved in critical functions of the HIV-1 envelope glycoproteins such binding to CD4 and subsequent fusion events. Regions of gp120 that appear to be involved in the binding of gp120 to CD4 have been identified in site-directed mutagenesis (47,48) and monoclonal antibody blocking studies (48,49). These residues are located in the C-terminal region of gp120 and do not overlap the major type-specific neutralization epitope (PB135) which is located closer to the center of the linear amino acid sequence of gp120. With respect to gp41, mutagenesis studies suggest that replacement of gp41 residues C-terminal to residue to the putative membrane spanning hydrophobic domain does not abolish syncytium forming ability of mutant envelope glycoprotein (47). It is therefore unclear why antibodies directed at epitopes C-terminal to the transmembrane domain cause neutralization. An important direction for future research is clearly a molecular analysis of how neutralizing antibodies actually inactivate HIV-1.

Another critical issue related to neutralizing antibodies is the question of why such antibodies, which are present in most seropositive individuals, fail to prevent disease progression. Neutralizing antibodies also appear to be insufficient to prevent HIV-1 infection in experimental animals. For example, preliminary experimental HIV-1 vaccines tested in primates have failed to prevent infection despite the fact that they induce neutralizing antibodies (50). Initial attempts to protect chimpanzees from HIV-1 infection by passive transfer of immune globulin from seropositive human donors with high titers of neutralizing antibodies have been also unsuccessful (51). There are several possible explanations for the failure of neutralizing antibodies to halt disease progression. One is that the virus can spread by direct cell-to-cell transmission. Recent work by Gupta and coworkers suggests HIV-1 infection can spread in an *in vitro* culture even in the presence of azidothymidine and neutralizing antibody (52). This process was inhibitable by anti-CD4 and may represent spread of genomic viral RNA from infected to uninfected cells during reversible CD4-dependent membrane fusion events. Another potential reason for the failure of neutralizing antibodies to eradicate HIV-1 infection is that the neutralization of viral infectivity *in vivo* is much more difficult than in *in vitro* assays because of the much higher concentrations of infectable target cells in *in vivo* sites such as within lymph nodes (Scott Layne, personal commun.) If this is the case, high titer neutralizing antibodies may be required to prevent infection and/or disease progression. A third explanation for the failure of neutralizing antibodies to prevent disease progression is the continuous emergence of neutralization-resistant variants. Although sera from infected individuals are not highly type-specific in neutralization, subtle changes in *env* sequence can dramatically affect neutralization by such sera (34). Late in the disease course, there is also most likely a declining capacity to mount *de novo* antibody responses to new antigenic determinants. Polyclonal B cell activation with concomitant hypergammaglobulinemia is a consistent part of the clinical picture of HIV-1 infection. At the same time, AIDS patients have greatly diminished antibody responses to novel and recall antigens both because of decreased T cell help and because of intrinsic B cell abnormalities (53). One might therefore expect that late in the disease course infected individuals would have a decreased capacity to generate antibodies to newly emergent, neutralization-resistant variants. Finally, the failure of anti-HIV-1 antibody to protect the host against disease progression may in part be due to the fact that such antibodies may in some cases facilitate the uptake of HIV-1 by

Fc-receptor positive cells such as monocytes and macrophages (54). Ennis and colleagues have shown that HIV-1 infection of the monocyte-like cell line U937 is enhanced by subneutralizing concentrations of sera from infected individuals. This effect appears to be mediated by the Fc receptors (FcRI and FcRII) expressed by these cells (54). Levy and colleagues have recently shown that antibody-mediated enhancement of HIV-1 infection of human peripheral blood macrophages involves the Fc receptor FcRIII but does not involve CD4 (55). These studies suggest that antibody-mediated enhancement is a deleterious consequence of the normal immune response to HIV-1. The *in vivo* significance of this phenomenom needs to be established.

Antibody-dependent effector mechanisms.

In addition to direct virus neutralization, certain classes and subclasses of anti-HIV-1 antibodies may in principle exert a beneficial anti-viral effect by activating cytotoxic mechanisms that destroy HIV-1-infected cells and thereby limit further virus production by those cells. Sera from HIV-1-infected humans do not mediate antibody-dependent, complement-mediated cytotoxicity (ACC) against HIV-1-infected target cells in the presence of rabbit or human complement, although under the same conditions sera from HIV-1-infected chimps will mediate ACC (56,57). However, sera from infected humans will mediate antibody-dependent, cell-mediated cytotoxicity (ADCC) against HIV-1-infected targets cells (57,58). In some (58,59) but not all (60) studies, the ADCC-promoting activity in sera from AIDS patients was lower than that in healthy seropositive individuals. The target antigen for this ADCC reaction appears to be gp120 since ADCC against $CD4^+$ targets cells that have been preincubated with purified gp120 is also observed (57). In addition, Koup and colleagues have recently used vaccinia vectors to express separately HIV-1 *env* and *gag* proteins in target cells. ADCC activity in sera from infected hemophiliacs was directly only at *env* determinants (60). Freshly isolated peripheral blood mononuclear cells (61) and purified $CD16^+$ cells (62) from infected individuals mediated gp120-specific, non-MHC restricted lysis of target cells expressing gp120, suggesting that *in vivo* arming of $CD16^+$ effector cells with anti-gp120 occurs. This gp120-specific ADCC provides a mechanism only for the destruction of HIV-1-infected cells *in vivo*, but also for the depletion of non-infected $CD4^+$ T cells that have taken up free gp120 by virtue of its high affinity binding to CD4 (57). The potential significance of this and other related cell-mediated cytolytic mechanisms for the destruction of $CD4^+$ cells is discussed below.

## HUMAN T CELL RESPONSES TO HIV

General aspects.

The analysis of human T lymphocyte responses to HIV-1 infection is complicated by the fact that infection produces not only a numerical depletion of one major subset of T lymphocytes, those which express CD4, but also a functional deficit in the remaining $CD4^+$ T cells. This functional deficit is particularly prominent in AIDS patients whose remaining $CD4^+$ T cells show poor proliferative responses to soluble protein antigens (63) and decreased *in vitro* cloning efficiency (64). Thus $CD4^+$ T cell responses to HIV-1 are difficult to evaluate in AIDS patients. This functional defect is present to a lesser extent in healthy seropositive individuals. Proliferative responses to disrupted HIV-1 virion preparations (65) and to purified gp120 (30) are observed in only a fraction of seropositive individuals. Some functional abnormalities in $CD8^+$ T cells are observed in infected individuals (64), but the $CD8^+$ T cell response to HIV-1 is very readily measured in the form of an HIV-1-specific, class I-restricted cytolytic activity (7,8). $CD4^+$ and $CD8^+$ T cell responses to HIV-1 are considered separately in the discussion below. Additional issues related to the human T cell response to HIV-1, including the delineation of T cell epitopes in HIV-1 proteins, the effects of HIV-1 sequence variability on T cell recognition, and the role of $CD4^+$ CTL in the pathogenesis of AIDS are discussed.

$CD4^+$ T cell responses to HIV-1: Proliferative responses and IL2 production.

Following antigen-induced activation, $CD4^+$ T cells release lymphokines which promote the proliferation and differentiation of antigen-activated B cells and $CD8^+$ CTL. The presence of an IgG antibody response and a $CD8^+$ CTL response to HIV-1 in infected individuals is presumptive evidence for the presence of activated HIV-1-specific $CD4^+$ T cells, at least early in the course of HIV infection. $CD4^+$ T cell responses are generally measured in proliferation assays. However, as discussed above, antigen-specific proliferative responses are often lost as the disease progresses. Nevertheless, HIV-1-specific proliferative responses can be observed in some infected individuals using as a stimulus HIV-1 virions (65), proteins (65) and synthetic peptides (46,66). For example, 24% of seropositive individuals have *in vitro* proliferative responses to a synthetic peptide representing amino acids 584-609 of gp41 of Z3 strain of HIV-1 (46). In a more extensive survey of proliferative responses to 21 synthetic peptides derived from the HIV-1 gp120, gp41, p24, and pol proteins, Schrier and colleagues found that 90% of infected individuals had proliferative responses to at least one of these HIV-1 peptides (66). No single peptide induced responses in more than 50% of those tested, perhaps reflecting the influence of HLA genotype on T cell responsiveness to particular epitopes.

Other methods have been used to detect $CD4^+$ T cell responses to HIV-1 in infected individuals. Clerici *et al.*, have shown that peripheral blood mononuclear cell (PBMC) cultures from about half of HIV-1-infected individuals will secrete IL2 in response to recall antigens such as influenza A virus or tetanus toxoid even though proliferative responses to the same antigens are undetectable in most of these donors (67). In donors whose $CD4^+$ T cell function is measurable by IL2 production, it is possible to assess IL2 production in response to various synthetic peptides from gp160. In this group, 85-90% of individuals were found to be responsive to at least one of four gp160 peptides tested.

The data cited above provide presumptive evidence for a $CD4^+$ T cell response to HIV-1 proteins in infected individuals. Definitive analysis of this response will require the isolation and characterization of HIV-1-specific human T cell clones. In an effort to understand the initial $CD4^+$ T cell response to HIV-1 infection, we have recently isolated a series of T cell clones specific for HIV-1 gp120 (68). These gp120-specific clones were derived by *in vitro* stimulation of PBMC from seronegative donors with recombinant gp120 followed by soft agar cloning. As expected from immunization with a soluble protein antigen, all of the clones we obtained belonged to the $CD4^+$ subset. As is discussed below, the characterization of these clones has provided insights into a number of aspects of the CD4$^+$ T cell response to HIV-1 infection including the location of T cell epitopes in HIV-1 proteins, the effects of HIV-1 sequence variability on T cell recognition, and the role of CD4$^+$ CTL in the pathogenesis of AIDS.

Delineation of T cell epitopes in HIV-1 proteins: identification of epitopes in gp120.

There is currently a great deal of interest in identifying T cell epitopes in HIV-1 proteins. Elegant studies in the murine system have demonstrated that the antigenic sites recognized by individual T cells specific for particular protein antigens consist of restricted regions of linear amino acid sequence (69,70,71). This finding reflects the fact that the association of peptide fragments of protein antigens with MHC restriction elements is necessary for T cell recognition (72-74). The stretches of linear sequence recognized by T cells (herein referred to as T cell epitopes) generally consist of a minimum of 7 but often 8-15 contiguous amino acids. T cell epitopes in a number of HIV-1 proteins have been identified (46,66-68,75,76). Rather than reviewing such studies in detail, the subsequent discussion will focus on the issue of whether T cell epitopes can be reliably predicted from primary amino acid sequence. Several methods for predicting T cell epitopes have been reported, including algorithms based on the presence of amphipathic $\alpha$-helical propensity (77), general sequence patterns (78), or MHC allele-specific consensus sequences (79). However the reliability of such methods for predicting human T cell epitopes is not yet well established since only a small number of human T cell epitopes have been identified.

Using gp120-specific human T cell clones, we have recently described epitopes within the HIV-1 gp120 (68). By fractionation of proteolytic digests of gp120 and further analysis with synthetic peptides, we have definitively identified a 14 amino acid epitope (residues 416-429) which is recognized by human T cell clones restricted to DR4. We have also shown that a spatially distinct gp120 epitope is recognized by DR2-restricted clones. Therefore, T cell clones restricted by different *allelic* forms of a given MHC restriction element can recognize spatially distinct regions of gp120. This finding is of obvious importance with respect to the design of synthetic peptide vaccines since it indicates that individuals with different MHC class II genotypes may respond differently to particular peptide fragments of gp120.

The precise delineation of the DR4-restricted epitope has allowed us to compare its structure with that of T cell epitopes defined for several model protein antigens in the murine system and for a small number of viral and bacterial proteins in the human system. Berzofsky *et al.* (77) have noted that many T cell epitopes have the appropriate periodicity of hydrophilic and hydrophobic residues to form an amphipathic $\alpha$-helix. It is of interest that the DR4-restricted gp120 epitope 416-429 partially overlaps a region of gp120 with a high degree of amphipathic $\alpha$-helical character shown by Cease *et al.* (75) to represent a class II-restricted murine T cell epitope (residues 421-436). However, the DR4-restricted gp120-specific clones were completely unresponsive to synthetic peptide 419-444 which encompasses this amphipathic region. We therefore feel that while predictive algorithms are potentially useful, definitive identification of T cell epitopes requires direct experimental delineation. Similar conclusions may be drawn from the analysis of proliferative responses to panels of HIV-1 peptides. Shrier *et al.*, have observed proliferative responses to a large number of such peptides, none of which conform to the amphipathic $\alpha$-helix or general sequence pattern algorithms (66).

Effect of HIV-1 genomic heterogeneity in $CD4^+$ T cell recognition of HIV-1.

The delineation of a DR4-restricted epitope also enabled us to examine in detail the effects of naturally occurring sequence variation in HIV-1 on the recognition of HIV-1 antigens by human T lymphocytes. The DR4-restricted epitope described above is not located in one of the more variable regions of gp120. Rather it is located near a component of the CD4 binding site (47,48) which is likely to be under strong selective pressure to maintain structural features essential for CD4 binding. Nevertheless, we found that the sequence variation that does occur at certain positions within this epitope generally has a strong negative effect on recognition by T cells and can often completely abrogate recognition. Since even a single amino acid substitution can dramatically reduce recognition by a given T cell clone, the minor degree of

HIV-1 sequence heterogeneity which occurs within a given infected individual (33) may be sufficient to allow the virus to evade an ongoing T cell response via the selection of variants with sequence changes in T cell epitopes. With regard to vaccine development, our results indicate that immunization with gp120 from a given viral strain is likely to give rise to T cell clones which recognize the gp120 molecules of most other strains poorly and which completely fail to recognize gp120 from some strains.

We have recently found that HIV-1 genomic heterogeneity can affect T cell recognition at two levels: (1) the binding of processed HIV-1 peptides to MHC molecules on antigen-presenting cells (APC) and (2) the subsequent recognition of peptide-MHC complexes by T cell receptors (Callahan, Fort, Obah and Siliciano, manuscript submitted). Using a peptide competition assay, we have surveyed gp120 sequences from a large number of HIV-1 strains for the capacity to be recognized in the context of a single human MHC molecule, DR4. For many strains, variation in the DR4-restricted epitope was sufficient to alter the interaction of antigen receptors on gp120-specific human T cell clones with peptide-DR4 complexes on APC. Only in the case of two highly divergent isolates was the natural variation sufficient to prevent the critical initial interaction between the relevent gp120 peptides and DR4.

CTL responses to HIV-1: $CD8^+$ CTL.

CTL play an important role in recovery from many viral infections (80), including retroviral infections (81). Thus there has been considerable interest in the nature of human CTL responses to HIV-1 and their potential role in controlling HIV-1 infection. HIV-1-specific CTL have been observed in healthy seropositive individuals and in AIDS patients (5,6,82). Cytolytic activity specific for the HIV-1 envelope proteins has been detected in freshly isolated, unfractionated PBMC from healthy seropositive individuals and AIDS patients (5,61,82). While a portion of this lytic activity may represent non-MHC restricted ADCC, there is clearly an MHC restricted component mediated by *env*-specific CTL (5,82). This is a remarkable finding given that virus-specific CTL responses are generally demonstrable only after an *in vitro* restimulation step which is needed to expand specific clones. Lymphocytes obtained from broncho-alveolar lavage of seropositive patients with lymphocytic alveolitis also show significant cytolytic activity towards autologous alveolar macrophages which were shown to be HIV-infected (8). These lung-derived CTL also lysed mouse mastocytoma cells transfected with the HIV-1 *env* genes and HLA-A2, indicating that at least a portion of the response is *env*-specific (8,83). Freshly isolated PBMC from eight of ten seropositive individuals were also shown to contain CTL activity directed at the HIV-1 reverse transcriptase (RT) (84). *Gag*-specific, class I-restricted CTL activity has been observed in PBMC from a high percentage of healthy seropositive individuals, although in some cases detection required an *in vitro* restimulation in the presence of autologous lymphoblasts treated with PHA to induce HIV-1 gene expression (61,85). Cytolytic activity towards targets expressing the HIV-1 *vif* and *nef* gene products has also been recently reported (61). In general, the HIV-1-specific CTL obtained from seropositive individuals belong to the $CD8^+$ subset and are MHC class I-restricted, although a set HIV-1-specific CTL cloned from the cerebrospinal fluid of AIDS patients with neurological involvement included both $CD4^+$ and $CD8^+$ clones (86). Human $CD8^+$ CTL clones characterized thus far recognize diverse strains of HIV-1, suggesting that a conserved epitope is involved (82). In contrast, murine $CD8^+$ CTL clones specific for HIV-1 gp120 in the context of $H\text{-}2D^d$ recognize an epitope contained within amino acids 308-322 (76). This epitope overlaps the major type-specific neutralizing antibody epitope described above. It has been suggested that the high degree of variability observed in this region of gp120 reflects selection of variants resistant to the effects of neutralizing antibodies and/or virus-specific CTL (76).

The presence of $CD8^+$, HIV-specific CTL in infected individuals raises the critical issue of whether such CTL exert a beneficial anti-viral effect. It is possible that such CTL play an important role in slowing down the spread of HIV-1 infection. CTL isolated from infected individuals have been shown to lyse infected macrophages (8,83). This is of particular importance given recent evidence suggesting that such cells represent a reservoir for HIV-1 in infected individuals (87,88). $CD8^+$ T cells from HIV-1 infected donors have been shown to suppress virus replication in infected cultures (89,90). The frequency of HIV-specific CTL has been shown in one study to decline as the clinical and immunological status of the patients deteriorated (91). Although further longitudinal studies will be needed to determine whether this correlation has pathophysiologic significance, it is quite likely that the $CD8^+$ HIV-specific CTL response exerts a significant anti-viral effect *in vivo* and is therefore beneficial to the host. The same cannot be said for the $CD4^+$ CTL response. As is discussed below, $CD4^+$ CTL may exert some anti-viral effect but may also contribute to $CD4^+$ cell depletion.

<u>$CD4^+$ CTL responses to HIV-1: Protective anti-viral effects and a role in $CD4^+$ T cell depletion.</u>

In considering the nature of CTL responses which may be generated in response to HIV-1 infection, it is important to note CTL may be derived from the $CD4^+$ subset as well as the $CD8^+$ subset. Although $CD4^+$ T cells are often referred to as belonging to the "helper-inducer" subset, this subset is actually comprised of cells which are heterogeneous in terms of effector functions such as lymphokine secretion and cytolysis, but which all share an MHC class II-restricted antigen recognition pattern (92,93). Virus-specific $CD4^+$ CTL have been observed in a number of human diseases. Kaplan *et al.*, isolated a series of $CD4^+$ CTL clones specific for influenza virus from PBMC of an influenza-immune donor (94). Jacobson and colleagues have cloned from PBMC of a patient with multiple sclerosis a series of $CD4^+$ CTL specific for measles virus (95). Yasukawa and Zarling have isolated from herpes simplex virus (HSV) Type-1 seropositive donors a series of $CD4^+$ CTL clones which lysed autologous EBV-transformed B cell lines infected with HSV-1 (96). Celis *et al.*, have isolated $CD4^+$ CTL specific for Hepatitis B surface antigen (HBSAg) from recipients of the HBSAg vaccine (97). In addition, a subset of the $CD4^+$ T cell clones specific for gp120 which we have recently described have antigen-specific cytolytic activity (68), and similar HIV-1-specific $CD4^+$ CTL have been observed in AIDS patients (86, S. Koenig, personal communication).

It is important to note that there are significant differences in the antigen processing pathways involved in antigen recognition by $CD4^+$ and $CD8^+$ CTL. Most virus-specific CTL are derived from the $CD8^+$ T cell subset, and they recognize processed peptide fragments of viral antigen in association with class I MHC gene products expressed on infected cells (98). The precise mechanisms by which viral proteins synthesized in infected cells are processed for association with class I MHC gene products is unclear. However, as a general rule, processing for association with class I MHC gene products occurs only when the protein is synthesized within the cytoplasm of an infected cell (99,100). Viral proteins taken up by an endocytic pathway are generally processed for association with class II MHC gene products, and only in very rare instances have $CD8^+$ CTL been generated in respones to soluble proteins antigens.

In light of the work cited above, it is not surprising that soluble gp120 elicits a $CD4^+$, gp120-specific CTL response (68), and it is important to consider the effects of such a response on the course of HIV-1 infection. The $CD4^+$ phenotype of such CTL and their associated class II MHC restriction specificity indicates that the lytic activity of such clones will be focused towards class II-expressing targets cells which in the human include monocytes, macrophages, B cells, and activated T cells. We have shown that $CD4^+$, gp120-specific CTL can lyse autologous monocytes, B lymphoblastoid cells, and activated T cells that have taken up free exogenous gp120 (68). Whether such CTL would serve a protective function *in vivo* depends upon whether HIV-1 infected monocytes, macrophages, and activated T cell would process endogenously synthesized gp120 for association with class II MHC gene products. Recently, we have shown that gp120-specific $CD4^+$ CTL clones can lyse autologous B lymphoblastoid targets infected with recombinant vaccinia vectors carrying the HIV-1 *env* gene (Koenig, Fauci, Moss, and Siliciano, unpublished results). We therefore expect that HIV-infected monocytes, macrophages, and activated T cells should display gp120 epitopes in association with their class II molecules permitting their elimination by gp120-specific CTL *in vivo*. Given that the activation of $CD4^+$ T cells induces both HIV-1 replication (101,102) and class II expression (103,104), these gp120-specific, class II-restricted CTL may be of substantial value in limiting the spread of infection. The lysis of infected macrophages would also be significant in light of the possibility that such cells serve as a reservoir for the virus (87,88). However, the lysis of infected macrophages may also decrease the number of cells capable of presenting HIV-1 antigens and may thereby limit the generation of additional anti-viral effector T cells.

While the lysis of infected T cells and macrophages may be important for host control of HIV-1 infection, we have also found that the same immune effector mechanism may also lead to the destruction of non-infected $CD4^+$ T cells that have taken up free gp120. The noncovalent association between gp120 and gp41 is readily disrupted, and gp120 can thus be shed from infected cells (105,106). Shed gp120 would then be expected to bind to $CD4^+$ T cells by virtue of the high affinity gp120-CD4 interaction. We have shown that free gp120 can bind to CD4 on activated human T cells and then be internalized and processed for association with class II molecules (68). We have also demonstrated that gp120-specific $CD4^+$ CTL can readily lyse any activated autologous $CD4^+$ T cells that have taken up and processed free gp120, regardless of whether or not the target cells are infected with HIV-1. This cytolytic reaction proceeds even when target cells are exposed to very low concentrations of gp120 ($<10^{-10}$M). Thus one of the most significant consequences of the existence of $CD4^+$, gp120-specific CTL may be the involvement of such cells in the fundamental pathophysiologic feature of AIDS, the severe depletion the $CD4^+$ subset of T lymphocytes.

The elucidation of this novel pathway for the depletion of non-infected $CD4^+$ T cells was dependent upon the discovery that T cells can process protein antigens that have bound with high affinity to a T cell surface structure (68). Similar observations have been recently made by Lanzavecchia and colleagues (107). Blocking studies with anti-CD4 and soluble CD4 have demonstrated that T cell uptake of gp120 is totally dependent upon an initial interaction between gp120 and CD4. Activated $CD8^+$ T cells can readily present peptide fragments of gp120 to gp120-specific CTL. However, $CD8^+$ cells lack a specific focusing mechanism for taking up intact gp120 and therefore are not generally susceptible to destruction by $CD4^+$, gp120-specific CTL.

The *in vivo* significance of this CTL-mediated pathway for $CD4^+$ cell destruction relative to other proposed pathways for $CD4^+$ cell destruction such as lytic infection and syncytium formation remains to be established. However, given the fact that only a very small fraction of T lymphocytes in AIDS patients express viral RNA (108) and the fact that syncytia have not been well demonstrated *in vivo*, this gp120-specific cytolytic mechanism warrants serious consideration. One important factor is clearly the amount of gp120 shed in infected individuals. In this regard, it should be noted that the autocytolytic mechanism described here can operate even when the concentration of gp120 is below the Kd of the gp120-CD4 interaction, indicating that uptake and processing of only a small number of gp120 molecules may be sufficient to render a T cell susceptible to lysis. Another factor is the level of circulating anti-gp120 antibodies and the effects of these antibodies on gp120 uptake by T cells. In this regard, it should be pointed out that these antibodies may also contribute to the destruction of $CD4^+$ through an ADCC mechanism (57). This would require the presence of intact gp120 on the surface of the $CD4^+$ T cell. Our data indicate that bound gp120 can be internalized and processed within the T cell to expose peptide epitopes, thereby rendering the cell susceptible to cytolysis by gp120-specific CTL for a prolonged period of time.

The pathway for $CD4^+$ cell depletion described above is of particular interest because it accounts for the depletion of precisely those clonal T cell specificities that are directed at ubiquitous opportunistic pathogens such as *Candida albicans* and at pathogens that coinfect HIV-1-infected individuals such as cytomegalovirus, Epstein-Barr virus, and herpes simplex virus. Given the findings presented above, one can imagine that activated $CD4^+$ T lymphocytes responding appropriately to such organisms may be selectively destroyed by gp120-specific, class II-restricted CTL as a consequence of gp120 uptake by their own surface CD4 molecules. This unique type of "innocent bystander" effect may delete critical helper T cell clonal specificities thus exposing the host to a myriad of opportunistic infections.

## REFERENCES


1. Barre-Sinoussi, F., Chermann, J.C., Rey, F., Nugeyre, M.T., Chamaret, S., Gruest, J., Dauguet, C., Axler-Blin, C., Vezinet-Brun, F., Rouzinouz, C., Rozenbaum, W. and Montagnier, L. Science 220:868-870, 1983.
2. Popovic, M., Sarngadharan, M.G., Read, E. and Gallo, R.C. Science 224:500-503, 1984.
3. Shaw, G.M., Wong-Staal, F. and Gallo, R.C. In: **AIDS, Second Edition** (Eds. V.T. DeVita, Jr., S. Hellman and S.A. Rosenburg), J.P. Lippincott, Philadelphia, 1988, pp 11-31.
4. Fauci, A.S. Science 239:617-622, 1988.
5. Barin, F., McLane, M.F., Allan, J.S., Lee, T.H., Groopman, J.E. and Essex, M. Science 228:1094-1096, 1985.
6. Allan, J.S., Coligan, J.E., Barin, F., McLane, M.F., Sodroski, J.G., Rosen, C.A., Haseltine, W.A., Lee, T.H. and Essex, M. Science 228:1091-1093, 1985.
7. Walker, B.D., Chakrabarti, S., Moss, B., Paradis, T.J., Flynn, T., Durno, A.G., Blumberg, R.S., Kaplan, J.C., Hirsch, M.S. and Schooley, R.T. Nature 328:345-348, 1987.
8. Plata, F., Autran, B., Martins, L.P., Wain-Hobson, S., Raphael, M., Mayaud, C., Denis, M., Guillon, J.M. and Debre, P. Nature 328:348-351, 1987.
9. Ho, D.D., Pomerantz, R.J. and Kaplan, J.C. New Engl. J. Med. 317:278-286, 1987.
10. Koenig, S. and Fauci, A.S. In: **AIDS, Second Edition** (eds. V.T. DeVita, Jr., S. Hellman and S.A. Rosenburg), J.P. Lippincott, Philadelphia, 1988, pp 61-77.
11. Pan, L.Z., Cheng-Mayer, C. and Levy, J.A. J. Infect. Dis. 155:626-632, 1987.
12. Sandler, S.G., Dodd, R.Y. and Fang, C.T. In: **AIDS, Second Edition** (eds. V.T. DeVita, Jr., S. Hellman and S.A. Rosenberg), J.B. Lippincott Co., Philadelphia, 1988, pp 121-136.
13. Franchini, G., Robert-Guroff, M., Aldovini, A., Kan, N.C. and Wong-Staal, F. Blood 69:437-441, 1987.
14. Imagawa, D.T., Lee, M.H., Wolinsky, S.M., Sano, K., Morales, F., Kwok, S., Sninsky, J.J., Nishanian, P.G., Giorgi, J., Fahey, J.L., Dudley, J., Visscher, B.R. and Detels, R. New Engl. J. Med. 320:1458-1462, 1989.
15. Cooper, D.A., Imrie, A.A. and Penny, R. J. Infect. Dis. 155:1113-1118, 1987.
16. Polk, B.F., Robin, F., Brookmeyer, R., Kanchanaraksa, S., Kaslow, R., Visscher, B., Rinaldo, C. and Phair, J. New Engl. J. Med. 316:61-66, 1987.
17. Moss, A.R., Bacchetti, P., Osmond, D., Krampf, W., Chaisson, R.E., Stites, D., Wilber, J., Allain, J.P. and Carlson, J. Brit. Med. J. 296:745-750, 1988.
18. Lifson, J.D., Reyes, G.R., McGrath, M.S., Stein, B.S. and Engleman, E.G. Science 232:1123-1127, 1986.
19. Sodroski, J., Goh, W.C., Rosen, C., Campbell, K. and Haseltine, W.A. Nature 322:470-474, 1986.

20. Farzadegan, H., Polis, M.A., Wolinsky, S.M., Rinaldo, C.R. Jr., Sninsky, J.J., Kwok, S., Griffith, R.L., Kaslow, R.A., Phair, J.P., Polk, B.F. et al., Ann. Intern. Med. 108:785-790, 1988.
21. Robert-Guroff, M., Brown, M. and Gallo, R.C. Nature 316:72-75, 1985.
22. Weiss, R.A., Clapham, P.R., Cheingsong-Popov, R., Dalgleish, A.G., Carne, C.A., Weller, I.V.D. and Tedder, R.S. Nature 316:69-72, 1985.
23. Ho, D.D., Sarngadharan, M.G., Hirsch, M.S., Schooley, R.T., Rota, T.R., Kennedy, R.C., Chanh, T.C. and Sato, V.L. J. Virol. 61:2024-2028, 1987.
24. Prince, A.M., Pascual, D., Kosolapov, A.B., Kurokawa, D., Baker, C. and Rubinstein, P. J. Infect. Dis. 156:268-272, 1987.
25. Sawyer, L.A., Katzenstein, D.A., Hendry, R.M., Boone, E.J., Vujcic, L.K., Williams, C.C., Quinnan, G.V. Jr., Zeger, S.L. and Saah, A.J. In: **Vaccines 89** (eds. R.A. Lerner, H. Ginsberg, R.M. Chanock and F. Brown), Cold Spring Harbor Laboratory, Cold Spring Harbor, NY, 1989, pp 163-166.
26. Weiss, R.A., Clapham, P.R., Weber, J.N., Dalgleish, A.G., Lasky, L.A. and Berman, P.W. Nature 324:572-574, 1986.
27. Lasky, L.A., Groopman, J.E., Fennie, C.W., Benz, P.M., Capon, D.J., Dowbenko, D.J., Nakamura, G.R., Nunes, W.M., Renz, M.E. and Berman, P.W. Science 233:209-212, 1986.
28. Robey, W.G., Arthur, L.O., Matthews, T.J., Langlois, A., Copeland, T.D., Lerche, N.W., Oroszlan, S., Bolognesi, D., Gilden, R.V. and Fischinger, P.J. Proc. Natl. Acad. Sci. USA 83:7023-7027, 1986.
29. Matthews, T.J., Langlois, A.J., Robey, W.G., Chang, N.T., Gallo, R.C., Fischinger, P.J. and Bolognesi, D.P. Proc. Natl. Acad. Sci. USA 83:9709-9713, 1986.
30. Krohn, K., Robey, W.G., Putney, S., Arthur, L., Nara, P., Fischinger, P., Gallo, R.C., Wong-Staal, F. and Ranki, A. Proc. Natl. Acad. Sci. USA 84:4994-4998, 1987.
31. Nara, P.L., Robey, W.G., Pyle, S.W., Hatch, W.C., Dunlop, N.M., Bess, J.W. Jr., Kelliher, J.C., Arthur, L.O. and Fischinger, P.J. J. Virol. 62:2622-2628, 1988.
32. Thomes, E.K., Weber, J.N., McClure, J., Clapham, P.R., Singhal, M.C., Shriver, M.K. and Weiss, R.A. AIDS 2:25-29, 1988.
33. Saag, M.S., Hahn, B.H., Gibbons, J., Li, Y., Parks, E.S., Parks, W.P. and Shaw, G.M. Nature 334:440-444, 1988.
34. Looney, D.J., Fischer, A.G., Putney, S.D., Rusche, J.R., Redfeild, R.R., Burke, D.S., Gallo, R.C. and Wong-Staal, F. Science 241:357-359, 1988.
35. Rushe, J.R., Javaherian, K., McDanal, C., Petro, J., Lynn, D.L., Grimaila, R., Langlois, A.J., Gallo, R.C., Arthur, L.O., Fischinger, P.J., Bolognesi, D.P., Putney, S.D. and Mathews, T.J. Proc. Natl. Acad. Sci. USA 85:3198-3202, 1988.
36. Palker, T.J., Clark, M.E., Langlois, A.J., Mathews, T.J., Weinhold, K.J., Randall, R.R., Bolognesi, D.P. and Haynes, B.F. Proc. Natl. Acad. Sci. USA 85:1932-1936, 1988.
37. Goudsmit, J., Debouk, C., Meloen, R.H., Smit, L., Bakker, M., Asher, D.M., Wolff, A.V., Gibbs, C.J. Jr. and Gajdusek, D.C. Proc. Natl. Acad.

Sci. USA 85:4478-4482, 1988.
38. Ho, D.D., Kaplan, J.C., Rackauskas, I.E. and Gurney, M.E. Science 239:1021-1023, 1988.
39. Willey, R.L., Smith, D.H., Lasky, L.A., Theodore, T.S., Earl, P.L., Moss, B., Capon, D.J. and Martin, M.A. J. Virol. 62:139-147, 1988.
40. Kennedy, R.C., Dreesman, G.R., Chanh, T.C., Boswell, R.N., Allan, J.S., Lee, T.H., Essex, M., Sparrow, J.T., Ho, D.D. and Kanda, P. J. Biol. Chem. 262:5769-5774, 1987.
41. Chanh, T.C., Dreesman, G.R., Kanada, P., Linette, G.P., Sparrow, J.T., Ho, D.D. and Kennedy, R.C. EMBO J. 5:3065-3071, 1986.
42. Dalgleish, A.G., Chanh, T.C., Kennedy, R.C., Kanda, P., Clapham, P.R. and Weiss, R.A. Virology 165:209-215, 1988.
43. Evans, D.J., McKeating, J., Meredith, J.M., Burke, K.L., Katrak, K., John, A., Ferguson, M., Minor, P.D., Weiss, R.A. and Almond, J.W. Nature 339:385-388, 1989.
44. Papsidero, L.D., Sheu, M. and Ruscetti, F. J. Virol. 63:267-272, 1989.
45. Reitz, M.S., Wilson, C., Naugle, C., Gallo, R.C. and Robert-Guroff, M. Cell 54:57-63, 1988.
46. Schrier, R.D., Gnann, J.W., Langlois, A.J., Shriver, K., Nelson, J.A. and Oldstone, M.B.A. J. Virol. 62:2531-2536, 1988.
47. Kowalski, M., Potz, J., Basiripour, L., Dorman, T., Goh, W.C., Terwilliger, E., Dayton, A., Rosen, C., Haseltine, W. and Sodroski, J. Science 237:1351-1355, 1987.
48. Lasky, L.A., Nakamura, G., Smith, D.H., Fennie, C., Shimasaki, C., Patzer, E., Berman, P., Gregory, T. and Capon, D.J. Cell 50:975-985, 1987.
49. Dowbenko, D., Nakamura, G., Feenie, C., Shimasaki, C., Riddle, L., Harris, R., Gregory, T. and Lasky, L. J. Virol. 62:4703-4711, 1988.
50. Berman, P.W., Groopman, J.E., Gregory, T., Clapham, P.R., Weiss, R.A., Ferriani, R., Riddle, L., Shimasaki, C., Lucas, C., Lasky, L. and Eichberg, J.W. Proc. Natl. Acad. Sci. USA 85:5200-5204, 1988.
51. Prince, A.M., Horowitz, B., Baker, L., Shulman, R.W., Ralph, H., Valinsky, J., Cundell, A., Brotman, B., Boehle, W., Rey, F. et al. Proc. Natl. Acad. Sci. USA 85:6944-6948, 1988.
52. Gupta, P., Balachandran, R., Ho, M., Enrico, A. and Rinaldo, C. J. Virol. 63:2361-2365, 1989.
53. Lane, H.C., Masur, H., Edgar, L.C., Whalen, G., Rook, A.H. and Fauci, A.S. New Engl. J. Med. 309:453-458, 1983.
54. Takeda, A., Tuazon, C.U. and Ennis, F.A. Science 242:580-583, 1988.
55. Homsy, J., Meyer, M., Tateno, M., Clarkson, S. and Levy, J.A. Science 244:1357-1360, 1989.
56. Nara, P.L., Robey, W.G., Gonda, M.A., Carter, S.G. and Fischinger, P.J. Proc. Natl. Acad. Sci. USA 84:3797-3801, 1987.
57. Lyerly, H.K., Matthews, T.J., Langlois, A.J., Bolognesi, D.P. and Weinhold, K.J. Proc. Natl. Acad. Sci. USA 84:4601-4605, 1987.
58. Rook, A.H., Lane, H.C., Folks, T., McCoy, S., Alter, H. and Fauci, A.S. J. Immunol. 138:1064-1067, 1987.
59. Ljunggren, K., Bottiger, B., Biberfeld, G., Karlson, A., Fenyo, E.M. and Jondal, M. J. Immunol. 139:2263-2267, 1987.
60. Koup, R.A., Sullivan, J.L., Levine, P.H., Brewster, F., Mahr, A., Mazzara,

G., McKenzie, S. and Panicali, D. J. Virol. 63:584-590, 1989.
61. Riviere, Y., Tanneau-Salvador, F., Regnault, A., Lopez, O., Sansonetti, P., Guy, B., Kieny, M.P., Fourel, J.J. and Montagnier, L. J. Virol. 63:2270-2277, 1989.
62. Tyler, D.S., Nastala, C.L., Stanley, S.D., Matthews, T.J., Lyerly, H.K., Bolognesi, D.P. and Weinhold, K.J. J. Immunol. 142:1177-1182, 1989.
63. Lane, H.C., Depper, J.M., Greene, W.C., Whalen, G., Waldmann, T.A. and Fauci, A.S. New Engl. J. Med. 313:79-84, 1985.
64. Margolick, J.B., Volkman, D.J., Lane, H.C. and Fauci, A.S. J. Clin. Invest. 76:706-715, 1985.
65. Wahren, B., Morfeldt-Mansson, L., Biberfeld, G., Moberg, L., Sonnerborg, A., Ljungman, P., Werner, A., Kurth, R., Gallo, R.C. and Bolognesi, D.P. J. Virol. 61:2017-2023, 1987.
66. Schrier, R.D., Gnann, J.W. Jr., Landes, R., Lockshin, C., Richman, D., McCutchan, A., Kennedy, C., Oldstone, M.B.A. and Nelson, J.A. J. Immunol. 142:1166-1176, 1988.
67. Clerici, M., Stocks, N.I., Zajac, R.A., Boswell, R.N., Berstein, D.C., Mann, D.L., Shearer, G.M. and Berzofsky, J.A. Nature 339:383-385, 1989.
68. Siliciano, R.F., Lawton, T., Knall, C., Karr, R.W., Berman, P., Gregory, T. and Reinherz, E.L. Cell 54:561-575, 1988.
69. Barcinski, M.A. and Rosenthal, A. J. Exp. Med. 145:726-736, 1977.
70. Berzofsky, J.A., Richman, L.K. and Killion, D.J. Proc. Natl. Acad. Sci. USA 76:4046-4050, 1979.
71. Livinstone, A.M. and Fathman, C.G. Annu. Rev. Immunol. 5:477-501, 1987.
72. Benacerraf, B. J. Immunol. 120:1809-1812, 1978.
73. Babbit, B., Allan, P., Matsueda, G., Haber, E. and Unanue, E.R. Nature 317:359-361, 1985.
74. Buus, S., Sette, A., Colon, S.M., Jenis, D.M. and Grey, H.M. Cell 47:1071-1077, 1986.
75. Cease, K.B., Margalit, H., Cornette, J.L., Putney, S.D., Robey, W.G., Ouyang, C., Streicher, H.Z., Fischinger, P.J., Gallo, R.C., DeLisi, C. and Berzofsky, J.A. Proc. Natl. Acad. Sci. USA 84:4249-4253, 1987.
76. Takahashi, H., Cohen, J., Hosmalin, A., Cease, K., Houghten, R., Cornett, J.L., DeLisi, C., Moss, B., Germain, R.N. and Berzofsky, J.A. Proc. Natl. Acad. Sci. USA 85:3105-3109, 1988.
77. Berzofsky, J.A., Cease, K.B., Cornette, J.L., Spouge, J.L., Margalit, H., Berkower, I.J., Good, M.F., Miller, L.H. and DeLisi, C. Immunol. Rev. 98:9-52, 1987.
78. Rothbard, J. Annu. Int. Past. 137E:518-526, 1986.
79. Guillet, J.G., Lai, M.Z., Briner, T.J., Buus, S., Sette, A., Grey, H.M., Smith, J.A. and Gefter, M.L. J. Exp. Med. 158:294-302.
80. Quinnan, G.V. Jr., Kirmani, N., Rook, A.H., Manischewitz, J.F., Jackson, L., Moreschi, G., Santos, G.W., Saral, R. and Burns, W.H. New Engl. J. Med. 307:7-13, 1982.
81. Earl, P.L., Moss, B., Morrison, R.P., Wehrly, K., Nishio, J. and Chesebro, B. Science 234:728-731, 1986.
82. Koenig, S., Earl, P., Powell, D., Pantaleo, G., Merli, S., Moss, B. and Fauci, A.S. Proc. Natl. Acad. Sci. USA 85:8638-8642, 1988.

83. Langlade-Demoyen, P., Michel, F., Hoffenback, A., Vilmer, E., Dadaglio, G., Garicia-Pons, F., Mayaud, C., Autran, B., Wain-Hobson, S. and Plata, F. J. Immunol. 141:1949-1957, 1988.
84. Walker, B.D., Flexner, C., Paradis, T.J., Fuller, T.C., Hirsch, M.S., Schooley, R.T. and Moss, B. Science 240:64-66, 1988.
85. Nixon, D.F., Townsend, A.R.M., Elvin, J.G., Rizza, C.R., Gallwey, J. and McMicheal, A.J. Nature 336:484-487, 1988.
86. Sethi, K.K., Naher, H. and Stroehmann, I. Nature 335:178-181, 1988.
87. Ho, D.D., Rota, T.R. and Hirsch, M.S. J. Clin. Invest. 77:1712-1729, 1986.
88. Gartner, S., Markovits, P., Markovitz, D.M., Kaplan, M.H., Gallo, R.C. and Popovic, M. Science 233:215, 1986.
89. Walker, C.M., Moody, D.J., Stites, D.P. and Levy, J.A. Science 234:543, 1986.
90. Trubota, H., Lord, C.I., Watkins, D.I., Moumoto, C. and Letvin, N.L. J. Exp. Med. 169:1421, 1989.
91. Hoffenbach, A., Langlade-Demoyen, P., Dadaglio, G., Vilmer, E., Michel, F., Mayaud, C., Autran, B. and Plata, F. J. Immunol. 142:452-461, 1989.
92. Meuer, S.C., Cooper, D.A., Hodgdon, J.C., Hussey, R.E., Fitzgerald, K.A., Schlossmann, S.F. and Reinherz, E.L. Science 222:1239-1241, 1982.
93. Swain, S.L. Immunol. Rev. 74:129-142, 1983.
94. Kaplan, D.R., Griffin, R., Braciale, V.L. and Braciale, T.J. Cell. Immunol. 88:193-206, 1984.
95. Jacobson, S., Richert, J.R., Biddison, W.E., Satinsky, A., Hartzman, R.J. and McFarland, H.F. J. Immunol. 133:757, 1984.
96. Zarling, J.M., Eichberg, J.W., Moran, P.A., McClure, J., Sridhar, P. and Hu, S.L. J. Immunol. 139:988-990, 1987.
97. Celis, E., Ou, D. and Otvos, L. Jr. J. Immunol. 140:1808-1815, 1988.
98. Townsend, A.R.M., Rothbard, J., Gotch, F.M., Bahadur, G., Wraith, D. and McMichael, A.J. Cell 44:959-968, 1986.
99. Braciale, T.J., Morrison, L.A., Sweester, M.T., Sambrook, T., Gething, M.J. and Braciale, V. Immunol. Rev. 98:95-114, 1987.
100. Moore, M.W., Carbone, F.R. and Bevan, M.J. Cell 54:777-785, 1988.
101. McDougal, J.S., Mawle, A., Cort, S.P., Nicolson, J.K.A., Cross, G.D., Sheppler-Campbell, J.A., Hicks, D. and Sligh, J. J. Immunol. 135:3151-3162, 1985.
102. Zagury, D., Bernard, J., Chenier, R., Feldman, M., Sarin, P. and Gallo, R.C. Science 231:850-853, 1986.
103. Fu, S.M., Chiorazzi, N., Wang, C.Y., Montazeri, G., Kunkel, H.G., Ko, H.S. and Gottlieb, A.B. J. Exp. Med. 148:1423-1431, 1978.
104. Evans, R.L., Faldetta, T.J., Humphreys, R.E., Pratt, D.M., Yunis, E.J. and Schlossman, S.F. J. Exp. Med. 148:1440-1445, 1978.
105. Kieny, M., Rautmann, G., Schmitt, D., Dott, K., Wain-Hobson, S., Alizon, M., Girard, M., Chamaret, S., Laurent, A., Montagnier, L. and Lecueg, J. BioTechnology 4:790-795, 1986.
106. Gelderbloom, H.R., Hausmann, E.H.S., Ozel, M., Pauli, G. and Koch, M.A. Virology 156:171-176, 1987.

107. Lanzavecchia, A., Roosnek, E., Gregory, T., Berman, P. and Abrignani, S. Nature 334:530-532, 1988.
108. Harper, M.E., Marselle, L.M., Gallo, R.C. and Wong-Staal, F. Proc. Natl. Acad. Sci. USA 83:772-776, 1986.

# 4

## THE CONTROL OF HUMAN HERPESVIRUS INFECTIONS BY CYTOTOXIC T LYMPHOCYTES (CTL): A COMPARISON TO AIDS VIRUS SPECIFIC CTL

Stephen Martin and Barry T. Rouse

**Department of Microbiology and Nutrition The Upjohn Company Kalamazoo, Michigan 49001 and University of Tennessee Department of Microbiology College of Veterinary Medicine Knoxville, Tennessee 37996-0845**

## ABSTRACT

Recovery from viral infection and subsequent protection against reinfection are intimately associated with T cell immunity. Cytotoxic T lymphocytes (CTL) which are able to recognize and destroy virus infected cells have been experimentally demonstrated to play a crucial role in this recovery process. In the first part of this review, some of the evidence which implicates anti-viral CTL as a principal controlling mechanism for infections with the human herpesviruses is discussed. Several aspects of the infections associated with some of the herpesviruses are remarkably similar to those seen after infection with the human immunodeficiency virus. Accordingly, in the second part of this review, the role of antiviral CTL in the control of AIDS is discussed in the light of our knowledge of the ability of the immune system, particularly CTL, to contain the herpesviruses.

## INTRODUCTION

The word herpes was originally used to describe a condition characterized by creeping lesions on the skin (1), but today the term herpes is used to describe the clinical consequence of infection with herpes simplex virus type 1 (or type 2; HSV), the prototype of a large family of herpesviruses, the herpesviridae. Most members of the herpesviridae are assigned to three sub-families (alpha, beta, and gamma) based upon properties such as *in vitro* host range, rate of multiplication, release from the host cells, and *in vivo* characteristics (2,3).

The alpha herpesviridae exhibit classical ganglionic latency *in vivo*. They are further divided into two subgroups based upon their ability to rapidly replicate in a variety of cultured cells, causing lysis and release of infectious progeny (alpha 1, typified by HSV), or to replicate more slowly, in a narrow range of tissue culture cells with infectious progeny remaining cell associated (alpha 2, typified by varicella zoster virus, VZV). The beta herpesviridae comprise the cytomegaloviruses (CMV) which demonstrate a very slow replication rate, a very narrow host range, and remain latent *in vivo* within lymphocytes. Finally, the gamma herpesviridae represent a group of lymphotropic viruses typified by Epstein-Barr Virus (EBV). They have an excessively narrow host cell range, the ability to immortalize cells, remain latent in T or B lymphocytes, and cause lymphoproliferative diseases in their natural hosts.

The unique feature which all herpesviruses share is their ability to persist during the life time of the individual host in a latent form. This non-infectious state may remain for years or may eventually reactivate whereupon infectious virus may be shed to infect a fresh host. The survival advantage that the virus derives from this life cycle is obvious and has been discussed previously (3). Reactivation may be clinically silent or it may be accompanied by the recurrence of clinical disease. What determines the difference in outcome is not known but resolution appears to depend on the efficiency of the adaptive immune response particularly aspects of cell mediated immunity (CMI) (reviewed in 4). The best evidence for this conclusion comes from observations of a far more severe and often lethal disease in situations of primary or secondary immunodeficiency especially those which affect CMI (5, 6). Herpesviruses become devastating disease producing agents in the acquired immune deficiency syndrome (5).

A review of the literature reveals that the host response to infection with a herpesvirus is very complex with involvement of both the innate natural resistance mechanisms as well as the adaptive immune response. Although innate immunity represents a formidable early barrier to herpesviruses (7), the eventual control of the infection and establishment of long-term immunity are mediated by the adaptive immune system. One important aspect of the adaptive immune system are cytotoxic T lymphocytes (CTL) which play a crucial role in the resolution and control of many viral infections (Reviewed in 8). These CTL usually recognize viral antigens in the context of the class I gene products of the major histocompatibility complex (MHC) (9). As a consequence of MHC restriction, the CTL usually respond to processed viral antigen presented in the cleft of a MHC element on the surface of an infected autologous cell (10). At least *in vitro*, CTL bind to such cells and subsequently deliver a lethal signal, a process usually measured by the release of bound $^{51}$Chromium (reviewed in 8,11). It is far from certain if cell lysis is the way CTL act *in vivo* or whether they subserve a protective effect by participating in another function such as lymphokine release or by causing detrimental metabolic changes in target cells. This topic was elegantly discussed recently by Martz and Howell (12). However, whatever the mechanism by which CTL act, the evidence that CTL serve to resolve many viral infections, including those caused by the herpesviruses, is unequivocal.

In the sections that follow, we review the characteristics, function, and specificity of antiviral CTL specific for each sub-family of herpesviridae and for the persistent retrovirus human immunodeficiency virus (HIV). In the final sections, we will compare our knowledge of the role of CTL in the control of the persistent and potentially lethal infections with herpesviruses to the devastating and currently uncontrolled infection with HIV.

**ALPHA 1 HERPESVIRIDAE**

These viruses, as exemplified by HSV-1 and HSV-2, cause natural disease usually in only one species, but a wide range of susceptible hosts and cell cultures may be detected experimentally. Primary disease caused by HSV-1 and HSV-2 may be subclinical and is usually confined to local cutaneous sites. However, dissemination, particularly to the nervous system, may occur especially if hosts are immunologically compromised or immature. It is also possible that neuropathic strains of virus exist although this notion has not been proven for human disease (13). Primary clinical infections are considered to have both a viropathic and later on, as immunity develops, an immunopathic mechanism of pathogenesis. Usually lesions resolve within 2 weeks, but not before the virus succeeds in invading via innervating sensory nerves the local ganglion where, after a few rounds of replication, latency is established. One consequence of replication in the ganglion is spread via the axons of motor nerves back to the periphery which may be the chief means of lateral lesion dissemination. This is called zosteriform spread (14). Latency rarely remains intact indefinitely and as a result of a variety of stimuli virus is reactivated and usually spreads down nerve fibers and replicates at or near the site of primary lesions. This recurrent event may be subclinical or clinical. Clinical recurrent disease is usually less severe and of shorter duration then with the primary lesions, because of a vigorous cellular immune response. However, such responses may have severe immunopathological consequences in some locations, the eye is an example (15).

Many animal models have been developed to study the pathogenesis of HSV and its control by the immune system, but none can be considered as fully satisfactory. Animal models are particularly poor at mimicking the spontaneous recrudescent disease that occurs in man. However, despite their deficiencies, animal models have provided our chief means of understanding the defense function of different components of the immune response. A variety of approaches, performed mainly in inbred mice, have implicated a principal role for T cell immunity in normal recovery form infection. This topic has been discussed in detail in other reviews (4,8) and will not be reiterated here. Basically, using adoptive transfer experiments, both $CD4^+$ and $CD8^+$ T lymphocytes appear involved in immunity (8,16,17), and in some instances immunopathology (18). However, for optimal immunity against significant infections, both $CD4^+$ and $CD8^+$ T cells need to function. The former cell type were assumed to function in delayed type hypersensitivity (19,20,21) and the latter to function as CTL (22-25).

Deciding if $CD8^+$ cells act *in vivo* as lytic cells has been especially difficult in murine model systems. Thus, even after acute infection, it is not possible to unequivocally demonstrate active CTL in lymphoid tissues taken directly from animals. Instead, it is necessary to culture cells for a few days either in the presence of antigen or under conditions in which accessory lymphokines could be produced (26,27). Presumably, during *in vitro* culture, CTL precursors could be differentiated into CTL as a result of their interaction with $CD4^+$ helper cells or factors released from other helper cells ($CD4^+$ or $CD8^+$). Alternatively, it has been suggested that the *in vitro* culture period permits otherwise active CTL to escape from the regulatory influence of suppressor cells (28). Whatever the mechanism by which class I restricted $CD8^+$ cells function *in vivo* the evidence that such cells are required for optimal immunity is substantial and consequently those interested in vaccine development have sought to define the antigens which are recognized by, and serve to induce CTL.

Early efforts mainly focused on the major glycoproteins since these are structural virion components that are abundantly expressed at the surface of infected cells. However, none have been universally shown to act as major CTL targets in all strains of mice investigated. There are several instances of contradictory observations in part because of studies performed in different mouse strains and using different recombinant gene constructs to measure responses. Thus, it is now widely accepted that epitope selection is dependent on the MHC elements that are involved, and that the type of vector that is used could markedly influence immunogenicity and immune recognition. Especially important is whether or nor not a given vector permits appropriate processing events to occur (reviewed in 8).

CTL induction depends on antigen processing via a non-lysosomal pathway of proteins that usually must be synthesized *de novo*. The time at which these proteins are synthesized in relation to the virus replication cycle may also play an important role (29). Currently, the consensus is that gD is not a CTL target in $H\text{-}2^d$, $H\text{-}2^k$, or $H\text{-}2^b$ mice (30,31), gB is at least a minor target in $H\text{-}2^k$, $H\text{-}2^d$, and $H\text{-}2^b$ mice (31-33) and gC is a target in $H\text{-}2^b$ mice (33; Martin and Rouse, unpublished observations). It is also evident that non-structural immediate early (IE) gene products represent significant targets for anti-HSV CTL responses in all strains of mice so far studied (34 and S. Jennings, personal communication). We have recently shown, using the recombinant vaccinia virus vector strategy, that ICP4 appears to be a target antigen in $H\text{-}2^k$ (but not $H\text{-}2^b$ nor $H\text{-}2^d$) mice (Martin and Rouse, submitted) and ICP27 acts as a target in only H-2 $^{d}$ mice (Banks *et al.*, in preparation). The ICP0 protein is apparently not a CTL target in any of the 3 mouse strains studied and ICP22 and ICP47 have yet to be investigated. There is considerable interest in the fact that IE proteins act as major CTL targets since this could permit the destruction of cells before new virus is replicated. We return to this topic when discussing CMV where far more data attest to the role of IE protein recognition, which appear to dominate the antiviral CTL responses.

Although mice respond to HSV infection by generating CTL responses that are mainly class I MHC restricted, there is a minor component of class II restricted reactivity mediated by $CD4^+$ T cells (30, Kolaitis and Rouse, in preparation). Such class II restricted $CD4^+$ T cells seemingly dominate CTL responses of human anti-HSV reactivity (35), although some older reports show class I restricted CTL (36) and more recent studies report both class I and class II restricted responses (37,38). Possibly as with other human herpesviruses, the apparent controversy of findings can be explained by the nature of antigens used for *in vitro* restimulation (39,40). Clearly it appears to be easier to detect class II MHC restricted CTL in humans than in mice and the precursors of such cells are present at high frequency in seropositive individuals (1/4000) (35). Moreover, all HSV specific CTL clones so far studied have turned out to be class II restricted (41-45) and recognize glycoproteins gB and gD. None were reported with reactivity to gC or IE proteins.

The role of antiviral CTL in human HSV-1 infections is much less clear than with the mouse model, although extrapolations from these studies would suggest a primary importance. A specific example is the demonstration that thymopentin treatment of mice increased both antiviral CTL activity and resistance to a lethal HSV-1 challenge (46). Similar treatment of human patients with frequently relapsing HSV infections markedly improved the relapse rate and was suggested to work via a CTL stimulatory mechanism (47).

The observation that class II restricted $CD4^+$ CTL responses may dominate reactivity in humans has important implications to vaccination strategies. Accordingly, inactivated vaccines may act as potent stimulants of such cells, whereas for class I restricted $CD8^+$ CTL induction, living vaccines are necessary (8). On this point it is interesting to note that individuals immunized with a HSV-2 glycoprotein subunit vaccine successfully induced a virus specific CTL response mediated by CD4+ T cells (48). It is not clear, however, if protection against HSV-2 infection was manifest.

**ALPHA 2 HERPESVIRIDAE**

Alpha 2 herpesviruses, as typified by VZV, show a narrow host cell range *in vitro*, a short replication cycle, predominantly cell associated progeny, and the development of ganglionic latency *in vivo*. VZV presents as two clinically distinct syndromes; chicken pox and shingles. Chicken pox is a widespread generalized exanthem that spreads rapidly through its susceptible population of children and adolescents. In contrast, shingles is less common and usually occurs in older or immunocompromised individuals, and is characterized by a vesicular eruption usually limited to a single dermatome.

The classical manifestation of infection with VZV is the varicella rash. The virus usually enters through the mucosa of the upper respiratory tract, undergoes a primary replication and subsequently disseminates via the bloodstream and lymphatics. The virus then infects cells of the reticuloendothelial system, replicates and produces a more extensive secondary viremia (49). The second viremia is associated with prodromal symptoms which precede the mucosal and cutaneous lesions. The focal lesions are initiated by infection of capillary endothelial cells (50-52) and appear in cycles which reflect successive bouts of viremia (53). After the third day, the immunocompetent host terminates subsequent viremias and new lesions fail to develop. Local lesions may continue to enlarge but will eventually be resolved by the developing immune system leaving only non-replicating VZV virus latent within the sensory ganglia.

The pathogenesis of secondary VZV, or zoster is not fully understood, but many of the precipitating conditions are immunosuppressive and this has led to the suggestion that the latent VZV escapes immunological containment and begins to actively replicate within the sensory ganglia (54). If the immune system is sufficiently suppressed virus replication proceeds, producing neuronal necrosis, intense inflammation. Antidromal spread down the nerve allows infection of the skin with subsequent development of the zosteriform rash (55). As with all of the human herpesvirus infections, the continual development, systemic spread, and high mortality of VZV is associated with depressed T cell mediated immunity (52), and apparently normal humoral responses (57,58).

The importance of CTL as mediating protection against VZV has not been intensively studied, but Bowden *et al.* (59) demonstrated a reduction of HLA-restricted VZV specific lysis in some immunocompromised patients. Several studies in different laboratories have revealed the presence of VZV specific CTL in the peripheral blood of seropositive individuals. However the T cell phenotype and HLA restricting element that is involved remain confused. Thus several groups (40,60,61) have demonstrated HLA class I, $CD8^+$ precursors (pCTL) in the blood of VZV infected individuals, but others report HLA-D or DR restricted $CD4^+$ cells (62,63). The discrepancy between these two findings may be due to the nature of the antigen used for *in vitro* restimulation, with soluble antigen inducing predominantly $CD4^+$ CTL and restimulation with infected cells inducing a majority of $CD8^+$ CTL (40). The contribution of $CD4^+$ and $CD8^+$ CTL *in vivo* remains unresolved and although few reports concern VZV specific CTL, several interesting findings are documented. Thus the frequency of pCTL specific for VZV appears low as compared to those for HCMV (66), but frequencies of VZV specific pCTL rapidly increase after clinical recrudescence. This demonstrates a long-term CTL memory response to VZV. The significant increase in VZV specific CTL numbers after secondary disease also indicates an important role in control of the virus. Only one study has addressed the specificity of VZV-specific CTL, and this demonstrated a recognition of glycoprotein gI by a proportion of the $CD4^+$ CTL population (63). Most significantly, perhaps, these gI specific and other CTL populations where detectable in VZV vaccinated individuals, indicating the ability of the vaccine Oka/Merck to stimulate a potentially protective immune response.

The persistence of VZV CTL long after primary VZV suggests that, as with other human herpesvirus infections, CTL play a pivotal role in disease control. As more cloned VZV genes become available a rapid increase in our knowledge of VZV CTL specificities and characteristics will emerge. Such knowledge will help in the understanding of the immunological control of herpesvirus infections.

**BETA HERPESVIRIDAE**

Human cytomegalovirus (HCMV) is the largest known herpesvirus with a linear double stranded DNA genome of approximately 235 Kb which has a capacity to code for about 150 proteins. As with the other human herpesviruses, primary CMV infection is often followed by persistent and/or recurrent infections. Typical of the beta herpesviridae, HCMV is highly species specific, grows very slowly in a select few tissue culture cell lines, and probably persists *in vivo* within monocytes or T lymphocytes (64). Clinically, reactivation is associated with immunosuppression, suggesting that CMI responses are important in maintaining the host/virus interaction.

HCMV possesses low pathogenicity in the immunocompetent host despite its cytolytic nature in infected cells. The primary infection occurs via oral or orogenital contact and gives rise to a generalized infection. Viremia is detectable for up to several months (65), with the polymorphonuclear leucocyte acting as the main source of virus (66). Virus is consistently detected in the urine and pharynx with viruria persisting for up to a year or more (65,67). Clinically, disease is characterized by malaise, myalgia, and lymphocytosis with an excess of atypical lymphocytes in usually the absence of lymphadenopathy, and reactive heterophile antibodies. It is these latter two features which distinguish CMV mononucleosis from the more commonly seen EBV mononucleosis (68).

Immunosuppressive properties have been retrospectively attributed to HCMV infection based upon clinical evidence of susceptibility to opportunistic infection. These clinical observations are substantiated by a wealth of literature documenting an inhibition of monocyte/macrophage function, including an inability to support mitogen induced T cell responses (69,70), abrogation of interleukin 1 production (71) and an inhibition of CTL activity (72). In general, all symptoms attributed to CMV infection pass with time and the majority of infections remain subclinical. As with the other herpesvirus, increased numbers of CD8$^+$ T lymphocytes are detectable in the peripheral blood during primary and recrudescent episodes (73). Although the function of the excess CD8$^+$ cells is unknown, the studies of Quinnan *et al.* (74), which demonstrated a positive correlation between CTL and recovery from CMV infection, would indicate a useful role for class I MHC restricted CTL populations. A positive correlation for CMV-specific-CTL and recovery from CMV infection has also been demonstrated in renal transplant patients (75), and in the early control of a primary CMV infection (76).

The characteristics of HCMV specific CTL have been independently studied in numerous laboratories. Both CD4$^+$, HLA-DR restricted and CD8$^+$, HLA, A, B, C, restricted CTL have been demonstrated. The CD4$^+$ phenotype is preferentially expanded upon restimulation of peripheral blood lymphocytes with soluble/cell free CMV (39,77,79). In contrast, the CD8$^+$ phenotype CTL is usually detected after restimulation with autologous CMV infected cells (39, 78,79). As with all of the human herpesvirus specific CTL, detection requires a period of *in vitro* culture with viral antigen to presumably expand memory CTL populations (39,74). Therefore, a definitive decision about the role of CD4$^+$ or CD8$^+$ CTL *in vivo* is impossible and both populations must currently be considered desirable for optimum immunity.

An interesting and fruitful model has been developed and intensively studied by Koszinowski and colleagues (reviewed in 80). Although this model uses murine CMV in its natural host, many striking similarities to HCMV have been reported. Thus, both MCMV (81) and HCMV (82) specific CTL were shown by limiting dilution analysis to recognize virus antigens specific for the replication stage. Furthermore, in the individuals studied, the majority of the CTL were specific for the non-structural IE nuclear antigens (pp89 in MCMV and p72 in HCMV). To date, despite the 150 possible viral protein candidates, only the IE antigens were shown to act as major CTL target antigens.

In the MCMV/mouse model, it has been possible to thoroughly characterize the anti-viral CTL response. By adoptive transfer experiments, it was shown that a protective effect against lethal MCMV disease was attributable to $CD8^+$ T lymphocytes specific for the IE antigen (83,84). These protective CTL were class I MHC restricted ($H\text{-}2L^d$), could be induced in the absence of $CD4^+$ helper T lymphocytes, were functional in the absence of $CD4^+$ cells, and recognized a single 19 amino acid epitope on the pp89 IE protein (80,85). There are many similarities between the MCMV/mouse model and the data obtained with HCMV infected patients. However, direct extrapolation of MCMV data to man is not advisable since obvious differences do exist. For instance, HCMV specific CTL appear not to be independent of $CD4^+$ cells, since the apparent lack of HCMV-CTL in AIDS patients is readily reversible *in vitro* by the addition of interleukin 2 (86). However, despite these reservations there is ample evidence from both HCMV and MCMV to indicate that virus specific CTL recognizing the IE gene product have a significant beneficial effect upon the course of infection.

## GAMMA HERPESVIRIDAE

Epstein Barr virus, the exemplar of this subfamily, has a 170 kb double stranded DNA genome which may encode up to 100 proteins (87). Since its discovery 25 years ago in cultured Burkitt lymphoma cells (88), EBV has been extensively studied. The virus is very common with more than 90% of the world's population thought to be infected (89). Infection usually causes infectious mononucleosis, a self limiting lymphoproliferative disease, but is also associated with the B cell tumour, Burkitt's lymphoma, and with nasopharyngeal carcinoma, an important cancer in Southeast Asia with an incidence of 1-2 per $10^4$ population a year.

The classical syndrome associated with primary infection is infectious mononucleosis (I.M.). After the initial infection in a variety of epithelial cells in the anterior oropharynx (90), circulating B lymphocytes in the vicinity become infected and then home to the marrow and various lymphoid organs. During the acute phase of the infection between 5 and 20% of the circulating B lymphocytes may express the Epstein-Barr nuclear antigen (EBNA), but by the second week of infection less than 2% of the circulating B lymphocytes are EBNA positive. However, the virus is not completely eliminated, since it is possible to recover EBV from small numbers of lymphocytes indefinitely (91).

It is likely that the majority of clinical manifestations of I.M. are associated with the immune response to the proliferating EBV infected B lymphocytes. The major locations involved in the body are the lymph nodes, peripheral blood, spleen and liver. Most of the infiltrates of the lymph nodes and spleen involve the paracortical areas of the organ indicative of the predominant antiviral T cell response (92). This is also reflected in the blood where the lymphocytosis and phenotypic analysis of these lymphocytes reveal an expansion of $CD8^+$ cells and thus a reversal in the T4:T8 ratio that persists for 4-8 weeks (93).

Early in the course of infection, a CTL response occurs which is judged to be specific for EBV-transformed cells, although not restricted to HLA antigens (94). It is evident, however, that the primary T cell response in I.M. although polyclonal, contains an EBV-specific HLA class I antigen restricted CTL population (95,96). During convalescence, a regression in the outgrowth of EBNA positive B lymphocytes occurs and this coincides with the development of an overt population of $CD8^+$, HLA class I restricted EBV specific CTL that are $CD4^+$ cell independent (97-99). A convincing *in vitro* demonstration of the functionality of these CTL is the "regression assay" in which the outgrowth of freshly EBV transformed B cells is prevented in a mixed culture (100).

The precise identification of all of the target antigens for EBV-specific CTL remains to be made. Thus, B lymphocytes immortalized by EBV infection *in vitro* and susceptible to CTL, consistently express at least eight viral proteins, six in the nucleus and two in the plasma membrane (101). Genetic evidence indicates that at least one, and maybe an additional two of the nuclear proteins are necessary for immortalization (102). The viral antigens identified as targets for CTL are the lymphocyte membrane protein (LMP/P 63) EBNA 2 and EBNA 3. The LMP was demonstrated as a target antigen for human CTL by transfection into B lymphocytes (103) and similarly as a target for murine CTL (104). The amino terminal 128 amino acids were implicated as containing the epitopes recognized by human CTL (103). The studies of Thorley-Lawson and

Israelsohn (105) used a different approach with synthetic peptides to demonstrate that at least one of the amino-terminal epitopes is present in the amino acids 43-53 of the LMP. Differential recognition of two EBV strains by strain specific CTL have implicated both EBNA 2 (106,107) and EBNA 3 (102) as target antigens.

The lack of expression of LMP, EBNA 2 and 3 in freshly isolated Burkitt's lymphoma cells (they only express EBNA 1; 108) may account for their resistance to lysis by EBV-specific CTL (106). However, this scenario is made more complex by the demonstration by Gregory *et. al.* (109) that down-regulation of the cell adhesion molecules LFA3 and ICAM I also contributes to the escape of the tumor cells from CTL surveillance.

Although it is clear that CTL play a pivotal role in the control of primary and latent EBV infection, the total spectrum of immunoregulatory effector populations involved during I.M. needs further investigation. In addition, the lag phase of several years between primary infection with EBV and the development of Burkitt's lymphoma, or NPC in susceptible individuals demonstrates a complex multi-hit mechanism for the escape from immunosurveillance.

## CTL AND HIV INFECTION

Although HIV infection seems ultimately to result in a profound immunosuppression, there are reasons to suspect, and indeed hope, that some components of the immune response play a protective role. Moreover, it is conceivable that methods could be devised that would stimulate preferential aspects of the immune response that serve a protective role. Such manipulations in infected individuals may postpone or even negate the inexorable immune destruction. To date, no data unequivocally identify any component of immunity as protective in HIV infection, but if some immune mechanism plays such a protective role, CTL represent a most likely candidate. Evidence for this notion is at present purely circumstantial and inferential. Thus, in other cell associated infections which are transmitted directly between cells, CMI and especially the CTL system is a crucial means of defense (8). This was discussed previously with respect to herpes viruses. The fact that neutralizing antibodies appear to have little influence on the outcome of HIV infection in man and lentivirus infections in experimental animals is taken to imply that protection, if it does occur, must proceed by a nonhumoral means (110,111).

Killing infected cells by CTL would obviously be of benefit should it succeed in destroying viral nucleic acids rather than release and disseminate virus or destroy infected yet still functioning cells. Some have shown that cells of the $CD8^+$ phenotype, including CTL clones of known specificity, may inhibit virus replication in permissive cells (112,113). Moreover, one group has demonstrated that virus replication in permissive $CD4^+$ lymphocytes in culture proceeds more effectively if $CD8^+$ cells are first removed from the cultures (114). This inhibition by $CD8^+$ cells occurs at least to a degree in the absence of cell contact (115). As discussed below, it is evident that CTL of multiple specificities and various phenotypes exist but thus far no convincing longitudinal data implicate CTL as participating in either a helpful or a harmful capacity. Most suspect that CTL could play a protective role (113-115,117), but others take the contrary viewpoint and indicate that CTL, particularly of anti-gp120 specificity, may participate harmfully by mediating an inflammatory response during the destruction of infected cells or even serve to destroy uninfected CD4+ cells that bind soluble gp120 because of their possession of the CD4 surface protein (118-121).

The many properties of anti-HIV CTL remain to be clarified, but it does appear that several characteristics are either unique or at least unusual in comparison to other well-studied viruses such as herpesviruses. In this brief discussion of the ever expanding and at times contradictory information on anti-HIV CTL, we emphasize features which seem peculiar to HIV.

In most well-studied antiviral CTL systems, a response, which is often feeble, can be demonstrated by direct assay for a relatively short period of time (8,9). Subsequent CTL detection requires *in vitro* restimulation with antigen or mitogen, a procedure necessary to generate effector CTL from primed precursors (122,123). In the case of HIV, however, substantial CTL activity of multiple specificities is sustained more or less indefinitely in seropositive individuals (117,118,124). Although so far too few patients have been followed longitudinally for adequate time periods, reliable correlates of CTL activity against some viral gene product with stage or outcome of HIV infections have not been reported. Some data indicate a disappearance of CTL in highly immunosuppressed individuals and a reappearance of such cells after AZT treatment (125, B. Walker personal commun.). One group following frequencies of CTL precursors at different stages showed a dramatic fall in frequency as the disease progressed (125).

Usually with antiviral CTL responses only a few viral gene products are recognized by $CD8^+$ class I restricted CTL and, more often than not, proteins recognized as immunodominant are nonstructural or internal proteins synthesized early during viral replication (8). Since our knowledge of antigen processing has improved, it is now possible to understand how peptides derived from nonstructural proteins can interact with MHC to be recognized by CTL (reviewed in 8 and 126). With HIV, all gene products so far investigated can be recognized by CTL. These include viral components coded for by the *env*, *gag* and *pol* genes as well as the *nef* and *vif* regulatory proteins (127-129). At the time of writing no HIV gene product has been reported as lacking in CTL recognizable epitopes. The consensus with anti-HIV CTL is that the *env* gene product, particularly the viral surface glycoprotein gp120, is the major target for the CTL response (128,129). Fine details of the recognition specificity still need to be resolved, but taken together the various studies seem to indicate that gp120 may contain a plethora of epitopes. The identity of the epitope recognized will vary according to the MHC restriction elements present (113). In man maybe all HLA restricting elements interact with some peptide of gp120 to become CTL recognizable, but the response in mice seems far more parsimonious (129). Thus, in the A strain of the four available class I MHC molecules, only the $H\text{-}2D^d$ protein was a restrictive element and the responsive CTL were directed at a single epitope that manifests in the highly variable segment of the envelope protein (129). Whether this means that the mouse is not such a good model to study the immunobiology of anti-HIV CTL remains to be seen.

At present studies with bulk CTL cultures as well as with clones, have provided conflicting data both with respect to anti-*env* specificity, restriction profile and the phenotypic nature of CTL. Some groups report that bulk culture CTL are $CD3^+$ $CD8^+$ $CD4^-$ and react specifically with *env* gene expressing targets in a class I restricted fashion (114). Others show that reactivity can occur with similar targets in the absence of MHC restriction (129), and that the cells involved express none of the classical CTL markers such as $CD3^+$ $CD8^+$ and $CD4^+$. Still others have reported $CD3^+$ $CD4^+$ effector cells (120). In one report, individuals found to possess nonrestricted CTL responses to gp160 additionally showed highly specific anti-*gag* activity mediated by $CD3^+$ $CD8^+$ $CD4^-$ cells (129). Such *gag* responses were directed against the HIV-1 *gag* but not HIV-2 *gag*, although the *gag* genes show a 60% level of homology (130). Falling between the extremes of nonrestriction and restriction are the studies of Koenig *et. al.* (117) which demonstrate recognition by bulk culture CTL of several variants of the *env* gene product by $CD8^+$ T cells in a nonrestricted fashion; however, clones derived from such cultures were always class I restricted $CD8^+$ cells.

Several investigators have generated clones of CTL which react with various HIV gene products, especially the *env* gene (113,117,119,120). Some investigators report the identity of such clones to be classical $CD8^+$ $CD4^-$ class I restricted effector cells (113), but others have demonstrated class II restricted effector cells (120) or cells of unusual phenotype (119). For instance, Sethi *et. al.* (119) generated class I and class II restricted clones from the cerebrospinal fluid, which expressed an array of phenotypes, including $CD4^-$ $CD8^+$, $CD4^+$ $CD8^-$, and $CD4^-$ $CD8^-$. but reacted specifically with gp120. Another particularly well studied anti-gp120 specific clone was studied in detail by the Reinherz group (120). This clone was generated from a seronegative individual and was class II restricted $CD4^+$ $CD8^-$. The clone showed precise recognition characteristics and interestingly could even recognize any normal $CD4^+$ cells after exposure to the appropriate gp120 protein. This finding inspired the cytolytic mechanism hypothesis designed to explain a possible means of destruction of $CD4^+$ cells *in vivo*. The hypothesis states that normal uninfected $CD4^+$ cells could become subject to killing by gp120 specific CTL (class II restricted) *in vivo* if they happen to bind soluble gp120 that is released abundantly from infected cells (131). Activated $CD4^+$ T cells express class II MHC proteins and, not only become subject to class II restricted gp120 specific CTL (if they indeed exist *in vivo*), but also may be highly efficient at trapping and processing gp120 for the stimulation of other gp120 specific cells (121). There is also some evidence that soluble gp120 may sensitize target cells for killing by class I restricted $CD8^+$ cells (119), although this observation appears to contradict the usual rules of antigen processing (126).

Characteristically, $CD8^+$ cells reactive with viral antigens are of very low frequency and are difficult to demonstrate in unprimed individuals (122, 123). Frequencies of less than 1/200,000 are commonplace. This contrasts with the frequency of pCTL reactive with alloantigens which are in the 1/100 to 1/1000 range (132). However, one group has reported frequencies of pCTL reactive with the HIV *env* gene product in normal seronegative individuals at levels that are similar to those reported for alloantigens (125). The frequency of such cells in seropositive individuals was of a similar order of magnitude except that the frequency progressively decreased as the disease developed became more severe, a trend that was reversed with AZT treatment. The demonstration of anti-gp120 specific CTL in seronegative individuals required that cells be cultured in an autologous mixed lymphocyte culture for as little as 3 days. Details of the specificity characteristics of such anti-HIV CTL need to be established, since it remains possible that some cross reactivity with a ubiquitous foreign antigen such as an infecting bacterium or endogenous retrovirus was being measured. Alternatively, it is conceivable that the anti-HIV antigen could cross-react with certain alloantigens. Support for this idea comes from the observations of sequence homology between HLA class II antigens and gp41 (*env*) of HIV-1 (133). The possibility that the assay used for study was measuring the fusion of some effector cells to target cell must also be discounted. The presence of pCTL in seronegative individuals has not been a general finding, although a high frequency of anti-*env* specific pCTL has been observed by others in seropositive individuals (117).

## HERPESVIRUS AND HIV-SPECIFIC CTL: A COMPARISON

Based upon our current knowledge of human CTL specific to herpesvirus antigens, there appears to be two distinct situations. In the first, as occurs with HSV and VZV, the response seems dominated by $CD4^+$ MHC class II restricted cells whose principal target antigens are the major glycoproteins. Against these viruses, $CD8^+$ MHC class I restricted CTL may represent a minor component and there is little evidence that such cells play a protective role in controlling infection. By way of contrast, in the second situation which represents infections with CMV and EBV, $CD8^+$ MHC Class I restricted cells are the major type of CTL and these cells mainly express their reactivity against nonstructural antigens which tend to be synthesized early during the viral replication cycle. Moreover, there is substantial evidence that these CTL play a crucial role in resolving and controlling infection. Possibly the reason $CD8^+$ CTL play such a useful role in EBV and CMV infection relates to the accessibility of cells which harbor virus. Thus, in these infections viral persistence occurs in lymphocytes or monocytes and partial gene expression occurs in such cells. In addition, EBV may persist in a permanently replicative form in a localized epithelial site (134,135). All locations are accessible to CTL and such cells could exert their controlling influence best if they acted early in the replication cycle. This contrasts with the alphaherpesviridae where virus persists in a nonreplicating form (true latency) at sites which may not be subject to constant surveillance by the immune system (e.g., the neuronal tissues).

There are parallels between the pathogenesis of $\beta$ and $\gamma$ herpesviridae and infection by HIV, but the outcome of infection and the ultimate role of CTL show striking differences. Both may begin with an infectious mononucleosis syndrome, associated lymphadenopathy and systemic immunosuppression, but in EBV and HCMV recovery occurs mediated in large part by $CD8^+$ CTL. However, virus persists indefinitely in a small minority (about 3 in $10^6$ B cells in the case of EBV, 134) of cells which express some, if not all, viral genes and presumably are subject to killing by the few effector CTL which remain active.

Accordingly, at the chronic stage although pCTL are abundant, large numbers of active CTL cannot be demonstrated. In contrast, in HIV infection $CD8^+$ CTL responses occur, but it is by no means clear if such cells play a useful role in clearing infection at any stage. Accordingly, virus persists in a small number of cells (about the same proportion as in EBV) but in such cells viral gene expression is florid and viral protein expression may account for 40% of the cells total production (124). Such proteins are readily released and one of them, gp120, happens to bind to any cell that expresses a functionally important cell membrane protein-the CD4 molecule. In HIV, in contrast to herpesvirus, CTL persist as active effectors and do not seem to destroy infected cells to the benefit of the host. The question arises if these directly detectable CTL are a consequence of, or the cause of the progressive immunosuppressive course of HIV infection? Evidence for this latter possibility exists.

The presence of directly detectable HIV-specific CTL may reflect the continuous exposure to antigen of follicular dendritic cells which are potent activators of a primary CTL response (136). Alternatively, it may reflect a lack of $CD4^+$ suppressor cell control such as has been demonstrated for the HSV-2 specific murine CTL response (17). An extension of this scenario would be that $CD8^+$ CTL are self perpetuating in the face of continual antigen stimulation and destruction of immunoregulatory $CD4^+$ T cells. Presumably, the immune lysis of infected $CD4^+$ cells would fail to destroy infectious progeny virus thereby ensuring a continual supply of antigen for the stimulatory antigen presenting cells (APC). This self-perpetuating cycle would continue until a critical minimal density of $CD4^+$ cells is reached which predisposes the individual to a lethal opportunistic infection.

Similar models of $CD8^+$ mediated AIDS have been postulated by others (137). A consequence of such an AIDS pathogenesis is that several testable hypotheses are generated (138). First, any non-cytopathic virus which infects both $CD4^+$ T lymphocytes and macrophages and induces a potent antiviral CTL response should give rise to a clinical immunosuppression. Second, the selective inhibition of the CTL response should prevent a fulminating immunosuppressive syndrome. Finally, congenital infection prior to immunocompetence should give rise to a non-clinical virus carrier state, whereas infection after the development of immunocompetence should give the classical CTL mediated immunosuppressive disease. The first hypothesis is in part substantiated by the studies of Zinkernagel and colleagues with the LCMV/mouse model system (138). This system demonstrates the induction of a virus initiated immune suppression as a result of the destruction of virus infected APC/and or helper T cells by LCMV-specific CTL. The second hypothesis is corroborated by the finding that the LCMV immunosuppression model is abrogated by the removal of $CD8^+$ cells prior to the adoptive transfer of LCMV immune spleen cells to infected nude (athymic) mice. The testing of any of these hypotheses in human subjects is problematic, but the selective depletion of HIV-specific CTL could not possibly worsen the prognosis and may provide to be of therapeutic value. The converse side of the argument suggests that the continually detectable presence of HIV-specific CTL is a consequence of HIV pathogenesis rather than a cause. If this is the case, then unlike CMV and EBV-specific CTL, the HIV-specific CTL are unable to control the virus infection.

Given the possible scenario of immunopathology mediated by CTL, it is difficult to advocate a vaccine strategy to stimulate the CTL response even if it were possible to easily achieve this goal. We suspect, however, that both in the case of HIV and herpesviruses, CTL responses to some proteins may be more beneficial than the response against others. With herpesviruses, the early nonstructural components may provide the most desired CTL specificity to achieve. In HIV, it is too early to predict, but the *pol* product may represent a good start since the reverse transcriptase is more conserved than many other proteins and potent responses to this target antigen can be induced (139). In contrast, there are some highly persuasive arguments not to use gp120 for vaccination purposes. Thus, apart from the fact that infected individuals usually have potent CTL activity against gp120 at all times yet fail to clear virus, the *env* proteins are highly variable (140). Moreover, as discussed previously if an *env* based vaccine resulted in the stimulation of type

II restricted CTL, this may accelerate the destruction of the immune system rather than protect the host. Clearly vaccines based on gp120 must only induce class I restricted CTL since such cells may confine their inhibitory effects to cells that synthesize antigen and ignore those which bind it passively (8). The challenge is not only to identify proteins or their relevant subcomponents which act as beneficial CTL targets, but also to find optimal means of inducing CTL *in vivo*.

## REFERENCES


1. Beswick, T.S.L. Med. Hist. 6:214-232, 1962.
2. Wildy, P. In: **The Herpes Viruses** (ed. A.S. Kaplan), Academic Press, N.Y. 1973, pp 1-25.
3. Wildy, P. Brit. Med. Bull. 41:339-344, 1985.
4. Wildy, P. and Gell, P.G.H. Brit. Med. Bull. 41:86-91, 1985.
5. Burns, W.H. and Saral, R. Brit. Med. Bull. 41:46-49, 1985.
6. Andiman, W.A., Katz, B.Z. and Miller, G. In: **Concepts in Viral Pathogenesis** (Eds. A.L. Notkins and M.B.A. Oldstone), Springer Verlag, 1986, pp 276-286.
7. Lopez, C. In: **Immune Mechanisms in Herpes Simplex Virus Infections** (eds. B.T. Rouse and C. Lopez), CRC Press, Boca Raton, Fla. 1983, pp 132-162.
8. Rouse, B.T., Norley, S. and Martin, S. Revs. Inf. Dis. 10:16-33, 1988.
9. Zinkernagel, R.M. and Doherty, P.C. Adv. Immunol. 27:51-177, 1979.
10. Bjorkman, P.J., Super, M.A., Sauwraoui, B., Bennett, W.S., Strominger, J.L. and Wiley, D.C. Nature 329: 512-518.
11. Rouse, B.T. and Horohov, D.W. Vet. Immunol. and Immunopathol. 6:35-66, 1984.
12. Martz, E. and Howell, D.M. Immunol. Today 10:79-86, 1989.
13. Sunstrum, J.C., Chrisp, C.E., Averill, D.R., Levine, M. and Glorioso, J.L. Virus Res. 11:17-32, 1988.
14. Teague, O. and Goodpasture, E.W. J. Med. Res. 44:185-200, 1923.
15. Rouse, B.T. In: **Immunobiology of Human Herpesviruses** (ed. B. Roizman and C. Lopez), Plenum Press, 1984, pp 103-119.
16. Howes, E.L., Taylor, W., Mitchison, N.A. and Simpson, E. Nature 277:67-68, 1979.
17. Nash, A.A., Jayasuinga, A., Phelan, J., Cobbold, S.P., Waldmann, H. and Prospero, T. J. Gen. Virol. 68:825-833, 1987.
18. Newell, C.K. and Martin, S. J. Virol. 63:769-775, 1989.
19. Nash, A.A., Field, H.J. and Quartey-Papafio, R. J. Gen. Virol. 48:351-357, 1980.
20. Nagafuchi, S., Hayashida, I., Higa, K., Wada, T. and Mori, R. Microbiol. Immunol. 26:359-362, 1982.
21. Martin, S. and Rouse, B.T. J. Immunol. 138:3441-3437, 1987.
22. Larsen, H.S., Russell, R.G. and Rouse, B.T. Infect. Immun. 41:197-204, 1983.

23. Sethi, K.K., Omata, Y. and Schneweiss, K.E. J. Gen. Virol. 64:443-447, 1983.
24. Larsen, H.S., Feng, M.F., Horohov, D.W., Moore, R.N. and Rouse, B.T. J. Virol. 50:56-59, 1984.
25. Bonneau, R.H. and Jenning, S.R. J. Virol. 63:1480-1484, 1989.
26. Pfizenmaier, K., Starzinski-Powitz, A., Rollinghoff, M., Falke, D. and Wagner, H. Nature 265:630-632, 1977.
27. Lawman, M.J.P., Rouse, B.T., Courtney, R.J. and Walker, R.D. Infect. Immun. 27:133-139, 1980.
28. Rouse, B.T. and Horohov, D.W. Rev. Inf. Dis. 8:850-873, 1986.
29. Coupar, B.E.H., Andrew, M.E., Both, G.W., and Boyle, D.B. Europ. J. Immunol. 16:1479-1487, 1986.
30. Martin, S., Moss, B., Berman, P.W., Lasky, L.. and Rouse, B.T. J. Virol. 61:726-734, 1987.
31. Blacklaws, B.A., Nash, A.A. and Darby, G. J. Gen. Virol. 68:1103-1114, 1987.
32. McLaughlin-Taylor, E., Willey, D.E., Cantin, E.M., Eberle, R., Moss, B. and Openshaw, H. J. Gen. Virol. 69:1731-1734, 1988.
33. Glorioso, J., Kees, U., Kumel, G., Kirschner, H. and Krammer, P. J. Immunol. 135:575-582, 1985.
34. Martin, S., Courtney, R.J., Fowler, G. and Rouse, B.T. J. Virol. 62:2265-2273, 1988.
35. Schmid, D.S. J. Immunol. 140:3610-3616, 1988.
36. Yasukawa, M., Shirogochi, T., Kobagash, Y. Infect. Immun. 40:190-197, 1983.
37. Cunningham, A.L. and Noble, J.R. J. Clin. Invest. 83:490-496, 1989.
38. Torpey, D.J., Lindsley, M.D. and Rinaldo, C.R. J. Immunol. 142:1325-1332.
39. Borysiewicz, L.K., Morris, S., Page, J.D. and Sissons, G. Eur. J. Immunol. 13:804-809, 1983.
40. Hickling, J.K., Borysiewicz, L.K. and Sissons, J.G. J. Virol. 61:3463-3469, 1987.
41. Yasukawa, M. and Zarling, J.M. J. Immunol. 133:422-427, 1984.
42. Yasukawa, M. and Zarling, J.M. J. Immunol. 133:2736-2742, 1984.
43. Yasukawa, M. and Zarling, J.M. J. Immunol. 134:2679-2682, 1985.
44. Zarling, J.M., Moran, P.A., Burke, P.L., Pachl, C., Berman, P.W. and Lasuy, L.A. J. Immunol. 136:4669-4673, 1986.
45. Zarling, J.M., Moran, P.A., Lasky, L.A. and Moss, B. J. Virol. 59:506-509, 1986.
46. Cappel, R., de-Cuyper, F., De-Neef, K., Hobel, W. and Bolla, K. Surv. Immunol. Res. 4:48-57, 1985.
47. Haneke, E., Bolla, K., Degreef, H., Djawari, D., Demaubeuge, J., Molin, L., Panconesi, E., Schopf, E., Stengel, R. and Stingl, G. Int. J. Clin. Pharmacol. Res. 4:439-443, 1984.
48. Zarling, J.M., Moran, P.A., Brewer, L., Ashley, R. and Corey, L. J. Virol. 62:4481-4485, 1988.
49. Feldman, S. and Epp, E. J. Pediatr. 94:746-748, 1979.
50. Bruusgaard, E. Br. J. Dermatol. 44:1-24, 1932.
51. Johnson, H.N. Arch. Pathol. 30:292-307, 1940.

52. Eisenbud, M. Am. J. Med. 12:740-746, 1952.
53. Gelb, L.D. Virology 591-627.
54. Hope-Simpson, R.E. Proc. R. Soc. Med. 58:9-20, 1965.
55. Oberg, G. and Svedmyr, A. Scand. J. Infect. Dis. 1:47-49, 1969.
56. Gershon, A.A. and Steinberg, S.P. Infect. Immun. 25:170-174, 1979.
57. Takahashi, M. In: **Concepts in Viral Pathogenesis II** (eds. Notkins, A.L. and Oldstone, M.B.A.), Springer-Verlag, 1986, pp 365-371.
58. Arvin, A.M., Koropchak, C., Williams, B.R.G., Grumet, F.C. and Foung, S.K.H. J. Infect. Dis. 154:422, 1986.
59. Bowden, R.A., Levin, M.J., Giller, R.H., Tubergen, D.G. and Hayward, A.R. Clin. Exp. Immunol. 60:387-395, 1985.
60. Cooper, E.C., Vujcilk, L.K., Quinnan, G.V. Jr., J. Infect. Dis. 158:780-788, 1988.
61. Borysiewicz, L.K., Graham, S., Hickling, J.K., Mason, P.D. and Sissons, J.G. Eur. J. Immunol. 18:269-275, 1988.
62. Hayward, A.R., Pontesilli, O., Herberger, M., Laszlo, M. and Levin, M. J. Virol. 58:179-184, 1986.
63. Diaz, P.S., Smith, S., Hunter, E. and Arvin, A.M. J. Immunol. 142:636-641, 1989.
64. Schrier, R.D., Nelson, J.A. and Oldstone, M.B.A. Science 230:1048.
65. Klemola, E. Ann. Intern. Med. 79:267-268, 1973.
66. Rinaldo, C.R.Jr., Stossel, T.P., Black, P.H. and Hirsch, M.S. J. Infect. Dis. 136:667-678.
67. Klemola, E., Von Essen, R. and Wagner, O. Ann. Intern. Med. 71:11-19, 1969.
68. Klemola, E., Von Essen, R., Henle, G. and Henle, W. J. Infect. Dis. 121:608-614, 1970.
69. Rinaldo, C.R. and Hirsch, M.S. J. Infect. Dis. 141:488-495, 1980.
70. Carney, W.P. and Hirsch, M.S. J. Infect. Dis. 144:47-54, 1981.
71. Rodgers, B.C., Scott, D.M., and Mundin, J. J. Virol. 55:527-532, 1985.
72. Schrier, R.D. and Oldstone, M.B.A. J. Virol. 59:127-131, 1986.
73. Schooley, R.T. N. Eng. J. Med. 308:307-313, 1983.
74. Quinnan, G.V. Jr., Kirmani, N., Rook, A.H., Manischewitz, J.F., Jackson, L., Moreschi, G., Santos, G.W., Saral, R. and Burns, W.H. N. Engl. J. Med. 307:6-13, 1982.
75. Rook, A.H., Quinnan, G.V. Jr., Frederick, W. Jr., Manischewitz, J.F., Kirmani, N., Dantzer, T., Lee, B.B. and Currier, C.B. Jr. Am. J. Med. 76:385-392, 1984.
76. Quinnan, G.V., Kirmani, N., Burns, W.H. Jr., Rook, A.H., Manischewitz, J., Jackson, L., Santos, G.W. and Saral, R. Rev. Infect. Dis. 6:156-163, 1984.
77. Lindsley, M.D., Torpey, D.J., Rinaldo, C.R. Jr., J. Immunol. 136:3045-3051, 1986.
78. Gehrz, R.C. and Rutzick, S.R. Clin. Exp. Immunol. 61:80-89, 1985.
79. Laubscher, A., Bluestein, H.E., Spector, S.A. and Zvaifler, N.J. J. Immunol. Methods 110:69-77, 1988.
80. Koszinowski, U.H., Reddehase, M.J., Keil, G.M., Volkmer, H., Jonjil, S., Messerle, M., del Val M., Mutter, W., Munch, K. and Buhler, B. Immunol. Lett. 16:185-192, 1987.

81. Reddehase, M.J., Keil, G.M. and Koszinowski, U.H. Eur. J. Immunol. 14:56-61, 1984.
82. Borysiewicz, L.K., Hickling, J.K., Graham, S., Sinclair, J., Cranage, M.P., Smith, G.L. and Sissons, J.G. J. Exp. Med. 168:919-931, 1988.
83. Reddehase, M.J. and Koszinowski, U.H. Nature 312:369-371, 1984.
84. Reddehase, M.J., Mutter, W., Munch, K., Buhring, H.J. and Koszinowski, U.H. J. Virol. 61:3102-3108, 1987.
85. del Val, M., Volkmer, H., Rothbard, J.B., Jonjic, S., Messerle, M., Schickedanz, J., Reddehase, M.J. and Koszinowski, U.H. J. Virol. 62:3965-3972, 1988.
86. Rook, A.H., Masur, H., Lane, H.C., Frederick, W., Kasahara, T., Macher, A.M., DJeu, J.Y., Manischewitz, J.F., Jackson, L., Fauci, A.S. and Quinnan, G.V. Jr. J. Clin. Invest. 72:398-403, 1983.
87. Farrell, P.J. Adv. Viral. Onc. 8:103-131, 1989.
88. Epstein, M.A., Achong, B.G. and Barr, Y.M. Lancet 1:702-703, 1964.
89. Galloway, J. Nature 338:463-464, 1989.
90. Niederman, J., Miller, G., Pearson, H. and Pagano, J. N. Engl. J. Med. 294:1355-1359, 1976.
91. Gerber, P. and Monroe, J.H. J. Natl. Cancer Inst. 40:855-866, 1968.
92. Miller, G. Virology 30:582, 1985.
93. Haynes, B.F., Schooley, R.T., Payling-Wright, C.R., Grouse, J.E., Dolin, R. and Fauci, A.S. J. Immunol. 123:2095-2101, 1979.
94. Svedmyr, E. and Jondal, M. Proc. Natl. Acad. Sci. USA 72:1622-1626, 1975.
95. Strang, G. and Rickinson, A.B. Immunology 62:647-654, 1987.
96. Strang, G. and Rickinson, A.B. Eur. J. Immunol. 17:1007-1013, 1987.
97. Fishwild, D.M., Benike, C.J. and Engleman, E.G. J. Immunol. 140:1994-1998, 1988.
98. Moss, D.J., Rickinson , A.B. and Pope, J.H. Int. J. Cancer 22:1355-1359, 1978.
99. Kontinnen, Y.T., Bluestein, H.G. and Zvaifler, N. J. Cell. Immunol. 103:84-95, 1986.
100. Rickinson, A.B., Moss, D.J. and Pope, J.H. Int. J. Cancer 23:610-617, 1979.
101. Knutson, J.C. and Sugden, B. Adv. Viral. Oncol. 8:151, 1989.
102. Sugden, B. Nature 339:179-180, 1989.
103. Murray, R.J., Wang, D., Young, L.S., Wang, F., Rowe, M., Keiff, E. and Rickinson, A.B. J. Virol. 62:3747-3755, 1988.
104. Reiss, C.S., Wang, D., Ghosh, D., Gaposchkin, C. and Kieff, E. J. Immunol. 39:711-714, 1987.
105. Thorley-Lawson, D.A. and Israelsohn, E.S. Proc. Natl. Acad. Sci. USA 84:5384-5388, 1987.
106. Moss, D.J., Misko, I.S., Burrows, S.R., Burman, K., McCarthey, R. and Sculley, T.B. Nature 331:719-721.
107. Murray, R.J., Young, L.S., Calender, A., Gregory, C.D., Rowe, M., Lenoir, G.M. and Rickinson, A.B. J. Virol. 62:894-901, 1988.
108. Rowe, M. EMBO J. 6:2743-2751, 1987.
109. Gregory, C.D., Murray, R.J., Edwards, C.F. and Rickinson, A.B. J. Exp. Med. 167:1811-1824, 1988.

110. Popovic, M., Sarngadharan, M.G., Read, E. and Gallo, R.C. Science 224:497-500, 1986.
111. Griffin, D.E., Narayan, O. and Adams, R.J. J. Inf. Dis. 138:340-349, 1978.
112. Walker, C.M., Moody, D.J., Daniel, P., Stites, and Levy, J.A. Science 234:1563-1566, 1986.
113. Walker, B.D., Flexner, C., Birch-Limberger, K., Fisher, L., Aldovini, A., Young, R., Moss, B. and Schooley, T. Science in press.
114. Kannagi, M., Chalifoux, L.V., Lord, C.I. and Letvin, N.L. J. Immunol. 140:2237-2242, 1988.
115. Walker, C.M. Res. Immunol. 140:115-118, 1989.
116. Walker, B.D., Chakrabarti, S., Moss, B., Paradis, T.J., Flynn, T., Durno, A.G., Blumberg, R.S., Kaplan, J.C., Hirsh, M.S. and Schooley, R.T. Nature 328:345-348, 1987.
117. Koenig, S., Earl, P., Powell, D., Pantaleo, G., Merli, S., Moss, B. and Fauci, A.S. Proc. Natl. Acad. Sci. USA 85:8638-8642, 1988.
118. Plata, F., Autran, B., Martins, L.P., Wain-Hobson, S., Raphael, M., Mayaud, C., Denis, M., Guillon, J.M. and Debre, P. Nature 328:348-351, 1987.
119. Sethi, K.K., Naher, H. and Stroehmann, I. Nature 335:178-181, 1988.
120. Siliciano, R.F., Lawton, T., Knall, C., Karr, R.W., Berman, P., Gregory, T. and Reinherz, E.L. Cell 54:561-575, 1988.
121. Lanzavecchia, A. Res. Immunol. 140:99-103, 1989.
122. Owen, J.A., Allouche, M. and Doherty, P.C. Cell Immunol. 67:49-59, 1982.
123. Rouse, B.T. and Wagner, H. Immunology 51:57-63, 1984.
124. Koup, R.A. and Sullivan, J.L. Res. Immunol. 140:92-95, 1989.
125. Hoffenbach, A., Langlade-Demoyen, P., Dadaglio, G., Vilmer, E., Michel, F., Mayaud, C., Autran, B. and Plata, F. J. Immunol. 142:452-462, 1989.
126. Long, E.O. and Jacoboson, S. Immunol. Today 10:45-48, 1989.
127. Nixon, D.F., Townsend, A.R.M., Elvin, J.G., Rizza, C.R., Gallewoy, J. and McMichael, A.J. Nature 335:484-487, 1988.
128. Walker, B.D., Flexner, C., Paradis, T.J., Fuller, T.C., Hirsch, M.S., Schooley, R.T. and Moss, B. Science 240:64-65, 1988.
129. Takahashi, H., Cohen, J., Hosmalin, A., Cease, K.B., Houghten, R., Cornette, J.L., DeLisi, C., Moss, B., Germain, R.N. and Berzofsky, J.A. Proc. Natl. Acad. Sci. USA 85:3105-3109, 1988.
130. Cuyader, M., Emerman, M., Sonigo, P., Clavel, F. and Montague, L. Nature 326:662-669, 1987.
131. Gelderbloom, H.R., Reupke, H. and Pauli, G. Lancet 11:1016-1017, 1985.
132. Kabelitz, D., Herzog, W.R., Zanker, B. and Wagner, H. Scand. J. Immunol. 22:329-339, 1985.
133. Golding, H., Robey, F.A., Gatesii, F.T., Linde, W., Beining, W., Hoffman, T. and Golding, B. J. Exp. Med. 167:914-923, 1988.
134. Rickinson, A.B., Yao, Q.Y. and Wallace, L.E. Br. Med. Bull. 41:75-79, 1985.
135. Venables, P.J., Teo, C.G., Baboonian, C., Griffin, B.E. and Hughes, R.A. Clin. Exp. Immunol. 75:359-364, 1989.

136. Hengel, H., Lindner, M., Wagner, H. and Heeg, K. J. Immunol. 139:4196-4202, 1987.
137. Fauci, A.S. Science 239:617-622, 1988.
138. Zinkernagel, R.M. Immunology Today 9:370-371, 1988.
139. Alizon, A., Wain-Hobson, S., Montagne, L. and Sonigo, P. Cell 46:63, 1986.
140. Fisher, A.G., Ensoli, B., Looney, D., Rose, A., Gallo, R.C., Saag, M.S., Show, G.M., Hahn, B.H. and Wong-Stahl, F. Nature 344:444-447, 1988.

# 5

## CMV INFECTIONS AND IMMUNE RESPONSES

W. H. BURNS

**Oncology Center, The Johns Hopkins University School of Medicine, Baltimore, Maryland 21205 USA**

### ABSTRACT

CMV is an important pathogen in immunocompromised persons, including neonates, AIDS patients, organ transplant patients and oncology patients. Immunologic studies in humans and mice indicate a critical role for cytotoxic lymphocytes in recovery from infection. Cytotoxic T cell responses to products of the immediate early genes may be particularly important and these findings may provide a basis for the development of effective vaccines and for strategies to adoptively transfer immunity to CMV.

### INTRODUCTION

Cytomegalovirus (CMV) is a herpesvirus that produces characteristic inclusion bodies in infected cells. These were first observed in the early twentieth century and mistaken for protozoa. In the early 1950's the virus was isolated by Weller in Boston (1), Rowe in Bethesda (2) and Smith in St. Louis (3). As a congenital infection it was soon identified as a cause of microencephaly, mental retardation, deafness and blindness. About 80% of adults in the USA are seropositive for the virus and in the normal adult population it is present as a generally benign, latent infection, but with occasional reactivation and shedding in urine, saliva, cervical secretions or semen, particularly if immunosuppression occurs. Cultures of urine, saliva and blood leukocytes of over 1000 normal bone marrow donors and bone marrow transplant (BMT) patients studied pre-transplant at our institution, have always been negative for CMV. However, following BMT, nearly half of the patients have documented CMV infections (4). Recently, using a sensitive immunoblot assay to detect IgM antibody to CMV-specific proteins, McVoy and Adler (5) reported evidence for an age and gender associated IgM seropositivity suggesting that over half of normal persons over age 60 have recently reactivated CMV. The sites of this unexpectedly high frequency of reactivation and the amount of produced virus are unknown.

CMV has been considered "the commonest cause of life- threatening opportunistic viral infection in patients with the acquired immunodeficiency syndrome (AIDS)" (6). Nearly 100% of homosexuals with HIV infection are seropositive for CMV, about 50% of AIDS patients have CMV viremia and at autopsy about 90% of AIDS patients have evidence for CMV infection. Despite this high incidence of CMV infection and apparent high virus load, most CMV infection in the AIDS population is subclinical or associated with other opportunistic infections that make it difficult to assess the pathogenetic role of CMV. Retinitis is the most striking infection in AIDS patients directly attributable to CMV and occurs in about 20% of patients. Assessments of the role of CMV in gastrointestinal disease, hepatitis and pneumonia associated with CMV are more difficult because of the presence of other viruses or pathogenic organisms. This review will concentrate on the immune responses to CMV in non-AIDS patients, particularly BMT patients, because there is clearer data from this population which also has a high incidence of active CMV infections. Elucidation of normal immune responses to CMV and their relative importances is necessary as a frame of reference for future studies of CMV infections in AIDS patients.

## LATENT INFECTION

Much remains to be learned about the biology and pathogenesis of CMV infections. Although many cell types and organs have been noted to be infected in humans, the sites and nature of the latent state are unclear. Sites of reactivation of the virus often include salivary gland, cervix, prostate, liver, spleen, kidneys and leukocytes. However, it is not clear that the virus is actually latent at these sites in cells other than leukocytes. Use of the polymerase chain reaction (PCR) assay has demonstrated the presence of the viral genome in leukocytes of most normal seropositive persons if 40 or 50 amplification cycles are used (7,8) and one generally does not detect infectious virus in the blood of such persons by usual culture techniques. Also, transfusion of blood products from normal seropositive donors is a known mode of virus transfer, preventable by removal of leukocytes (9-11). Thus, in humans, leukocytes are the one proven site of latency in the sense of transmissible, reactivatable virus that cannot be cultured by usual techniques. However, until the latent form and expression of the viral genome is characterized molecularly, it is impossible to definitively differentiate a latent infection from the presence of the viral genome after its reactivation or after a new infection of a cell. It is probable that there are many sites of latency in addition to leukocytes. Interestingly, the data from PCR studies indicate that the granulocyte is the probable latency site in leukocytes, suggesting that myeloid or other hematopoietic precursors may actually be sites of latency. These observations differ from the findings of Olding *et al.* (12) implicating the B lymphocyte as a site of latency for mouse cytomegalovirus (MCMV).

**REINFECTION**

Immunocompromised patients (organ transplants and AIDS patients) often have reactivation of endogeneou CMV or are reinfected by new strains of the virus. Early studies using endonuclease restriction analysis of multiple CMV isolates from the same or different persons indicated that isolates from epidemiologically unrelated persons differ, and a single strain was isolated from one person. Thus pre-transplant isolates from the urines of 4 BMT patients differed from each other but were identical to the post-BMT isolates from the respective patients (13). However, more recently, we have compared endonuclease restriction patterns of isolates from 3 or more sites from individual BMT patients and in 6 of 11 patients we concluded that the individual harbored more than one strain of CMV (14). Similar data have been published on CMV isolates from renal transplant recipients (15-18), homosexuals and AIDS patients (19-21). It therefore appears that in immunocompromised patients, infection by multiple strains of CMV and their reactivation may be more common than previously thought. At present the implications of these findings for the pathogenesis, latency, and immunological responses to possible CMV antigenic variants are not known. The latter possibility may have important consequences. Isolates obtained from one of our patients evidenced differences in the endonuclease restriction patterns within the immediate early (IE) gene and the IE gene product(s) that is known to contain important epitopes for cytotoxic T cell recognition in the mouse-MCMV model (22).

## CELL-MEDIATED IMMUNITY TO CMV IN HUMANS

Studies of cell-mediated responses to CMV originally examined lymphocyte proliferation (LP) to CMV antigens following organ or bone marrow transplantation. These studies generally showed transplant recipients were capable of making LP responses but the assays were not very predictive if performed early after transplant (23). The LP assay is a measure of a mixture of cell types responding to the recognition of viral antigens and the manner of antigen presentation and also reflecting culture conditions. It is not a direct measure of effector cell function. In collaboration with Quinnan and coworkers, we prospectively and longitudinally studied BMT patients for direct (unamplified) lymphocyte cytotoxic activities against CMV-infected fibroblasts derived from skin biopsies from the bone marrow donors (24,25). In a given

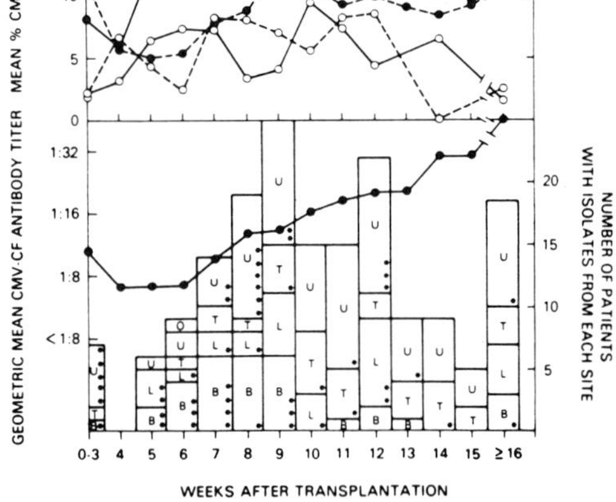


**Figure 1.** Cytomegaloviral (CMV)-specific cytotoxic responses of recipients of bone marrow transplants. Mean results of assays using HLA-matched (●——●) and mismatched (●- - - - -●) target cells and lymphocytes from infected (●) and uninfected (O) patients are shown by week after transplantation in the upper panel. In the lower panel, geometric mean antibody responses (●——●) are shown for all patients compared with numbers of virus isolates obtained from buffy coat or bone marrow (B), lung or liver (L), throat (T), urine (U), or other (O) sites. Asterisks (*) indicate the first CMV isolates obtained from individual patients.

**Table 1**. Effector cells mediating cytomegalovirus (CMV)-specific cytotoxic lymphocyte activity in peripheral blood of patients receiving bone marrow transplants.

| Patient Group | Type of CMV-specific cytotoxicity: HLA-Restricted | Not HLA-Restricted | Restriction Not Determined | None |
|---|---|---|---|---|
| CMV-infected | 25 | 3 | 7 | 16 |
| Uninfected | 0 | 18 | 0 | 9 |

experiment a patient's lymphocytes could be assayed against targets matched and mismatched at HLA loci. Results of an investigation of 78 patients studied in the first two months following BMT are summarized in Fig. 1 and Table 1 (adapted with permission from ref. 25).

We found that 35 of 51 infected patients were capable by 4 to 8 weeks after BMT of making significant CMV-specific responses (> 15% virus-specific lysis). Twenty-five of these 35 responding patients made HLA-restricted responses, 3 made unrestricted responses and restriction could not be determined for 7 responders because too few targets with differing HLA antigens were available in their assays. We were able to show through enrichment and depletion experiments that the HLA-restricted activity was primarily mediated by T cells. Although many patients had cytotoxicity mediated by both T cells and non-T lymphocytes, the presence of HLA-restriction always indicated that cytotoxicity was at least partly, and usually mostly, mediated by T cells. There was a highly significant correlation between the capacity to make a virus-specific cytotoxic response and survival from CMV infection. At least half of the 16 non-responding infected patients died directly of CMV infection (10 with CMV interstitial pneumonitis). Although the virus-specific HLA-unrestricted cytotoxic activity was not significantly increased over that found before infection, both it and the general natural killer (NK) cell activity (as measured against K562 targets) were significantly increased over that measured for the patients having fatal CMV infections (24). None of the uninfected patients exhibited HLA-restricted virus-specific cytotoxicity.

These results must be applied with caution to AIDS or other immunosuppressed patients. The BMT patients are unique because of their allogeneic setting, an immune system developing anew, and possible adoptive transfer of donor immunity. These findings must be corroborated by other groups and for other patient populations. Similar virus-specific cytotoxic responses mediated by a combination of cytotoxic T cells and NK cells were found after inoculation of normal seropositive volunteers with the Toledo strain of CMV, a not very attenuated strain, and lower responses were observed after inoculation with the more highly attenuated Towne strain (26).

Impressive as the unamplified cytotoxic T cell responses to CMV were in these patients, based on these studies we cannot separate their significance from that of virus-specific NK cell activity virus specific or general. Observations of the severity of herpesvirus and vaccinia infections in patients with congenital conditions of T cell deficiencies demonstrate the importance of T cells for recovery from these infections (see ref. 27). Until recently a human condition in which a similar degree of NK cell deficiency occurred was unknown. Sullivan and coworkers recently described a girl almost totally lacking lymphocytes possessing NK cell phenotype or function (28). She presented with life-threatening varicella and a few years later developed severe, protracted CMV infection and pneumonia requiring ganciclovir treatment. Various phenotype and functional studies indicated she had normal T cell numbers and functions and was able to make appropriate antibody responses. These human observations suggest important roles for both T cells and NK cells in controlling CMV infections and herpesvirus infections in general.

## CELL-MEDIATED IMMUNITY TO MCMV

Because of the extreme tropism of the cytomegaloviruses appropriate animal models must consist of the particular species and the natural cytomegalovirus of that species. Fortunately, mouse cytomegalovirus (MCMV) has been available for many years and investigations of immune responses to MCMV abound. There is considerable genetic variation in susceptibility to

MCMV with evidence for H-2-associated and non-H-2 associated control of resistance. For example, H-2$^k$ strains such as C3H and CBA mice are highly resistant while Balb/c (H-2$^d$) mice are susceptible and the susceptibility associated with H-2$^d$ is dominant to the H-2$^k$ resistance and segregates in backcross mice as a single gene. C57BL mice are also resistant but on a non-H-2 basis and the B10.BR strain, having H-2$^k$ on a C57BL background, are the most resistant mice. Early studies indicated that resistance to MCMV correlated with activation of NK cells during infection but an exception to the general finding was the B10.BR strain, the most resistant strain yet a low NK cell responder (29). Beige mice congenitally have depressed NK cell function with comparatively normal T cell functions and were noted to have increased susceptibility to MCMV infection (30). Further studies emphasized the importance of early interferon production after infection for resistance to MCMV but again there were too many exceptions for this to be the critical or sole basis for genetic resistance.

The situation is more complex than originally thought and Quinnan and Manischewitz (31) have recently published an extensive investigation in which they examined NK cell responses to infection, interferon production, inherent ability of adherent spleen cells or primary embryo cultures of different mouse strains to replicate MCMV, and the ability of strains to respond with cytotoxic T cells to infection. Their conclusions are that although there are consistent correlations between NK cell responses to MCMV infection and resistance for many mouse strains, there are important non-immunological factors determining genetic resistance to lethal MCMV infection. In their experiments these consisted of an H-2-linked interferon response gene and a non-H-2-linked gene with interferon independent effects. These genes influenced replication of the virus at the cellular level, independent of functions of the intact organism.

Grundy and Melief (32) found that Nu/Nu mice were more susceptible than Nu/+ littermates but that the increase in susceptibility was much greater for the more susceptible strains, suggesting that T cell immunity was more important in the latter strains, which had lesser NK cell responses to infection.

Several other investigators had demonstrated the presence and a protective role of T cells, particularly cytotoxic T cells, in MCMV infections (33-35). However, the most revealing investigations of the role of T cells in MCMV infections are the elegant studies of Koszinowski and coworkers. This group used X-irradiation to immunosuppress mice and showed enhanced virus growth and increased mortality from MCMV in such mice. Resistance could be restored by transfer of spleen cells from immune mice infected earlier and the necessary cells were shown to be T cells (36). This group then demonstrated that an important target antigen for cytotoxic T cells is the IE antigen(s) (22,37-39).

The IE locus codes for at least 3 proteins in MCMV (40-42). These proteins are synthesized first after infection and serve important regulatory roles for viral transcription and replication and are again synthesized late in infection. When DNA from the IE region was transfected into L cells along with a host gene coding for a class I MHC antigen, the resulting cell could serve as a target for virus specific T cells raised in mice of the appropriate MHC (43). Frequency analysis experiments to determine the relative frequency of precursor T cells with IE specificity or specificity for all MCMV expressed antigens indicated that IE antigen(s) are T cell immunodominant antigen(s) with the majority of precursor cytotoxic T cells specific for it (37). A clone of cytotoxic T cells specific for the IE antigen was selected and confirms that the IE protein in some form has accessible epitopes for T cells (43). This is somewhat surprising since IE products are phosphoproteins largely confined to the nucleus where they carry out their regulatory functions for replicating virus. It has been impossible to detect the IE antigens on the cell surface using antibody-based techniques. That this nuclear protein may have an important cell surface immunological role is consistent with recent findings that influenza nucleocapsid protein may play a similar role in cell-mediated responses in influenza infections (44,45). Thus the hypothesis that glycoproteins and other proteins clearly localizing to the cell surface are the most important in cell-mediated immunity must be reexamined.

The finding that mouse IE protein may elicit the largest number of precursor cytotoxic T cells does not necessarily mean that such a response is sufficient or critical for control of MCMV infection. The expanded clone of cytotoxic T cells specific for IE antigen was not capable of transferring protection to infected X-irradiated mice, although such failure could be due to inadequate cell numbers or inadequate local IL-2 or other lymphokines that helper T cells might provide. Reactivity with more than a single epitope might also be needed. The importance of responses to IE antigen(s), however, is shown by experiments in which mice were protected from lethal MCMV infection by immunization with a recombinant vaccinia virus into which the continuous coding sequence of the major MCMV IE (IE1) had been inserted (46). Furthermore, transfer of cytotoxic ($CD8^+$), but not helper ($CD4^+$) T cells from these immunized mice could protect X-irradiated recipients from lethal infection (47). This series of experiments strongly suggest that cytotoxic T cell responses to the IE antigen(s) of MCMV are sufficient for recovery from the infection. They do not negate an important role for responses to other MCMV antigens in the natural infection. Nor do they negate an important role for antibody, particularly in prevention of infection, although humoral responses probably play a secondary role to that of cell-mediated immunity in recovery from herpesvirus infections (see ref. 27).

## CONCLUSIONS

The role of CMV in the pathogenesis of AIDS and AIDS- related syndromes is still unclear except for retinitis. The presence of multiple simultaneous infections in AIDS patients makes it difficult to assign definite pathogenetic roles to CMV infections. Since virtually 100% of individuals in most groups at high risk for AIDS are infected with CMV, one is deprived of a CMV uninfected group for epidemiological comparisons. An exception are hemophiliacs. A recent comparison of HIV infected hemophiliacs seropositive or seronegative for CMV suggested a more rapid progression to AIDS in seropositive individuals, although confounding variables weakened the association (48). There are also obvious difficulties in measuring immune responses in the highly immunodeficient AIDS patients and a reference framework must be established of what is to be expected based on studies in other patient populations and normal persons.

In this chapter, therefore, studies in humans and mice that have produced clear or useful data have been reviewed with emphasis on cell-mediated immune response since they appear of primary importance in herpesvirus infections. From these studies it is probable that both cytotoxic T cell and NK cell activities play important roles in controlling CMV infections. The latter seem most important in early phases of primary infections while the T cell responses may be more important in the ultimate and long-term control of the infection.

New techniques make it possible to examine responses to specific antigens and to evaluate their relative importances in animal models. These techniques are now being applied to the study of infections in humans (49) and should provide a rational basis for the construction of effective vaccines and for strategies to adoptively transfer effector cells with defined specificities to patients unable to mount effective responses. These lymphocytes can be designed so as not to be infected by HIV, provide their own growth factors, or be administered along with a protective antiviral to AIDS patients. Along with the growing basic knowledge of CMV at the molecular level and the tools this knowledge provides, there is a critical need for their application in experimental pathogenesis for we are still largely ignorant of the basic biology of CMV latency, sites of latency, mechanisms of reactivation and details of interaction of the virus with the immune system.

## REFERENCES

1. Weller, T.H., Macaulay, J.C., Craig, J.M. and Wirth, P. Proc. Soc. Exp. Biol. Med 94:4-12, 1957.
2. Rowe, W.P., Huebner, R.J., Gilmore, L.K., Parrot, R.H. and Ward, T.G. Proc. Soc. Exp. Biol. Med 84:570-573, 1953.
3. Smith, M.G. Proc. Soc. Exp. Biol. Med. 92:424-430, 1956.
4. Wingard, J.R., Chen, D. Y. H., Burns, W. H., Fuller, D. J., Braine, H.G., Yeager, A.M. Kaiser, H., Burke, P.J., Graham, M.L., Santos, G.W. and Saral, R. Blood 71:1432-1437, 1988.
5. McVoy, M.A. and Adler, S.P. J. Infect. Dis. 160:1-10, 1989.
6. Jacobson, M.A. and Mills, J. Annals of Int. Med. 108:585-594, 1988.

7. Spector, S.A., Merrill, R., Meyers, J.D., Dankner, W.M. and Hsia, K. Second International Cytomegalovirus Workshop, Univ. of California, San Diego, CA, p. 60, 1989.
8. Street, A.C., Hamilton, J.D. and Klotman, M.E. Second International Cytomegalovirus Workshop, Univ. of California, San Diego, CA, p. 109, 1989.
9. Adler, S.P. Rev. of Infect. Dis. 5:977-993, 1983.
10. Bowden, R.A., Sayers, M., Slichter, S.J. and Meyers, J.D. Second International Cytomegalovirus Workshop, Univ. of California, San Diego, CA, p. 67, 1989.
11. de Gast, G.C., Gratama, J.W., Brand, A., Verdonck, L.F. and Mudde, G.C. Second International Cytomegalovirus Workshop, Univ. of California, San Diego, CA, p. 13, 1989.
12. Olding, L.B., Jensen, F.C. and Oldstone, M.B.A. J. Exp. Med. 141:561-572, 1975.
13. Winston, D.J., Huang, E.S.,. Miller, M.J., Lin, C.H., Ho, W.G., Gale, R.P. and Champlin, R.E. Annals of Int. Med. 102: 16-20, 1985.
14. Burns, W.H. and Sandford, G.R. Second International Cytomegalovirus Workshop, Univ. of California, San Diego, CA, p. 6, 1989.
15. Chou, S. N. Eng. J. Med. 314:1418-1423, 1986.
16. Grillner, L., Hokeberg, I., Eriksson, B.M. and Olding-Stenkvist, E. Second International Cytomegalovirus Workshop, Univ. of California, San Diego, CA, p. 8, 1989.
17. Chou, S. J. Infect. Dis. 160:11-15, 1989.
18. Chou, S. J. Infect. Dis. 160:16-21, 1989.
19. Drew, W.L., Sweet, E.S., Miner, R.C. and Mocarski, E.S. J. Infect. Dis. 150: 952-953, 1984.
20. Spector, S.A., Kirata, K.K. and Neuman, T.R. J. Infect. Dis. 150:953-956, 1984.
21. Collier, A.C., Chandler, S.H., Handsfield, H.H., Corey, L. and McDougall, J.K. J. Infect. Dis. 159:123-126, 1989.
22. Reddehase, M.J. and Koszinowski, U.H. Nature (London) 312:269-371, 1984.
23. Meyers, J.D. and Flournoy, N.T. J. Infect. Dis. 142:816-824, 1980.
24. Quinnan, G.V., Kirmani, N., Rook, A.H., Manischewitz, J.F., Jackson, L., Moreschi, G., Santos, G.W., Saral, R. and Burns, W.H. New Engl. J.Med. 307:7-13, 1982.
25. Quinnan, G.V., Burns, W.H., Kirmani, N., Rook, A.H., Manischewitz, J., Jackson, L., Santos, G.W. and Saral, R. Rev. Infect. Dis. 6:156-163, 1984.
26. Quinnan, G.V., Delery, M., Rook, A.H., Frederick, W.R., Epstein, J.S., Manischewitz, J.F., Jackson, L., Ramsey, K.M., Mittal, K., Plotkin, S.A. and Hilleman, M.R. Annals of Int. Med. 101:478-483, 1984.
27. Burns, W.H. and Allison, A.C. Virus Infections and the Immune Responses They Elicit. In: **Antigens, Vol. III** (Sela, M., Ed.). Academic Press, New York, pp. 479-575, 1975.
28. Biron, C.A., Byron, K.S. and Sullivan, J.L. New Engl. J. Med. 320:1731-1735, 1989.

29. Bancroft, G.J., Shellam, G.R. and Chalmer, J.E. J. Immunol. 126:988-994, 1981.
30. Shellam, G.R., Flexman, J.P., Farrell, H.E. and Papadimitriou, J.M. Scand. J. Immunol. 22:147-155, 1985.
31. Quinnan, G.V. and Manischewitz, J.F. J. Virol. 61:1875-1881, 1987.
32. Grundy (Chalmer), J.E. and Melief, C.J.M. J. Gen. Virol. 61:133-136, 1982.
33. Starr, S.E. and Allison, A.C. Infec. Immun. 17:348-361, 1977.
34. Quinnan, G.V., Manischewitz, J.E. and Ennis, F.A. Nature 273:541-543, 1978.
35. Quinnan, G.V., Manischewitz, J.E. and Ennis, F.A. J. Gen Virol. 47:503-508, 1980.
36. Reddehase, M.J., Weiland, F., Munch, K., Jonmic, S., Luske, A. and Koszinowski, U.H. J. Virol. 55:264-273, 1985.
37. Reddehase, M.J., Keil, G.M. and Koszinowski, U.H. Eur. J. Immunol. 14:56-61, 1984.
38. Koszinowski, U.H., Keil, G.M., Schwarz, H., Schickendanz, J. and Reddehase, M.J. J. Exp. Med. 166:289-294, 1987.
39. Reddehase, M.J., Mutter, W., Munch, K., Buhring, H.J. and Koszinowski, U.H. J. Virol. 61:3102-3108, 1987.
40. Keil, G.M., Ebeling-Keil, A. and Koszinowski, U.H. J. Virol. 61:526-533, 1987.
41. Keil, G.M., Ebeling-Keil, A. and Koszinowski, U.H. J. Virol. 61:1901-1908, 1987.
42. Keil, G.M., Fibi, M.R. and Koszinowski, U.H. J. Virol. 54:422-428, 1985.
43. Koszinowski, U.H., Reddehase, M.J., Keil, G.M. and Schickedanz, J. J. Virol. 61:2054-2058, 1987.
44. Andrew, M.E., Coupar, B.E.H., Boule, D.B. and Ada, G.L. Scand. J. Immunol. 25:21-28, 1987.
45. Taylor, P.M. and Askonas, B.A. Immunology 58:417-420, 1986.
46. Volkmer, H., Bertholet, C., Jonjic, S., Wittek, R. and Koszinowski, U.H. J. Exp. Med. 166:668-677, 1987.
47. Jonjic, S., del Val, M., Keil, G.M., Reddhase, M.J. and Koszinowski, U.H. J. Virol. 62:1653-1658, 1988.
48. Webster, A., Lee, C.A., Cook, D.G., Grundy, J.E., Emery, V.C., Kernoff, P.B.A. and Griffiths, P.D. Lancet, II:63-65, 1989.
49. Borysiewicz, L.K., Hickling, J.K., Graham, S., Sinclair, J., Cranage, M.P., Smith, G. L. and Sissons, J.G.P. J. Exp. Med. 168:919-931, 1988.

# 6

## IMMUNOBIOLOGY OF EBV INFECTIONS AND POTENTIAL INTERACTIONS WITH HIV

GARY R. PEARSON AND JOSE MENEZES

**Department of Microbiology, Georgetown University Medical School, Washington, D.C. and Laboratory of Immunovirology, Hospital Sainte-Justine and University of Montreal, Montreal, Canada.**

## ABSTRACT

Human Immunodeficiency Virus (HIV) is the causative agent of AIDS. However, numerous observations by various investigators indicate that co-factors may also be important in the pathogenesis of this disease. Herpesviruses are considered as potentially major factors in the etiology of AIDS because of their immunosuppressive activities and their role in HIV transactivation. This is particularly true for the Epstein-Barr Virus (EBV), a virus which exerts dramatic effects on immunoregulatory events during an active infection. The purpose of this chapter is to review the immunobiology of this virus and to identify those immunoregulatory events that are mediated by EBV which might be relevant to the pathogenesis of AIDS.

## INTRODUCTION

Acquired Immunodeficiency Syndrome (AIDS) is caused by an RNA virus designated Human Immunodeficiency Virus (HIV) and, because of its high morbidity and mortality rates, AIDS has rapidly become one of the most feared diseases of mankind. AIDS is characterized by a profound defect or suppression of cell-mediated immunity which predisposes the infected host to progressive opportunistic infections and unusual malignancies such as Kaposi's sarcoma and non-Hodgkin's lymphomas. Additionally, it has been shown in recent years that a considerable proportion of AIDS patients also develop neurological complications, the most common being a subacute encephalopathy with progressive dementia.

There is, in general, little doubt among most scientists that HIV is the primary cause of AIDS. It is also clear however that the majority of individuals with serologic evidence for HIV infection are asymptomatic for long periods of time. This suggests that co-factors may be critical for the progression of HIV infection to full-blown AIDS. Among the candidate co-factors are members of the *Herpesviridae* family which are immunosuppressive in themselves and for which there is good evidence for active infections in AIDS patients. Whether these infections function in a causal manner by acting as co-factors to HIV in the pathogenesis of AIDS or whether the infections are a consequence of the disease process which, by disturbing the cellular immune system, activate latent herpesvirus infections, is unknown. The purpose of this article is to examine one of these herpesviruses, Epstein-Barr Virus (EBV), and to discuss how this virus might function as a co-factor in the AIDS process as well as a major causative factor of the non-Hodgkin's lymphomas that occur at a high frequency in AIDS patients.

EBV, discovered in 1964 by Epstein and co-workers (1), is the causative agent of heterophil-positive infectious mononucleosis (IM) (2). It is also etiologically associated with nasopharyngeal carcinoma (NPC) and African Burkitt's lymphoma (BL) (3). More recently, the virus has been associated with polyclonal proliferations and malignant lymphomas in immunocompromised patients such as transplant patients and patients with AIDS and other immunodeficiency conditions (4-7).

The behaviour of the virus following primary infection appears to vary depending on the genetic makeup of the infected individual, the age of infection and the environment (2,3). In developing countries, infection generally occurs in early childhood without any apparent disease symptoms. In more developed countries and in higher socioeconomic populations, primary infection is generally delayed until adolescence and results in the development of IM. Once infection takes place, the virus is carried for life generally in a quiescent state. Evidence for persistent infection in the asymptomatic host is found in the nasopharynx where virus is shed intermittently (8,9), in peripheral blood lymphocytes where a small percentage of B-lymphocytes can normally be identified that express viral proteins (10) and by the presence of persisting antibodies directed against a number of EBV antigens (3,11). Environmental disturbances which result in disruption of the cellular immune system can activate the latent virus with clinical consequences which remain largely undefined. In general, reactivated infections are diagnosed by increased shedding of infectious virus in the saliva and changes in the antibody response patterns to different EBV antigens (11,12). The profound effects that EBV is capable of exerting on the immune system, as will be discussed later in this chapter, raise the possibility that active replication of EBV in the presence of an existing cellular immune defect could contribute to the rapid loss of immunoregulation observed in patients with immunodeficiency diseases such as AIDS and thereby accelerate the disease process. Therefore, should co-factors be of importance in the development of AIDS, EBV has to be considered as a major candidate.

**EPSTEIN-BARR VIRUS (EBV)**

EBV is an enveloped DNA virus that belongs to the herpesvirus family. The native virus particle measures between 150-180 nm in diameter, and the nucleocapsid is comprised of 162 capsomeres and has icosahedral symmetry (13). It has a very restricted host range and is therefore subclassified as a "gamma" herpesvirus. In the new nomenclature for herpesviruses, EBV is designated Human Herpesivurs 4 or HHV-4.

The EBV genome consists of a double-stranded DNA that is approximately 172 kb long and exhibits a buoyant density of 1.718 $g/cm^3$ corresponding to a G + C content of 58%. It consists of unique domains (U), major interanl repeats (IR) and tandem terminal repeat (TR) sequences of DNA. The genome is organized as TR-U1-IR1-U2-IR2-U3-IR3-U4-IR4-U5-TR (14). The TR or terminal repeat sequences are about 550 bp in length and are involved in the circularization of DNA to form episomes (14). Among EBV isolates, most of the polymorphism results from: (a) different numbers of IR elements, (b) minor repeats in the largely unique DNA domains and (c) differences in specific restriction endonuclease sites. The entire genome of the B95-8 strain of EBV has been cloned and sequenced and entire regions encoding for specific proteins have been mapped (15). In the productive EBV cycle, over 30 mRNAs have been isolated and translated *in vitro* to give a variety of polypeptides that react with human sera and monoclonal antibodies (11,16). As discussed later, functions have now been assigned to many of these proteins.

## VIRUS-CELL INTERACTIONS

It is known that the primary target for EBV infection is the mature resting B-lymphocyte population (17). Within this population, EBV is capable of infecting cells that synthesize immunoglobulins (18). In addition, the virus replicates in epithelial cells present in the upper respiratory tract (19,20) and this is now considered the probable major source of infectious virus isolated in the saliva of seropositive individuals and is probably the major source of the antigenic stimulus responsible for maintaining persistent antibody levels to EBV antigens.

Two types of interactions between B cells and EBV have been described. The first is the productive infection cycle whereby the virus replicates in the cell leading to the production of progeny virion and cell death. During this cycle, there is the expression of a variety of virus-specified antigens that are required for the production of new virus particles. The second type of infection leads to a non-productive proliferation of B-lymphocytes which is referred to as "immortalization". In the sequence of events that lead to immortalization *in vitro*, the virus binds to B-cells through a specific receptor. This attachment of the virus to the cell activates the resting B-cell and induces the expression of a B-cell activation marker designated CD23 (21). Sixteen to 20 hours post infection, the linear EBV genome reaches the nucleus of the cell and is circularized as a result of the covalent binding of the viral terminal tandem repeats to each other forming an episome (22). In addition, a series of virally encoded proteins, designated the Epstein-Barr Virus Nuclear Antigens (or EBNA), are transcribed from an 84 kb transcript whose expression is under the control of a promoter located in the U1 region of the viral genome (14). This transcript, which extends uninterrupted over half of the entire viral genome, is differentially spliced to generate the different EBNA proteins (23). These proteins are then transported to the nucleus of the infected cell. Expression of these proteins appears to be necessary to push the cell into DNA synthesis

(S phase), an event usually occuring 48 to 72 hours post infection. At this time a second viral-encoded protein designated latent membrane protein (or LMP) is also detected in the mebranes of infected cells (24). The resting B cell now has become a proliferating lymphoblast expressing cell surface markers characteristic of activated B cells and expressing all of the EBV functions required to maintain the viral genome. These are the events that lead to what is referred to as "polyclonal proliferation" of B-cells, a major characteristic of EBV infections. Despite the presence of the complete viral genome, only a few genes are expressed in immortalized cells.

Initially, episomal EBV DNA replicates in a coordinate fashion with the cellular genome in the infected cell. However, at about 1-2 weeks post infection, the viral genome begins to replicate faster than the cell DNA resulting in an amplification of the number of EBV episomes per cell (25). As a result, cells immortalized by EBV contain multiple copies of the EBV genome which are found mostly in non-integrated episomal forms (26,27). Integrated DNA however, has been detected in some cell lines (28,29). In cells latently infected with EBV, the viral genome is likely to be replicated via a cellular DNA polymerase since treatment of cells with drugs that have an effect on the viral DNA polymerase, such as acyclovir, does not alter episomal EBV DNA replication (30).

**THE EBV RECEPTOR**

EBV infects primarily B-lymphocytes. However, the viral genome has also been detected by cytohybridization (19,20) in epithelial cells and in T-cell lymphomas of patients with Kawasaki-like disease (31). Early data suggested that the host range for the virus was restricted to cells that expressed the receptor that binds the C3d component of complement (also referred to as CR2). Thus, all EBV susceptible cell lines were CR2 receptor positive and vice-versa (32,33). Moreover, two color immunofluorescence staining for EBV and CR2 receptors revealed completely overlapping staining patterns on lymphoblastoid cell lines and both receptors were found to co-cap (34). Finally, EBV adsorption was blocked with antibodies to CR2 and purified CR2 bound to EBV indicating that this type 2 complement receptor was the main receptor for EBV on B cells (33,35-37).

CR2 is expressed on mature B cells and more recently has been detected on human epithelial cells (38). This could account for the susceptibility of oropharyngeal epithelial cells to EBV. However, there is also evidence for receptors structurally different from CR2. It has been reported that EBV will also bind to some T-cell lines such as Molt-4 and that the binding of virus to these cells is not blocked by antibody to CR2 (39,40). In contrast to B cells, EBV binding to such T cells was not followed by internalization of the bound virus. This indicated that structure(s) other than CR2 could also serve as receptor(s) for EBV on certain cell lines. The exact identity of this receptor on non-B cells still needs to be determined.

**EBV ANTIGENS**

A variety of EBV-associated antigen complexes have been identified in EBV genome-positive cells using primarily immunofluorescence (IF) techniques (11,41). These antigens have been designated *VCA* for viral capsid antigens; *MA* for EBV-induced membrane antigens which are also expressed in the viral envelope; *EA* for EBV-induced early antigens which are expressed in the presence of inhibitors of viral DNA synthesis such as phosphonoacetic acid or cytosine arabinoside; and *EBNA* for the EBV-associated nuclear antigens which are detected in the nuclei of all EBV genome-positive cells. In addition, an antigen designated *LYDMA* (lymphocyte-detected membrane antigen) has been described which was believed to be the target for T-cell immunity directed against EBV genome-positive cells (41).

Until recently relatively little was known about the composition or functions associated with each complex. This was partly the result of the difficulty in purifying sufficient quantities of each protein from infected cells until the advent of monoclonal antibody technology and the development of sophisticated gene cloning techniques. Many of the proteins associated with the different antigen complexes have now been identified and biochemically and biologically characterized (11). In addition, the viral gene fragments encoding for many of these proteins have now been mapped. The current composition of the different antigen complexes is presented in Table 1.

**Table 1.** Proteins associated with different EBV antigen complexes.

| |
|---|
| Membrane Antigen (MA) |
| gp 350/300 |
| gp 250/200 |
| gp 85 |
| Viral Capsid Antigens (VCA) |
| p 160 |
| gp 125 |
| Early Antigen-Diffuse (EA-D) |
| p 140 |
| p 60 |
| p 47-56 |
| Early Antigen-Restricted (EA-R) |
| p 85 |
| p 17 |
| EBV-Induced Nuclear Antigens (EBNA) |
| p 78/70 |
| p 84/81 |
| p 140/130 |
| p 160/150 |
| p 70/22 |

**EPSTEIN-BARR VIRUS NUCLEAR ANTIGENS (EBNA).**

EBNA was first detected by Reedman and Klein in EBV genome-positive producer and non-producer lymphoblastoid cell lines by anti-complement IF (42). It was also demonstrated to be expressed in most African Burkitt's lymphoma biopsies, nasopharyngeal carcinoma biopsies and, more recently, in many of the polyclonal proliferations and monoclonal non-Hodgkins lymphomas occurring in immunodeficient patients (4-7). In fact EBNA is expressed in every cell that contains the EBV genome. This finding suggested that this antigen complex, which also binds to both double stranded and single stranded DNA (43), was of importance to the immortalization of infected cells. Because of this, the EBNA complex has been characterized extensively. It is now known that it is composed of at least 5 components designated EBNA 1 through EBNA 5 and biological functions have now been assigned to some of these components.

EBNA 1 was first described by Strnad *et al.* who identified this major EBV protein with the use of immunoblotting (44). This now appears to be the major EBNA component detected by anti-complement immunofluorescence and has a molecular weight between 65-80 kDa depending on the cell line used as the source of antigen. The gene encoding for EBNA 1 has been mapped to the BamHI K fragment (45) and the variation in the size of EBNA 1 proteins isolated from different cell lines is apparently related to the number of repeat glycine-alanine copolymers encoded by the third internal repeat segment (IR3) located in the BamHI K fragment (44,46). EBNA 1 binds to the origin of replication (Ori P) located in the BamHI C fragment and appears to be responsible for maintaining the EBV genome in episomal form (47). More recently, it has been reported that EBNA 1 can also bind to sequences present on the BamHI Q fragment and therefore might also possess other regulatory functions (48). This component however does not appear to be involved in the immortalization cycle since transfection of the BamHI K fragment into various cell lines has not resulted in altered growth potential.

EBNA 2 was also originally described by Strnad *et al.* (44). Using certain human sera, this group detected a protein with a molecular weight of approximately 81 kDa which was expressed in some but not all EBV genome-positive cell lines. Of particular interest was the absence of this antigen in extracts prepared from the $P_3$HR-1 and Daudi cell lines since virus isolates from these lines did not immortalize B-lymphocytes (49). This raised the possibility that EBNA 2 was an immortalizing protein. This now appears to be true. Following the identification of the BamHI WYH fragment as containing the gene encoding for EBNA 2 (50,51), this gene fragment was transfected into various cell lines with results which indeed suggested that this protein was important in the immortalization cycle (52).

EBNA 3 was discovered by Hennessy *et al.* (53) and Dillner *et al.* (54) who identified a 140 kDa protein that reacted with sera from certain patient populations such as those with rheumatoid arthritis or chronic infectious mononucleosis. This protein was immunologically distinct from EBNA 1 and EBNA 2. Subsequent immunoblotting experiments resulted in the identification of three high molecular weight antigens associated with EBNA 3 ranging in size from 132-150 kDa. More recently, the genes encoding for the EBNA 3 polypeptides were mapped to different open reading frames located on the BamHI E fragment of EBV DNA. Following transfection of these open reading frames into baby hamster kidney cells, Shimuzu *et al.*(55) identified three proteins with molecular weights of 136, 142 and 147 kDa. Thus EBNA 3 appears to be composed of a family of proteins. The protein with a molecular weight of 135 kDa is currently designated EBNA 3 while the protein with a molecular weight ranging from 150-180 kDa has been designated EBNA 4. The third band often noted in the EBNA 3 preparations is thought to be the result of post-translational modification of EBNA 4. No functions have as yet been assigned to EBNA 3 and 4.

EBNA 5 or EBNA-LP was identified from a cDNA clone that had been isolated from the Raji cell line (56,57). This is the product of a highly spliced transcript from the BamHI WYH fragment. The protein encoded by this region has a molecular weight between 22-70 kDa. Again, the function of this protein in one of the virus cycles remains to be determined.

## EPSTEIN-BARR VIRUS-INDUCED EARLY ANTIGEN COMPLEX.

The EBV-induced early antigens (EA) were defined as those antigens that were expressed early in the virus replication cycle and whose synthesis was inhibited in the presence of inhibitors of viral DNA synthesis (11,41). These antigens were further subdivided into diffuse (D) and restricted (R) components based on IF staining patterns and sensitivity to methanol fixation (58). The D components were localized in both the nuclei and cytoplasm of infected cells by IF and were resistant to denaturation by methanol. In contrast, the R components were localized primarily in the cytoplasm and were denatured by methanol fixation. Immunoprecipitation with human sera revealed a large number of proteins with molecular weights ranging from 30 to 165 kDa which were tentatively classified as components of the EA complex (11,41). The development of monoclonal antibody technology however, has resulted in further classification and characterization of these antigens. Based on current information, EA-D is composed of polypeptides ranging from 47 to 140 kDa (11). The major components are DNA-binding proteins with molecular weights of 50-52 kDa and 60 kDa (59,60). *In vitro* translation and immunoprecipitation with the monoclonal antibody to the 50/52 kDa proteins identified a 47 kDa protein encoded by the BamHI M fragment. The gene encoding for the proteins precipitated by this antibody has now been mapped to the BamHI MRF1 open reading frame by gene transfer experiments (61). The biological function associated with this complex of proteins has not been definitively determined but recent evidence identifed a topoisomerase activity associated with the 50/52 kDa proteins (62).

The second EA-D component was identifed by transfection experiments using a DNA fragment containing the BamHI MLF1 open reading frame (60). This is also a nuclear protein with a molecular weight of 60 kDa and has been reported to be a generalized trans-activator (63). It has been speculated that this protein is involved early in the activation of the virus replication cycle.

Two polypeptides have also been identified that are associated with the EA-R complex. These polypeptides have molecular weights of 85 kDa and 17 kDa (59,64). The 85 kDa polypeptide which is associated with cytoplasmic filaments (65) has been characterized extensively. When the amino acid sequences of peptides cleaved from the 85 kDa protein by cyanogen bromide were determined, a significant homology was noted between this protein and the viral-specified ribonucleotide reductase encoded by the BamHI ORF2 reading frame (66). This relationship was confirmed when it was noted that the monoclonal antibodies to the 85 kDa protein immunoprecipitated the large subunit of this enzyme and neutralized the viral enzyme activity *in vitro* (67). Thus, this 85 kDa EA-R component is associated with the large subunit of this virus specified enzyme.

The second EA-R polypeptide was more recently described. Using a monoclonal antibody prepared against the fusion product isolated from *E. coli* containing the BHRF1 open reading frame of EBV DNA, a 17 kDa protein was precipitated from virus-producing cell lines (64). Predicted amino acid sequences showed that there was significant homology between this protein and polyoma middle T antigen (47%) and the product of the bcl-2 oncogene (27%) expressed in follicular lymphoma cells. This protein was also detected in nasopharyngeal carcinoma cells by immunoblotting (68). More recently it was observed that the protein was expressed in a cell-cycle-dependent manner, which was compatible with cell survival, and was associated with the internal cytoplasmic and nuclear membranes (69). In addition, a threonine-serine protein kinase activity was consistently detected in immunoprecipitates containing the 17 kDa protein (69). This suggested that this protein could be involved in cellular immortalization as well as in the virus replication cycle.

Other EA proteins are also expressed in infected cells as determined by immunoprecipitation with human sera (11,41). These include a major 140 kDa polypeptide that appears to be methanol resistant and therefore a member of the EA-D complex. Currently there is little information about the characteristics of this protein.

ELISA assays have been developed for measuring antibody responses to each of these individual polypeptides in individuals infected with EBV (70). Measurement of antibodies that separate components of the EA complex instead of antibodies against the whole complex could reveal more specific markers for active virus replication and specific disease associations.

**VIRAL CAPSID ANTIGENS.**

Viral capsid antigens (VCA) are produced late in the virus replication cycle and are dependent upon the synthesis of new viral DNA for their expression (11,41). They were originally defined by IF and shown to be different from other antigen complexes by absorption and blocking experiments. Based on the use of monoclonal antibodies, it is now evident that the VCA complex, as defined by IF, is composed of polypeptides associated with the capsid and the viral envelope of the complete virus particle.

Through the use of monoclonal antibodies, two major polypeptides have now been identified which compose the VCA complex (71-73). These polypeptides have molecular weights of 160 and 125 kDa. The 160 kDa protein was previously recognized by Dolyniuk *et al.* as the major polypeptide composing the viral capsid (74). Antibody directed against this component reacted with the naked virus particle and also with soluble antigen in the nuclei of infected cells (73). The 160 kDa protein is encoded by the BamHI C fragment of the EBV genome (74).

The 125 kDa protein is expressed in all producer cell lines but not in non-producer cultures. It was conclusively identified as a VCA protein present in virus particles as well as in the infected cells through the use of monoclonal antibodies (71,72). In most immunoprecipitation experiments, the 125 kDa polypeptide was usually slightly larger in immunoprecipitates prepared from the $P_3$HR-1 cell line as opposed to the B95-8 culture (72). This suggested that the protein underwent post-translation modifications. The conclusion was substantiated by experimental findings which demonstrated that this protein was highly glycosylated and rich in mannose residues (72). More recent findings indicated that this protein was the antigenic homologue of herpes simplex virus glycoprotein B (gB) and varicella-zoster virus glycoprotein II (gp II) because antibodies produced against these two glycoproteins reacted with the 125 kDa protein of EBV and vice-versa (75). Whether this protein serves a function similar to HSV-1 gB is unknown. The BamHI A genome fragment has been shown to contain the gene that encodes for this major VCA glycoprotein (76).

All sera tested to date from EBV-infected individuals contain IgG antibodies reactive with both p 160 and gp 125 (11,73). However, while all sera from IM patients also contain IgM antibodies to the 125 kDa glycoprotein, this is not true for IgM antibodies to the 160 kDa VCA protein (11). These results indicate that the 125 kDa protein is the major immunogen following primary infection with EBV. ELISA tests have now been developed that specifically measure IgM antibody to this component as a method for diagnosing IM (11).

## EPSTEIN-BARR VIRUS-INDUCED CELL MEMBRANE ANTIGENS.

The membrane antigen (MA) complex is found on the surface of infected lymphoblastoid cells and in the envelopes of viral particles (77-80). These are late viral antigens that are responsible for the induction of neutralizing and cytotoxic antibodies and therefore have been the antigens of interest for the development of a vaccine against this virus (11). Accordingly, this antigen complex has been well characterized.

Three major membrane glycoproteins have now been identified by a number of investigators (11,41). These have recently been designated gp 350/300, gp 250/200 and gp 85 to take into account molecular weight variations reported by different laboratories (81). The genes encoding for gp 350/300 and gp 250/200 have been mapped to the BamHI L DNA fragment (82). This fragment encodes for 3.4 and 2.8 kb RNAs which, when translated, encode for proteins of 135 and 100 kDa respectively. The 135 kDa protein is similar in size to the protein precursor of gp 350/300, and the 100 kDa protein is the expected size of the precursor for gp 250/200. This observation indicated that it was very likely that these glycoproteins were encoded by overlapping messages from the same DNA region.

The membrane glycoprotein designated gp 85 is the least abundant of the different membrane glycoproteins. The gene encoding for gp 85 has been mapped to the BamHI X region of EBV DNA (83). Because of its DNA homology with the gH protein of herpes simplex virus and to gIII of varicella zoster virus, it has been postulated that gp 85 might be involved in fusion of the virus envelope to the host cell membrane (84,85). This still needs to be resolved.

Recent studies have shown that gp 350/300 mediates attachment of EBV to the C3d receptor on B cells (86,87). The purified glycoprotein was able to attach to the isolated C3d receptor in a dose-dependent manner as shown by dot blot immunoassays. Moreover, the binding was inhibited by monoclonal antibodies directed against both CR2 and gp 350/300. Computer analysis and comparison of the amino acid sequences of both gp 350/300 and the C3d complement fragment showed two regions of significant primary sequence homology suggesting that a common region of these two proteins is involved in CR2 binding (88).

Several monoclonal antibodies have been produced and shown to react with gp 350/300 from B95-8 cells and gp 250/200 from $P_3$HR-1 cells (11,41). These monoclonal antibodies significantly reduced the infectivity of these two strains of virus *in vitro* thereby showing that neutralizing antibodies were directed against these glycoproteins. Similarly neutralizing antibodies have been detected with monoclonal antibodies to gp 85 (89). Antibodies to gp 350/300 but not to gp 85 also mediate antibody dependent cellular cytotoxicity (ADCC) (90). gp 350/300 and gp 250/200 share a variety of antigenic determinants whereas gp 85 is antigenically distinct from the others. However, monoclonal antibodies to gp 350/300 have been produced which react only with epitopes on the B95-8 cell line (1). Thus, these high molecular weight glycoproteins isolated from different cell lines are not antigenically identical.

Major efforts have been directed at employing gp 350/300 as a vaccine for EBV infections with varying degrees of success. In the cotton top tamarin, vaccination with native gp 350/300 has been used successfully for preventing the induction of EBV-induced lymphoma (92). However, other EBV antigen preparations have not been so successful. Recombinant gp 350/300 is now available as a secreted protein (93) or expressed through different viral vectors such as vaccinia or varicella-zoster virus (94,95). Such preparations are currently being evaluated as appropriate vaccines for administering to human populations.

## IMMUNE RESPONSE TO EBV ANTIGENS

### Humoral immune responses.

A variety of antibodies are produced against EBV following primary or reactivated infections. These antibodies have been detected in different immunoglobulin classes and their presence, in some circumstances, can be of diagnostic or prognostic importance (11,41). In addition, humoral factors have been identified in some situations which block immune control of EBV infected cells (96). These factors could contribute to the development of EBV-associated diseases as will be discussed below.

The humoral response to EBV structural and non-structural antigens following primary infection has certain characteristics that are different from those generally noted with responses to other infectious agents. These characteristics result from the fact that EBV, in addition to being an immunogen, is also a B lymphocyte polyclonal activator. For this reason, EBV is capable of inducing B cells to produce, in addition to EBV-specific antibodies, a variety of immunodulatory factors. This is illustrated in Figure 1.

**Figure 1**. B-lymphocyte response to EBV infection.

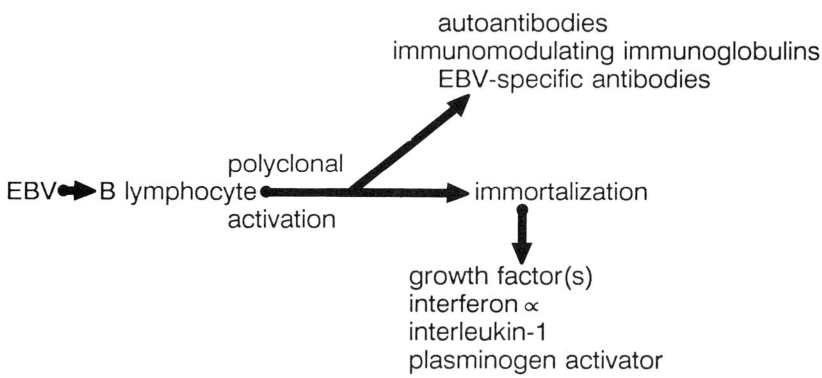


Following primary infection with EBV, antibodies to EBV antigens can be detected in the IgG, IgM and sometimes, but less frequently, in the IgA immunoglobulin fractions (97). The IgM antibody response to VCA, detected during the acute phase of IM, has developed into a useful diagnostic test for this disease. Following the subsidence of this disease, IgM antibodies disappear but IgG antibodies to VCA and MA persist for life.

During a primary EBV infection, a transient antibody response to EA is also frequently noted. This antibody, which appears to be a reliable indicator of an <u>active</u> EBV infection, usually decreases to undetectable levels during the convalescent stage of this disease (11,41,97). Antibodies to EA are also frequently detected at high titers in patients with defects in cellular immunity, including AIDS patients, suggesting that these patients have an active infection with EBV. This has been substantiated by the detection of increasing amounts of infectious virus in saliva of immunocompromised individuals (4-7).

Antibodies to EBNA and those antibodies detected by ADCC generally appear late in infection by comparison to other antibodies (7). The antibody response to EBNA was recently dissected by the Henles and co-workers and it was shown that the majority of the antibodies produced against EBNA are directed against either EBNA 1 or EBNA 2 (98). In the case of IM, antibodies to EBNA 2 emerged first and then declined to undetectable levels. Anti-EBNA 1 antibodies emerged with the disappearance of antibodies to EBNA 2 and then persisted for life. Therefore, in acute phase sera, the ratio of EBNA 1 to EBNA 2 antibodies was generally less than 1 while the ratio was greater than 1 in convalescent sera. Interestingly, in some instances the EBNA 1 to EBNA 2 antibody ratio also remained less than 1 in patients with immunodeficiencies (124). This result needs to be substantiated but indicates that the antibody response to EBNA 1 is depedent on a functional cellular immune system.

The delay in the antibody response to MA as measured by the ADCC assay reflects a delay in the antibody response to gp 350/300, the target antigen for this cytotoxic antibody (90). Antibodies to gp 85 can however be detected during the acute phase of EBV infection. The delayed antibody response to gp 350/300 might also be indicative of a cellular immune defect since unpublished results from some preliminary studies indicate that patients with diseases such as the chronic fatigue syndrome, which appears to be at least partially related to immunodeficiency, also have low levels of both ADCC antibodies and antibodies to gp 350/300.

In certain EBV-associated malignancies and in immunocompromised individuals, antibody titers to different EBV antigens change with disease course suggesting that such antibodies might be useful prognostic markers in patients with EBV associated malignancies (11,41). This is particularly true for the anti-EA antibody response noted in patients with nasopharyngeal carcinoma or African Burkitt's Lymphoma. In these patients, decreases in antibody titers to different EA components are generally good prognostic markers while increases denote the presence of recurrent or progressive disease.

Another unusual antibody response noted in patients with nasopharyngeal carcinoma is the presence of IgA antibodies to different EBV antigens (99). IgA antibodies are detected at a high frequency in the sera of patients with this disease compared to a variety of control populations. Because of this specificity, the IgA test has developed into a useful technique for the diagnosis of nasopharyngeal carcinoma and for the detection of individuals at risk for the development of this disease (100). The reason for high IgA antibody levels specifically in this patient population is not clear. Interestingly, a second population has recently been identified that also has an unusual high frequency of serum IgA antibodies. These are patients with AIDS as illustrated in Table 2. Why sera from AIDS patients are also frequently positive for IgA antibodies to EBV is unknown but could indicate that their appearance is related to a specific cellular immune defect to EBV-infected or transformed cells.

The importance of the IgA antibody response to the progression of EBV-associated diseases is not clear. However, it is now known that IgA antibodies or IgA antibody-antigen complexes are capable of blocking some of the immune responses that are considered of importance in controlling the growth of EBV-transformed cells and the spread of virus. For example, evidence has been obtained that ADCC antibodies might be of importance in immunity to the EBV-infected or -transformed cells. Of particular importance was the observation that progressive disease was observed more frequently in patients with low ADCC antibody titers in comparison with those with high antibody titers (100,101). Subsequent studies revealed that the low ADCC titers in patients with nasopharyngeal carcinoma were the result of IgA antibodies blocking this IgG-mediated cellular cytotoxic reaction (102). Similarly, it has been known for a long time that T lymphocytes from EBV-infected individuals will respond to EBV in *in vitro* proliferation assays demonstrating the presence of EBV-specific T memory cells in these individuals (103). Sera from patients with some diseases such as nasopharyngeal carcinoma frequently are capable of inhibiting the recognition of EBV antigens in these proliferation assays (96,104). This was subsequently demonstrated to be the result of blocking of immune recognition by IgA antibody-antigen complexes present in the circulation of these individuals (105). These results therefore suggest that IgA antibodies contribute to the progression of EBV-associated malignancies by inhibiting certain specific immune effector functions. Whether the presence of similar IgA or IgA-complexes in the sera of patients with AIDS would have a

**Table 2**. Typical patterns of antibodies to EBV in AIDS patients compared to patients with nasopharyngeal carcinoma and normal individuals persistently infected with EBV.

Antibody Titers

| IgG anti-VCA | | | IgG anti-EA | | | IgA anti-VCA | | |
|---|---|---|---|---|---|---|---|---|
| Norm. | NPC | AIDS | Norm. | NPC | AIDS | Norm. | NPC | AIDS |
| 80 | 640 | 1280 | <10 | 320 | 40 | <10 | 160 | <10 |
| 160 | 640 | 640 | <10 | 320 | 80 | <10 | 80 | 20 |
| 80 | 1280 | 640 | 20 | 640 | 80 | <10 | 320 | 20 |
| 40 | 640 | 1280 | <10 | 160 | 20 | <10 | 320 | 40 |
| 40 | 320 | 320 | <10 | 160 | <10 | <10 | 80 | <10 |
| 80 | 1280 | 640 | <10 | 320 | 40 | <10 | 640 | <10 |
| 20 | 1280 | 320 | <10 | 640 | <10 | <10 | 320 | 20 |
| 160 | 640 | 1280 | 20 | 320 | 160 | <10 | 320 | <10 |
| 160 | 640 | 640 | <10 | 320 | 40 | <10 | 160 | 40 |
| 80 | 320 | 640 | <10 | 80 | 40 | <10 | 80 | 20 |

more generalized immunosuppressive effect and thereby contribute to the progression of this disease, is worth investigating.

Cellular immunity.

There is substantial evidence in the literature for the existence of both non-specific and specific types of cellular immunity that function in controlling the growth of EBV transformed cells (106). For example, a cytotoxic lymphocyte response, which initially appeared to be specific for IM, was described and the target antigen for this response was designated LYDMA (107). It is now known that this cytotoxicity mediated by IM lymphocytes is not HLA-restricted and is mediated by activated T cells with properties similar to natural killer and lymphokine-activated killer (LAK) cells. These killer cells lyse EBV genome-negative cell lines as readily as EBV genome-positive cell lines so that there is no viral specificity in this cellular immunity which develops following primary EBV infection. The lack of viral specificity of this cytotoxic reaction has led to the hypothesis that cytotoxic cells from IM patients represent activated killer cells similar to LAK cells activated by IL-2 (108).

In contrast to what has been identified following primary EBV infection, HLA-restricted $CD8^+$ virus-specific cytotoxic cells have been identified in the blood of individuals persistently infected with EBV (106,108). These were recognized initially by the fact that, whereas spontaneous immortalization of EBNA-positive B cells occurs *in vitro* when B cells from a seropositive donor are cultured alone, the addition of autologous T cells inhibited this immortalization process. The T cell nature of this inhibitory effect was determined in inactivation studies using cyclosporin A and anti-T cell antibody. Moreover, patients treated with cyclosporin A or anti-T cell antibody exhibited a progressive increase of EBV-infected B cells *in vivo* further suggesting the existence of EBV-specific effector T cells. The T cells capable of inhibiting the outgrowth of autologous EBV-infected B cells mediated this inhibition by killing, as demonstrated by the chromium release assay, and this cytotoxicity was restricted by HLA class I antigens. These observations indicated that specific memory T cells become activated upon recognition of virus determinant(s) exposed on B cells. The identity of the target viral protein still has not been conclusively identified.

Of potentially more importance in regards to the co-factor question are the findings from a number of groups that EBV also induces the appearance of suppressor cells of the $CD8^+$ phenotype. During IM, it has been reported that suppressor T cells are activated that inhibit EBV-induced B-cell proliferation, pokeweed mitogen-driven B cell stimulation, and immunoglobulin production by cultured normal cells (109). The suppressor cells activated during IM are not HLA-restricted and do not mediate suppression through a soluble factor since supernatants of IM T cell cultures did not show any inhibitory effects (110). These suppressor cells were also not cytotoxic. The general depression in IM is therefore attributed to the activation of suppressor T cells with a broad range of inhibitory functions following an active EBV infection (96,106,108,110).

Besides the HLA non-restricted suppressor cells produced in the course of IM, HLA-restricted suppressor T cells have also been identified by Tosato *et al.* (111,112). Since EBV infects mature immunoglobulin secreting B cell populations, the number of B cells transformed *in vitro* can be quantitated by determining the number of immunoglobulin-producing cells using an immunoglobulin plaque-forming assay. With such an assay system, it has been shown that B cells infected with EBV *in vitro* yield increasing numbers of immunoglobulin producing cells. However, addition of T cells from an EBV-infected individual results in a marked decrease in the number of immunoglobulin secreting cells. Other findings indicated that this decrease was not mediated by cytotoxic T-cells and was HLA-restricted. Results from another study revealed that EBV antigen-specific $CD8^+$ suppressor cells could also be generated in the autologous system under conditions of antigen excess using lymphocytes from high responder EBV seropositive donors (113). Thus a virus-specific suppressor cell population could also be activated by EBV infection. Again, the nature of the target antigen remains unknown. Taken together, these results indicate that the immune response to primary EBV infection includes essentially the following components: production of EBV-specific antibodies, lymphocyte activation, an increase in NK and LAK-type cellular cytotoxicity, an elevation of suppressor ($CD8^+$) lymphocytes, and the development of EBV-specific cytotoxic cells. During both acute and chronic EBV infections, suppressor and cytotoxic cells are induced which can inhibit not only the proliferation of EBV-transformed cells but also the immune response to other antigens. This could be highly significant to the progression of AIDS in HIV-infected individuals with an active EBV infection.

**EBV AND AIDS.**

Most AIDS patients have serological and virological evidence for an active EBV infection. Moreover, it has been observed that HIV-infected patients have an increased risk of developing EBV-associated lymphoproliferative disorders, especially lymphomas (114,115). In addition, EBV-positive hairy leukoplakia lesions have been identified in patients with AIDS or at high risk for developing this disease (116). It is clear from the above discussion that EBV has profound effects on immunoregulation that could contribute significantly to the progression of AIDS as well as the other AIDS-related disease manifestations. These effects include the induction of both humoral and cellular elements with both specific and non-specific immunosuppressive activities. Of particular potential importance to the AIDS process is the induction of suppressor cells. These cells are capable of altering, not only the immune response to EBV itself, but also to non-related antigens thereby potentially contributing to the overall immunosuppression noted in patients with AIDS. In addition, it has been reported that EBV-immortalized cells are capable of producing soluble immunosuppressive factors such as plasminogen activator which could also play important role in the progression of this disease (96). Thus, because of the major effects that EBV has on immunoregulation, it should be considered as a major candidate co-factor in the pathogenesis of AIDS.

It is also possible that EBV can contribute to the progression of AIDS in other important ways. For example, it has been reported that following the conversion of B cell lines with EBV, these cells become susceptible to HIV infection (117,118). This is an extremely important observation and it suggests that EBV infection of B cells might result in the production of another population of cells *in vivo* that is capable of supporting the replication of the causative agent of AIDS. Similarly, as with other herpesviruses (119,120), it has been reported that an EBV immediate early gene product is capable of transactivating the HIV-LTR *in vitro* (121). This suggests another mechanism by which EBV might activate HIV expression if both viruses are present in the same cells. More recently, Clouse *et al.* (122) suggested a third mechanism by which an active EBV infection might activate HIV replication. They reported that EBV was capable of activating the replication of HIV by inducing cytokine production by monocytes or T cells, presumably through a specific immunological recognition mechanism. It was previously reported by this same group that other antigens were also capable of inducing a variety of cytokines capable of activating HIV (123). This therefore offers another avenue by which an active EBV infection might serve as a co-factor in the pathogenesis of AIDS.

In summary, the following scenario could be proposed to implicate EBV as a co-factor in the progression of HIV disease (Figure 2). First of all, as discussed above, EBV by itself could exert an immunosuppressive effect in the HIV-infected individual. Further, following HIV infection and initial alterations in

**Figure 2**. Scheme identifying possible pathways of interaction between EBV and HIV in the pathogenesis of AIDS.

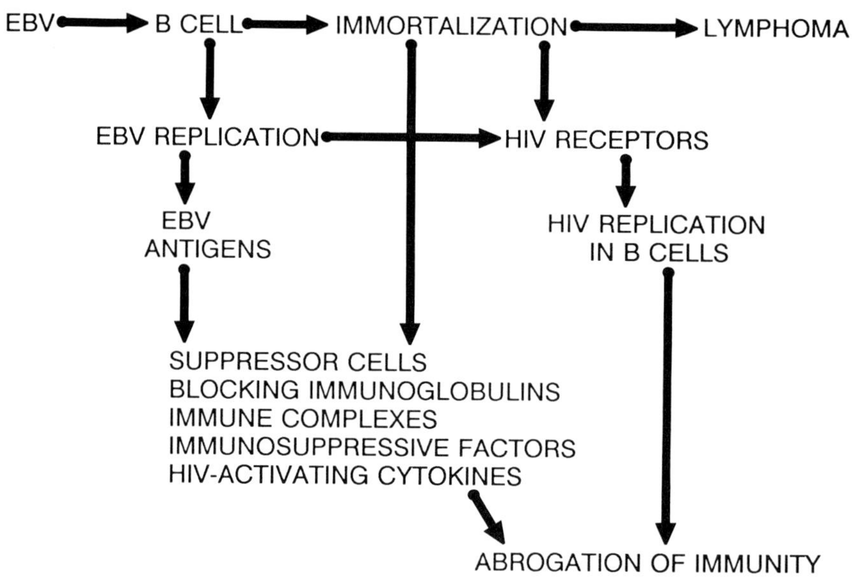


cellular immunity in the infected host, latent viruses such as EBV could be activated to replicate and to infect new B cells. The infection of new B lymphocytes by EBV would result in the expression of new HIV receptors thereby converting these cells into another potential reservoir for the production of HIV, thus increasing the HIV load in the host. The replication of HIV could be further augmented by cytokines produced by macrophages or T cells in response to EBV antigens. In addition, as discussed above, active

EBV replication could result in the reappearance of suppressor cells that have previously been identified in patients with IM as well as the production of immunosuppressive factors. The eventual emergence of EBV genome-positive lymphomas would also contribute to the immunosuppressive load in HIV-infected individuals. The overall effect would be a dramatic suppression of the immune effector mechanisms when considered in the context of the existing defects already induced by HIV, thereby accelerating the progression of HIV-associated disease.

**REFERENCES.**


1. Epstein, M.A., Achong, B.G. and Barr, Y.M. Lancet 1:702-703, 1964.
2. Henle, G. and Henle, W. In: **The Epstein-Barr Virus** (Eds. M.A. Epstein and B.G. Achong), Springer Verlag, Berlin, 1979, pp 297-320.
3. Henle, W. and Henle, G. In: **The Epstein-Barr Virus** (Eds. M.A. Epstein and B.G. Achong), Springer Verlag, Berlin, 1979, pp 61-78.
4. Purtilo, D.T. and Klein, G. Cancer Res. 41:4209-4304, 1981.
5. Cleary, M.L., Dorfman, R.F. and Sklar, J. In: **The Epstein-Barr Virus: Recent Advances** (Eds. M.A. Epstein and B.G. Achong), Heinemann, London, 1986, pp 163-182.
6. Hamilton,-Dutoit, S.J., Pallesen, G., Karkov, J., Skinhj, P., Franzmann, M.B. and Pedersen, C. Lancet 1:554, 1989.
7. Knowles, D.M., Inghirami, G., Ubriaco, A. and Dalla-Favera, R. Blood 73:792-799, 1989.
8. Gerber, P., Nonoyama, M., Lucas, S., Perlin, E. and Goldstein, L.I. Lancet 2:988-989, 1972.
9. Miller, G., Niederman, J.C. and Andrews, L. N. Engl. J. Med. 288:229-232, 1973.
10. Nilsson, K., Klein, G., Henle, N. and Henle, G. Int. J. Cancer 8:443-450, 1971.
11. Pearson, G.R. and Luka, J. In: **The Epstein-Barr Virus: Recent Advances** (Eds. M.A. Epstein and B.G. Achong), Heinemann, London, 1986, pp 48-74.
12. Yao, Q., Rickinson, A.B. and Epstein, M.A. Int. J. Cancer 35:35-42, 1985.
13. Miller, G. In: **Fields Virology** (Ed. B.N. Field), Raven Press, N.Y., 1985, pp 565-589.
14. Dambaugh, T., Hennessy, K., Fennewald, S. and Kieff, E. In: **The Epstein-Barr Virus: Recent Advances** (Eds. M.A. Epstein and B.G. Achong), Heinemann, London, 1986, pp 13-46.
15. Baer, B., Banker, A., Biggin, M.D., Deninger, P.L., Farrell, P.J., Gibson, T.J., Hatfield, G., Hudson, G.S., Satchnell, S.C., Seguin, C. and Barrell, B.G. Nature 310:207-211, 1984.
16. Dillmer, J. and Kallin, B. Adv. Cancer Res. 50:95-158, 1988.
17. Aman, P., Ehlin, B. and Klein, G. J. Exp. Med. 159:208-220, 1984.

18. Brown, N., Smith, D., Miller, G., Niederman, J., Liu, C. and Robinson, J. J. Infect. Dis. 150:517-522, 1984.
19. Sixby, J.W., Vesterinen, E.H., Nedrud, J.G., Raab-Traub, N., Walton, L.A. and Pagano, J.S. Nature 306:480-483, 1983.
20. Sixby, J.W., Nedrud, J.G., Raab-Traub, N., Hanes, R.A. and Pagano, J.S. N. Engl. J. Med. 310:1225-1230, 1984.
21. Thorley-Lawson, D.A. and Mann, K.P. J. Exp. Med. 162:45-59, 1985.
22. Hurley, E.A. and Thorley-Lawson, D.A. J. Exp. Med. 168:2059-2075, 1988.
23. Bodescot, M. and Perricaudet, M. Nucleic Acid Res. 14:7103-7164, 1986.
24. Mann, K.P., Stanton, D. and Thorley-Lawson, D. J. Virol. 55:710-720, 1985.
25. Sugden, B., Phelps, M. and Domoradzki, J. J. Virol. 31:590-595, 1979.
26. Nonoyama, M. and Pagano, J.S. Nature 238:169-171, 1972.
27. Lindahl, T., Adams, A., Bjursell, G., Borkamm, G.W., Kaschkadierich, G. and Jehn, U. J. Mol. Biol. 102:511-530, 1976.
28. Matsuo, T., Heller, M., Petti, L., O'Shira, E. and Kieff, E. Science 226:1322-1325, 1984.
29. Kieff, E., Dambaugh, T. and Hennessy, K. In: **The Role of Viruses in Human Cancer** (Eds. G. Giraldo and E. Beth), Elseiver, N.Y. 1984, pp 103-118.
30. Sixbey, J.W. and Pagano, J. J. Virol. 53:299-301, 1985.
31. Kikuta, H., Taguchi, Y., Tomizawa, K., Kojima, K., Kawamura, N., Ishizaka, A., Sakiyama, Y., Matsumoto, S., Imai, S., Kinoshita, T., Koizumi, S., Osato, T., Kobayashi, I., Hamada, I. and Hirai, K. Nature 333:455-457, 1988.
32. Jondal, M., Klein, G., Oldstone, M., Bokish, V. and Yefenof, E. Scand. J. Immunol. 5:401-410, 1976.
33. Fingeroth, J., Weiss, J., Tedder, T., Strominger, J., Bird, P. and Fearon, D. Proc. Natl. Acad. Sci. 81:4510-4514, 1984.
34. Yefenof, E., Klein, G., Jondal, M. and Oldstone, M.B.A. Int. J. Cancer 17:693-700, 1976.
35. Nemerow, G., Wolfert, R., McNaughton, M. and Cooper, N. J. Virol. 55:347-351, 1985.
36. Nemerow, G., Siaw, M.F.E. and Cooper, N.R. J. Virol. 58:709-712, 1986.
37. Mold, C., Cooper, N.R. and Nemerow, G. J. Immunol. 136:4140-4151.
38. Young, L.S., Clark, D., Sixbey, J.W. and Rickinson, A.B. Lancet 1:240-241, 1986.
39. Menezes, J., Seigneurin, J.M., Patel, P., Bourkas, A. and Lenoir, G. Virol. 22:816-821, 1977.
40. Stocco, R., Sauvageau, G. and Menezes, J. Virus Research 11:209-225, 1988.
41. Pearson, G.R. In: **Viral Oncology** (Ed. G. Klein), Raven Press, N.Y., 1980, pp 739-767.
42. Reedman, B.M. and Klein, G. Int. J. Cancer 11:499-520, 1973.
43. Sculley, T.B., Spelsberg, T.C. and Pearson, G.R. Intervirol. 22:191-200, 1984.
44. Strnad, B.C., Schuster, T.C., Hopkins, R.F., Neubauer, R.H. and Rabin,

H. J. Virol. 38:996-1004, 1981.
45. Summers, W., Grogan, E., Sheed, D., Robert, M., Lui, C. and Miller, G. Proc. Natl. Acad. Sci. 79:5688-5692, 1982.
46. Gergely, L., Sternas, L., Dillner, J. and Klein, G. Intervirol. 22:85-89, 1984.
47. Yates, J.L., Warren, N. and Sugden, B. Nature 313:812-815, 1985.
48. Ernberg, I., Kallin, B., Dillner, J., Falk, K., Ehlin-Henriksson, B. and Klein, G. Int. J. Cancer 38:729-737, 1986.
49. King, W., Dambaugh, T., Heller, M., Dowling, J. and Kieff, E. J. Virol. 43:979-986, 1982.
50. Dambaugh, T., Hennessy, K., Chammanankit, L. and Kieff, E. Proc. Natl. Acad. Sci. 81:7632-7636, 1984.
51. Dillner, J., Kallin, B., Ehlin-Henriksson, B., Timman, L. and Klein, G. Int. J. Cancer 35:359-366, 1985.
52. Dambaugh, T., Wong, F., Hennessy, K., Woodland, E., Rickinson, A. and Kieff, E. J. Virol. 59:435-462, 1986.
53. Hennessy, K., Fennewald, S. and Kieff, E. Proc. Natl. Acad. Sci. 82:5944-5948, 1985.
54. Dillner, J., Kallin, B., Ehlin-Henriksson, B., Rymo, L., Henle, W., Henle, G. and Klein, G. Int. J. Cancer 37:195-200, 1986.
55. Shimizu, N., Yamaki, M., Sakuma, S., Ono, Y. and Takada, K. Int. J. Cancer 41:744-751, 1988.
56. Dillner, J., Kallin, B., Alexander, H., Ernberg, I., Uno, M., Ono, Y., Klein, G. and Lerner, R. Proc. Natl. Acad. Sci. 83:190-197, 1986.
57. Wang, F., Pettit, L., Braun, D., Seung, S. and Kieff, E. J. Virol. 61:945-954, 1987.
58. Henle, G., Henle, W. and Klein, G. Int. J. Cancer 8:272-278, 1971.
59. Pearson, G.R., Vroman, B., Chase, B., Sculley, T., Hummel, M. and Kieff, E. J. Virol. 47:193-201, 1983.
60. Cho, M., Jeang, K. and Hayward, D. J. Virol. 56:852-859, 1985.
61. Cho, M., Milman, G. and Hayward, D. J. Virol. 56:860-866, 1985.
62. Takoada, Y., Luka, J., Raab-Traub, N. and Pearson, G.R. Unpublished results.
63. Lieberman, P.M., O'Hare, P., Hayward, G.S. and Hayward, S.D. J. Virol. 60:140-148, 1986.
64. Pearson, G.R., Luka, J., Pettit, L., Sample, J., Birkenbach, M., Braun, D. and Kieff, E. Virol. 160:151-161, 1987.
65. Luka, J., Miller, G., Jornvall, H. and Pearson, G.R. J. Virol. 58:748-756, 1986.
66. Goldschmitdts, W., Luka, J. and Pearson, G.R. Virol. 157:220-226, 1987.
67. Goldschmidts, W., Ginsburg, M. and Pearson, G.R. Virol. 170:330-333, 1989.
68. Lika, J., Deeb, Z.E., Hartmann, D.P., Jenen, B. and Pearson, G.R. J. Natl. Cancer Inst. 80:1164-1167, 1988.
69. Kocache, M. and Pearson, G.R. Unpublished results.
70. Luka, J., Chase, R.C. and Pearson, G.R. J. Immunol. Methods 67:145-156, 1984.
71. Takada, K., Fujiwara, S., Yano, S. and Osato, T. Med. Microbiol. Immunol. 171:225-231, 1983.

72. Kishishita, M., Luka, J., Vroman, B., Poduslo, J.F. and Pearson, G.R. Virol. 133:363-375, 1984.
73. Vroman, B., Luka, J., Rodriguez, M. and Pearson, G.R. J. Virol. 53:107-113, 1985.
74. Dolyniuk, M., Wolff, E. and Kieff, E. J. Virol. 18:289-297, 1976.
75. Emini, E.A., Luka, J., Armstrong, M.E., Keller, P.M., Ellis, R.W. and Pearson, G.R. Virol. 157:552-555, 1987.
76. Gong, M., Ooka, T., Matsuo, T. and Kieff, E. J. Virol. 61:499-508, 1987.
77. Klein, G., Clifford, P., Klein, E., Smith, R.T., Minowada, J., Kourilsky, F.M. and Burchenal, J.H. J. Natl. Cancer Inst. 39:1027-1044, 1967.
78. Pearson, G.R., Dewey, F., Klein, G., Henle, G. and Henle, W. J. Natl. Cancer Inst. 45:989-995, 1970.
79. Pearson, G.R., Henle, G. and Henle, W. J. Natl. Cancer Inst. 46:1243-1250, 1971.
80. Pearson, G.R. and Qualtiere, L.F. J. Virol. 28:344-351, 1978.
81. Thorley-Lawson, D.A., Edson, C.M. and Geilinger, K. Adv. Cancer. Res. 36:295-348, 1982.
82. Hummel, M., Thorley-Lawson, D.A. and Kieff, E. J. Virol. 49:413-417, 1984.
83. Heineman, T., Gong, M., Sample, J. and Kieff, E. J. Virol. 62:1101-1107, 1988.
84. Oba, D. and Hutt-Fletcher, L.M. J. Virol. 62:1108-1114, 1988.
85. Miller, N. and Hutt-Fletcher, L.M. J. Virol. 62:2366-2372, 1988.
86. Nemerow, G., Mold, C., Schwend, V.K., Tollefson, V. and Cooper, N. J. Virol. 61:1416-1420, 1987.
87. Tanner, J., Weis, J., Fearon, D., Whang, Y. and Kieff, E. Cell 50:203-213, 1987.
88. Nemerow, G., Houghten, R.A., Moore, M.D. and Cooper, N.R. Cell 56:369-377, 1989.
89. Strnad, B.C., Schuster, T., Klein, R. and Rabin, H. J. Virol. 41:258-264, 1982.
90. Qualtiere, L.F., Chase, R. and Pearson, G.R. J. Immunol. 129:814-818, 1982.
91. Qualtiere, L.F., Chase, R., Vroman, B. and Pearson, G.R. Proc. Natl. Acad. Sci. 79:616-620, 1982.
92. Epstein, M.A. and Morgan, A.J. In: **The Epstein-Barr Virus: Recent Advances** (Eds. M.A. Epstein and B.G. Achong), Heinemann, London, 1986, pp 272-289.
93. Whong, Y., Silberklong, M., Morgan, A., Munski, S., Lenny, A. and Kieff, E. J. Virol. 61:1796-1807, 1987.
94. Mackett, M. and Arrand, J.R. EMBO J. 4:3229-3231, 1985.
95. Lowe, R.S., Keller, P.M., Keech, B.J., Davidson, A.J., Whong, Y., Morgan, A.J., Kieff, E. and Ellis, R.W. Proc. Natl. Acad. Sci. 84:3896-3900, 1987.
96. Menezes, J., Sundar, S.K. and Ahronheim, G.A. In: **Viral Mechanisms of Immunosuppression**, Allan R. Liss, Inc., 1985, pp 115-134.
97. Pearson, G.R. In: **Infection mononucleosis: humoral response** (Ed. O. Schlossberg), Praeger Monographs in Infectious Diseases, vol. 1, N.Y., 1983, pp 141-158.
98. Henle, W., Henle, G., Andersson, J., Ernberg, I., Klein, and Straus, S.E.

Proc. Natl. Acad. Sci. 84:570-574, 1987.
99. Henle, G. and Henle, W. Int. J. Cancer 17:1-7, 1976.
100. Pearson, G.R. In: **Current Hematology and Oncology** (Ed. V.F. Fairbanks), Year Book Medical Publishers, Inc., 1986, pp 123-138.
101. Neel, H.B., Pearson, G.R. and Taylor, W.F. Arch. Otoloryngol. 110:742-747, 1984.
102. Mathew, G.D., Qualtiere, L.F., Neel, H.B. and Pearson, G.R. Int. J. Cancer 27:175-180, 1981.
103. Gerber, P. and Lucas, S.J. Cell. Immunol. 5:318-324, 1972.
104. Sundar, K.S., Ablashi, D.V., Kamaraju, L.S., Levine, P.H., Faggioni, A., Armstrong, G.R., Pearson, G.R., Krueger, G.R.F., Hewetson, J.F., Bertram, G., Sesterhenm, K. and Menezes, J. Int. J. Cancer 29:407-412, 1982.
105. Sundar, K.S., Menezes, J., Levine, P.H., Ablashi, D.V., Kamaraju, L.S., Faggioni, A. and Prasa, U. In: **Nasopharyngeal Carcinoma** (Ed. U. Prasad, D.V. Ablashi, P.H. Levine, G.R. Pearson), University of Malaya Press, 1983, pp 293-298.
106. Rickinson, A.B. In: **Epstein-Barr Virus: Recent Advances** (Ed. M.A. Epstein and B.G. Achong), Heinemann, 1985, pp 75-126.
107. Svedmyr, E. and Jondal, M. Proc. Natl. Acad. Sci. 72:1622-1626, 1975.
108. Tosato, G. Adv. Cancer Res. 49:75-125, 1987.
109. Tosato, G., Magrath, I., Koski, I., Dooley, W. and Blaese, M. N. Engl. J. Med. 301:1133-1137, 1979.
110. Tosato, G. and Blaese, R.M. Adv. Immunol. 37:99-149, 1985.
111. Tosato, G., Magrath, I. and Blaese, R.M. J. Immunol. 128:575-579, 1982.
112. Tosato, G., Pike, S.E., Koski, I.R. and Blaese, R.M. J. Immunol. 128:1986-1991, 1982.
113. Sundar, S.K. and Menezes, J. Int. J. Cancer 35:351-357, 1985.
114. Hanto, D.W., Frizzera, G., Gajl-Pezalske, K.J. and Simmons, R.L. Transplant. 39:461-472, 1985.
115. Ziegler, J.L., Beckstead, J.A., Volberding, P.A. , Abrams, D.I., Levine, A.M., Lukes, R.J., Gill, P.S., Burkes, R.L., Meyer, P.R., Metroka, C.E. et al. N. Engl. J. Med. 311:565-570, 1984.
116. Greenspan, J.S., Greenspan, D., Lennette, E.T., Abrams, D.I., Conant, M.A., Peterson, V. and Freese, U.K. N. Engl. J. Med. 313:1564-1571, 1985.
117. Montagnier, L., Gruest, J., Chamaret, S., Dougreet, C., Axler, C., Geustard, D., Nugeyre, M.T., Senoussi, F.B., Brunet, J.B., Klatzmann, D. and Gluckman, J.C. Science 225:63-66, 1984.
118. Salahuddin, S.Z., Ablashi, D.V., Hunter, E.A., Gonda, M.A., Sturzenegger, S., Markham, P.D. and Gallo, R.C. Int. J. Cancer 39:198-202, 1987.
119. Gendelman, H.E., Phelps, W., Feigenbaum, L., Ostrove, J.M., Adachi, A., Howley, P.M., Khoury, G., Ginsberg, H.S. and Martin, M.A. Proc. Natl. Acad. Sci. 83:9759-9763, 1986.
120. Mosca, J.D., Bednarik, D.P., Raj, N.B., Rosen, B.C., Sadroski, J.G., Haseltine, W.A. and Pitha, P.M. Nature 325:67-70, 1987.
121. Pagano, J.S., Kenney, S., Markovitz, D. and Kamine, J. J. Virol. Methods 21:229-239, 1988.

122. Clouse, K.A., Robbins, P.B., Fernie, B., Ostrove, J.M. and Fauci, A.S. J. Immunol. 143:470-475, 1989.
123. Clouse, K.A., Powell, D., Washington, I., Poli, G., Strebel, K., Farrar, W., Barstad, P., Kovacs, J., Fauci, A.S. and Folks, T.M. J. Immunol. 142:431-438, 1989.
124. Masucci, G., Mellstedt, H., Henle, G., Henle, W., Rymo, L., Masucci, M.G., Ernberg, I. and Klein, G. Int. J. Cancer 43:1017-1021, 1989.

# 7

## THE ROLE OF IMMUNOSUPPRESSION IN HERPES SIMPLEX VIRUS PATHOGENESIS AND ITS RELATIONSHIP TO AIDS

LAURE AURELIAN

**Virology Immunology Laboratories, The University of Maryland School of Medicine and Divisions of Biophysics and Comparative Medicine, The Johns Hopkins Medical Institutions, Baltimore, Maryland 21201 USA**

### ABSTRACT

Latency, the hallmark of Herpes simplex virus (HSV) infection is defined as long-term virus persistence in a clinically silent state with periodic episodes of overt disease caused by reactivated virus. The ability of the reactivated virus to replicate in the presence of apparently normal HSV-specific immunity has intrigued the students of HSV latency for many generations. This chapter reviews HSV immunity focusing on immunoregulatory events during latency and recurrent disease. Particular emphasis is placed on the role of immunosuppression and the definition of factors that govern its generation in response to secondary exposure to HSV antigen. HSV-mediated immunoregulatory events that might impact on infection with human immunodeficiency virus (HIV) and in the pathogenesis of Acquired Immunodeficiency Syndrome (AIDS), are discussed.

### INTRODUCTION

Two HSV types infect man. HSV type 1 (HSV-1) that primarily causes facial lesions, and HSV type 2 (HSV-2) that is a leading cause of ulcerative lesions of the genital tract. The number of patients consulting for genital HSV infections in the US is currently estimated at 5-9 million/year. In immunosuppressed patients, HSV infections are severe and even fatal. Genital infection in early pregnancy has been linked with the development of fetal malformations and increased abortion rates (1). Severe infections of newborn infants acquired by contact with virus at the time of delivery, have been reported, with mortality rates as high as 90% (2).

Recovery from infection is associated with the development of virus-specific antibody and antigen-reactive lymphocytes. Despite this apparently normal sequence of immunologic events, approximately one-half of the individuals infected with HSV develop clinical recurrences and reinfection is relatively common (3). This raises major questions pertaining to the role of immunity in the control of HSV infections.

The immune response to the herpesviruses includes an early non-specific containment phase characterized primarily by local host defense factors, and a later virus-specific acquired phase that consists of humoral and cell mediated (CMI) functions. These responses do not prevent HSV latency, recurrent disease and reinfection. Accordingly, consideration of the role that immunity plays in the pathogenesis of HSV disease and the potential impact of HSV mediated immunoregulatory events on the progression of HIV infection must address the function, protein-specificity, and regulation of protective immunity. This chapter will consider the role of immunity in HSV recurrent disease and its potential contribution to the pathogenesis of AIDS.

**HSV LATENCY**

Primary HSV disease may be subclinical and is usually confined to local cutaneous sites. It is invariably followed by the development of virus-specific immunity and the establishment of latency, operationally defined as a condition in which virus cannot be isolated from homogenates of infected tissues but it can be isolated by tissue cocultivation. Using this approach latent virus was shown to persist in the ganglia that supply sensory innervation to the skin areas in which HSV lesions recur. Presumably virus spreads to the PNS along the peripheral nerves by retrogade axonal transport. Latent infection of the central nervous system (CNS) has also been reported and it has been suggested that outflow of virus from CNS may help establish latency in neurons that do not supply the site of primary infection. Despite some evidence for virus persistence in extra- neural tissues, it is widely accepted that recurrent disease is due to reactivated ganglionic virus (6).

Productive infection of the neuron does not seem to be required for the establishment of latency (7). However, the exact mechanism whereby latency is established and maintained is still poorly understood. In one study that addressed the state of the persistent viral DNA, terminal sequences were not detected, suggesting that HSV DNA molecules persist in a circular, episome-like structure (8). Others found terminal fragments of uncharacteristic size (9,10), consistent with the interpretation that linear HSV DNA sequences had integrated into cellular DNA at a limited number of sites. Persistent HSV DNA was detected in only 0.4-5% of the neurons (11,12), suggesting that reinfection and/or periodic virus reactivation and replication at the periphery are necessary in order to retain the store of latently infected neurons (13). For a comprehensive review of latency see references 4 and 5.

HSV genes fall into three classes: immediate early (IE or alpha), delayed early (DE or beta) and late. IE genes are expressed in the absence of prior viral protein synthesis. At least three of them (IE175, IE110 and IE63 also designated ICP4, ICP0 and ICP27) have regulatory functions. Using temperature sensitive (*ts*) and deletion mutants, the product of the IE175 gene was shown to be essential for DE and late gene expression (14-16). The product of the IE110 gene has strong trans-activating activity on IE, DE and late promoters (17-20). According to some studies, the two genes function independently (17,18). Others concluded that they act synergistically in a transcriptional complex (16,19). IE63 can both enhance and repress the individual or combined abilities of IE110 and IE175 to stimulate expression from other viral gene promoters (21). Deletion and *ts* mutants in IE175 and IE63 are not viable under nonpermissive conditions demonstrating an essential role for these proteins in virus replication (14,22,23). *ts* mutants in IE110 are growth competent at high, but not low multiplicities of infection consistent with the interpretation that IE110 is not absolutely required for virus replication. The DE genes include enzymes involved in DNA replication, and most structural virion proteins are encoded by late genes (24).

All or nearly all viral genes are expressed during the replicative phase. However, it is still unclear whether expression is required for the establishment and maintenance of latency. Using mutants defective in their production, IE175 was implicated in the establishment of latency (25) and IE110 in the reactivation of latent virus (26). A novel transcript that overlaps the 3'-end of the IE110 gene by 1000 bases but is antisense to the IE110 transcript (LAT) was found in latently infected ganglia (27). Early studies suggested that LAT is required for the maintenance of latency because it acts as an antisense transcript that inhibits IE110 expression (28). More recently, transcripts arising from the LAT region were implicated in determining the frequency of virus reactivation from a latent state (29). However, LAT does not seem to be involved in either one of these aspects since a mutant deleted in the 5'-end of LAT and its control sequences was shown to establish latency while failing to express LAT or any other viral transcript (30). Furthermore, another mutant in which the LAT gene was disrupted by insertion of the *E.coli* lac Z gene was also shown to establish latency and become reactivated with kinetics and efficiency identical to those of the parental wild type virus (31). Presumably, changes in neuronal metabolism lead to reactivation of latent HSV by causing it to become permissive for virus replication at an as yet undefined stage.

Collectively, the data indicate that virus persists in privileged tissue, thus avoiding immune recognition. In order to maintain the store of latently infected neurons virus must replicate at the periphery, thereby risking eradication by the immune system. Therefore, the induction of transient immunosuppression has survival advantage.

## HSV-SPECIFIC IMMUNITY

Within the context of host-parasite relationships, immunity depends to a great extent on the host's ability to detect the replicating virus earlier (and therefore in lower concentration) than in the primary infection. Indeed the hallmark of secondary immunity includes a rapid and significant increase in serum antibody titers as compared to those obtained during the primary response, a shorter interval between immunization and peak antibody titers, and increased antibody affinity (32). Presumably the failure of a reinfecting virus to establish clinical disease (if not subclinical infection) is to a major extent determined by these factors, defined as *"relevant immunity"*. Therefore, when taken to its logical conclusion, this interpretation argues that the failure of HSV-specific immunity to protect from reinfection and recurrent disease is related to regulatory defects in virus-specific secondary immunity.

As already stated, *"relevant"* virus-specific immunity means effective secondary responses in which positive regulatory cell circuitries dominate. Such responses effectively recognize minute amounts of viral antigen and rapidly reach those levels (or differentiate into those cell types) that are required in order to eliminate the incoming virus. There remains the question of identifying such *"relevant"* responses and defining those factors that determine their induction. Can we safely assume that increased titers or unique protein/epitope specificity of virus-induced neutralizing antibody or immune memory suffice in order to achieve this efficient i.e. *"relevant"* response? Since recurrent disease and reinfection occur in patients with high titers of neutralizing antibody and an apparently normal immune memory, the answer to this question is apparently negative. On the other hand, if the problem is one of immunoregulation, the obvious question is what are the significant regulatory aspects and how can they be modified? What viral antigenic determinants are responsible for inducing *"relevant immunity"*? What is the role of antigen presentation in its generation?

Presently available data on HSV-specific immunity come mostly from studies of animal models, all of which suffer from certain limitations. In the mouse, the most commonly used model, HSV causes a lethal disease that is different from the cutaneous recurrent lesions seen in humans. The guinea pig is a valid model of recurrent cutaneous disease, but it lacks the markers required for immune cell dissection. The data reviewed below, the inferences and the conclusions draw freely on information acquired from all models as well as human studies.

Humoral responses.

Kinetic and quantitative details of serum antibody detection depend on the nature of the antigen (protein and epitope specificity), the method of antigen presentation (as part of the replicating virus or as a potentially soluble subunit protein), the site of antigen localization (local or systemic) and the animal species used in the particular study. In mice, antibody was first detected on day 5 post labial HSV-1 infection when using the enzyme immunosorbant

(ELISA), radioimmunopreciptation (RIA), or antibody dependent cellular cytotoxicity (ADCC) assays, and on day 10 when using the neutralization assay. Neutralizing antibody was not induced by intravaginal HSV-2 infection (33,34).

In humans, almost all subjects with primary genital infections develop IgG1 neutralizing antibodies. IgG2, IgG3 and IgG4 antibodies are detected in acute phase sera from patients with recurrent genital HSV, but in none of those with primary infection. IgG4 antibodies are significantly more frequent in sera from men than women with recurrent genital infections (35), consistent with the involvement of various regulatory factors.

HSV antigenic determinants appear to differ in their antibody inducing potential. The antibody repertoire predominant in primary infection is directed against the virion envelope (presumably glycoproteins) and antibodies against capsid proteins predominate in patients with recurrent disease. Antibodies against soluble antigens, readily detectable in sera from animal models, are only rarely found in human sera (36) reflecting differences in species or routes of exposure to the viral antigen. The antibody response is heterogeneous (23-31 HSV-1 and 10-27 HSV-2 proteins are precipitated by hyperimmune sera) and characterized by individual patient variability (37,38).

By virtue of their location on the surface of the virus-infected cells and the virion envelope, HSV glycoproteins [gB, gC (gF for HSV-2), gD, gE and gG (HSV-2 specific)] have been the subject of intense investigation as targets (elicitors) of *"relevant immunity"*. The picture that seems to emerge from these studies is that all glycoproteins can induce neutralizing antibody, although not all antigenic determinants are equally effective. The quality of the antibodies that are induced by the various determinants, in terms of binding affinity and/or neutralizing potential, is not yet understood and conflicting results were obtained as to their protective ability. gB, gE and gC were shown to induce neutralizing antibody, in some but not all studies (39,40) and monoclonal antibodies (MAbs) specific for gB, gC, gE and gD exhibited neutralizing activity (41-43). gD is generally accepted as the major elicitor of neutralizing antibody. Of 33 MAbs selected for their ability to bind to HSV virions, only 6 had neutralizing activity and all were specific for gD. However only some of the gD antigenic determinants were involved (43).

The potential involvement in *"relevant immunity"* of antibody to unique antigenic determinants, has not been excluded. However, the repertoire of determinants in neutralizing hyperimmune sera does not seem to hold much promise in this context. Qualitative or quantitative defects in production of antibodies that recognize specific viral proteins were not observed in patients with frequent recurrences (38). Neutralizing antibody titers are significantly higher in patients with a history of recurrent disease than in those without such a history (seropositive controls) (44,45). In animal models, vaccination or hyperimmunization produced high titers of neutralizing antibodies without reducing the incidence of spontaneous (guinea pigs) or induced (mice) recurrent disease (46-48).

Animal studies support the interpretation that antibody contributes to the outcome of fatal HSV infection. Thus, passive transfer of hyperimmune serum and leukocytes given intraperitoneally (i.p.) to newborn mice protected against lethal challenge with HSV-1 (49). Immunization of female mice, or oral feeding of newborns with human hyperimmune IgG given 24 hrs prior to i.p. infection ($10^2$ PFU, day 2 after birth), protected from fatal neonatal HSV disease (50).

The identity of the proteins that induce this response and their antigenic determinants are still poorly understood. MAbs to gC and gD injected i.p. protected mice from fatal challenge with HSV-1 or HSV-2 (51). In one study, protection was associated with neutralizing potential. Antibody against a major type-common antigenic determinant located within residues 8-23 on the mature gD glycoprotein neutralized the infectivity of HSV-1 and HSV-2 and protected mice from lethal challenge by either HSV type (52,53). In other studies protection was mediated by non-neutralizing MAbs to gD, gB or gC (against HSV-1) or gD, gB, gC, gE or gF (against HSV-2) (54,55) and was associated with ADCC activity (54). However, in humans, there is no apparent correlation between HSV antibodies and the development of neonatal (fatal) HSV (56).

The role of antibody, if any, in protection from the establishment of latency is even less well understood. Neutralizing antibody has been implicated in preventing infection of neurons, or in restricting HSV replication once the virus is in the ganglia. Thus, transfer of gD antibody interfered with HSV transport to the CNS. When injected intraveneously before day 3 post-infection (p.i.), antibody prevented the establishment of ganglionic latency (57), presumably by blocking virus replication at the periphery. Non-neutralizing MAbs to gB or gE suppressed HSV-1 replication in trigeminal ganglia in the absence of complement or immune cells, but virus replication resumed upon antibody removal (55).

The most reasonable conclusion seems to be that antibody of broad protein specificity and functional diversity plays an active role in decreasing the virus load. If the virus levels are reduced below the threshold required for lesion development or latency establishment, antibody may be considered as having played a significant role in resolving the acute phase of infection. This may occur in the neurotropic fatal disease of the mouse model. If on the other hand, virus levels are not decreased below this critical threshold, such as might occur in cutaneous human disease, antibody would seem to play no role in protection.

Cellular immunity.

The fundamental question is what determines whether HSV infection or exposure to specific viral antigens gives rise to a *"relevant"* immune response. In this context, *"relevant"* applies to the component or class of the immune response that efficiently recognizes viral antigenic determinants that presumably are displayed on the virion or the cell surface, thereby causing their elimination from the site of infection and/or preventing the virus from establishing ganglionic latency and recurrent disease. It is commonly believed that, at least as relates recovery from infection, immunity to HSV primarily involves the T cell system. This is based on the severity of the disease in the immunosupressed and on adoptive transfer studies that have established a major role for T cells in recovery from lethal HSV infection in mice (58,59).

Effector mechanisms recognized in cellular immunity to HSV include cytotoxic T lymphocytes (CTL) (60-62), delayed type hypersensitivity (DTH) (63,64) and effector lymphokines (65). The relative role of these components of the immune system in protection from infection and/or the establishment of ganglionic latency is not clear. In general it has been difficult to induce CTL to HSV. In mice, HSV induces CTL precursors with cytotoxic activity detectable only after cyclophosphamide treatment of mice prior to HSV immunization or by extended (3 days) *in vitro* culture of the effector cells before assay (60,61,66,67). However, their involvement in *"relevant immunity"* remains controversial. In contrast to other viruses, CTL to HSV were demonstrated only in draining lymph nodes and were no longer demonstrable at 12 days p.i. (63). Adoptive transfer studies indicate that HSV-specific $CD8^+$ CTL ($Lyt2^+$ in the mouse) present at the time of HSV infection, can participate in the control of the acute infection and the subsequent establishment of ganglionic latency by reducing HSV replication within ganglia (68,69). T-cell recognition of both I-coded [helper T (Th) cells] and K/D-coded (CTL) self antigens (H-2 complex in the mouse) was shown to be important in immune protection against fatal HSV-1 challenge, with long-lasting immunity being conferred by I-region compatibility (Th cells) (59, 70). Recently, CTL directed against IE antigens were described. However, in view of the intracellular location of the IE antigens it is difficult to understand how such CTL might be involved in *"relevant immunity"* (for review of CTL see chapter 4).

CTL clones established from HSV-1 or HSV-2 stimulated human peripheral blood monocytes (PBM) have a helper phenotype ($CD4^+$) and are restricted by HLA class II MB or DR antigens (71). HSV-1 specific CTL clones are directed against gB, gD or gE; HSV-2 specific clones, against gC, gD, gE or gG (72). Significantly, these human clones are bifunctional, providing both cytotoxic and helper activity. They proliferate in response to HSV antigens and in the absence of exogeneous IL-2, and the HSV type specificity and histocompatibility restriction of the proliferative response is identical to that of the CTL activity. HSV-specific stimulation of these clones results in the production of a factor (presumably IL-2) which induces proliferation of an IL-2 dependent T cell line (77). This raises the interesting possibility that cells of the same lineage might fulfill different functions under various selection conditions. It remains to be determined what conditions favour one or another function and whether this is significant in the pathogenesis of HSV disease. The possibility might be considered that in HSV infections, CTL play a regulatory rather than effector role.

DTH is an early response first detected in mice at 4-7 days after intradermal (i.d.) or subcutaneous (s.c.) HSV-1 infection, prior to the detection of neutralizing antibody (33,63). The DTH response is HSV type-common (74) and remains inducible for at least 2 years after infection (63). The lymphocytes responsible for DTH and lymphokine release (which can be provisionally included in the Th subset) show I region compatibility requirements. DTH can be adoptively transferred with draining lymph node cells (LNC) only at 6-10 days p.i., suggesting that DTH mediating cells are absent at late times after infection or that they are under the control of suppressor T (Ts) cells (63).

At 5 days p.i. the frequency of HSV-specific IL-2 producing Th cells ($CD4^+$; $L3T4^+$ in the mouse) present in lymph node suspensions was estimated at 1:2,470-1:5,800. Their detection required that mice be primed with infectious virus and that LNC be stimulated *in vitro* by exposure to viral antigen for 9 days. These findings suggest that infection induces Th-cell precursors that are strictly antigen dependent. Priming with non-infectious virus failed to induce these cells. Additionally, the frequency of Th-cell precursors was increased 2-3 fold by deletion of the $CD8^+$ cells prior to limiting dilution assays (75), consistent with the interpretation that infection also induced the generation of $CD8^+$ suppressor cells (Ts).

HSV-mediated blastogenesis coincides with DTH (33) and is primarily HSV type-common (65). It is first observed in cells from lymph nodes draining the site of infection followed by the spleen cells (SC) and then the PBM (76). By analogy to the progression toward high affinity antibody seen as the hallmark of secondary responses there is a progressive increase in the proliferation of SC from HSV-2 immunized guinea pigs that reflects an increase in the number of cells responding to low antigen concentrations (immune maturation) (77). However their function (help or suppression) is unknown. The finding of a significantly higher proportion of Ts cells during virus reactivation from a latent state (see below) supports the interpretation that the bulk of the cells responding to low antigen concentrations are Ts.

The antigenic determinants recognized in DTH and blastogenesis and their role in *"relevant immunity"* are not yet clearly established. Recent studies indicate that gB, although not possessing epitopes that induce neutralizing antibodies, contains antigenic determinants that activate Th cells capable of adoptively protecting mice from lethal HSV challenge (78). Our studies with vaccinia recombinants that express gD, suggest that protection is mediated by type-common determinants on gD. Long-term protection imparted by these determinants is conferred by I-region compatibility and is transferable with T cells that mediate DTH (and blastogenesis) (79-82). However, short-term protection does not appear to depend on the generation of this T cell subset, and another, as yet unidentified cell from the recipient animals, is also required for protection (Wachsman *et al.*, in preparation).

## HSV-INDUCED IMMUNOSUPPRESSION.

*In vitro*, HSV-1 suppresses the induction of antibody to an unrelated antigen due to nonproductive infection of a small proportion of Th cells (83,84). Additionally, inactivated HSV-1 or HSV-2 inhibit mitogen or antigen induced proliferative responses, presumably by complexing with IL-2. Addition of IL-2 to the cultures restores the proliferative responses (85,86). *In vivo*, HSV infection has been associated with an increase in Ts cell populations at the time of recurrent disease or re-infection (secondary exposure to antigen). In this section we shall review these data and discuss our present understanding of the cell circuitry involved in suppression.

Secondary exposure to other viruses results in virus specific immune responses comprised of positive cell circuitries that are capable of rapidly and effectively recognizing and clearing minute amounts of viral antigen (*"relevant immunity"*). In contrast, secondary exposure to HSV causes suppression of HSV-specific immunity accompanied by the failure to recognize and clear the secondary viral antigen, be it of exogenous (reinfection) or endogenous (reactivated virus) origin.

SC obtained 3 days after secondary i.d. infection of HSV immune guineapigs evidenced depressed HSV-specific proliferative responses and were negative for lymphokine production. This was due to antigen-dependent Ts cell precursors (pTs) as determined by cell mixing experiments (Fig. 1). Down-regulation of virus specific immunity coincided with the growth of the secondary virus at the site of infection (skin) and its ability to establish ganglionic latency (Aurelian *et al.*, in preparation). In mice and rats, secondary HSV infection suppressed antibody formation. The response was restored by administration of macrophage/monocyte toxicants (87), consistent with the interpretation that macrophages are involved in suppression. In humans (45,88-92) [and in the guinea pig model (47,93)], HSV-specific CMI responses are significantly impaired in subjects with recurrent disease. Impairment is reflected in the need to use higher antigen concentrations (up to 114-fold) and longer *in vitro* culture periods (up to 2.5 days) in order to mount responses comparable in magnitude to those observed in seropositive controls (Fig. 2). This may reflect fewer HSV-specific immune memory cells or a damping effect that reduces proliferation. Patients studied at recrudescence (days 0-3 after lesion onset) and prodrome (24-48 hrs. before overt lesions) had significantly higher proportions of macrophages (OKM2$^+$) and CD8$^+$ Ts cells that inhibited HSV-induced prolipheration in autologous cell mixing experiments. Depletion of these cell populations restored HSV-specific CMI responses. The proportion of Ts and pTs returned to normal levels at convalescence (days 3-9) when immune response was restored. Ts cells were not increased at quiescence (>14 days post lesion onset) and in seropositive controls (88). The possibility

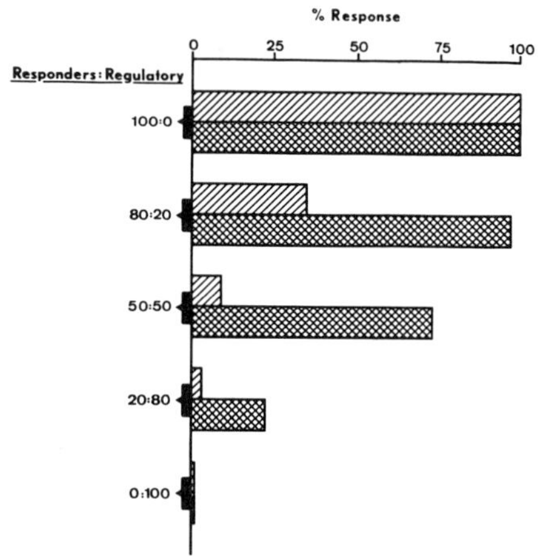


Fig.1 SC from HSV-2 seropositive guinea pigs (responders) were mixed with various proportions (0-100%) of regulatory cells consisting of SC from HSV immune animals at 3 days after secondary infection (▨) or non-immune animals after secondary infection (▩). Regulatory cells were pre-cultured (3 days) with HSV antigen. Mixed cultures were assayed for HSV-induced proliferation.

that the CD8⁺ cells are bifunctional has not been excluded, but it seems rather unlikely since human CTL clones were all CD4⁺ (71-73). Recrudescence was also associated with the appearance of a significant proportion of cells with the triple positive phenotype CD3⁺CD4⁺CD8⁺ (88). This cell subset differs from the single T cell clones CD4⁺ or CD8⁺ in that the cells have a diminished capacity to produce IL-2 and IFN $\gamma$ and their major function involves immunoregulation and surveillance (94). Additionally, suppression involves prostaglandin $E_2$ ($PGE_2$) [presumably made by the macrophages] and soluble suppressive factors (SFs) (presumably made by the Ts). Thus, supernatants from HSV-stimulated cultures of human PBM obtained at recrudescence, failed to display lymphokine activity [LIF or natural killer (NK) enhancing activity] unless cultured in presence of an inhibitor of PG synthesis (indomethacin) or depleted of macrophages by complement mediated lysis (90). However suppression was not mediated directly by $PGE_2$ since the levels of $PGE_2$ were similar throughout the disease cycle and in seropositive controls. Lymphokine activity was inhibited by a 8.2kDa SF that apparently blocks binding of the lymphokine to its target. SF induction was HSV-specific. However its effect was not antigen specific (90).

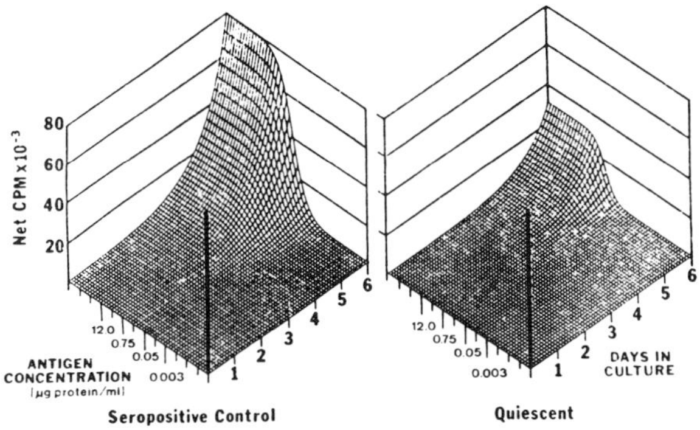


Fig. 2 Mathematical modeling of HSV induced proliferative responses in subjects with recurrent disease (quiescent) and seropositive controls as a function of time in culture and antigen dose.

Supporting the conclusion that HSV-induced immune responses are regulated differently than those induced by other antigens, the frequency of precursor cells to tetanus toxoid increased significantly on day 10 after injection of human volunteers with the tetanus toxoid (Fig. 3). This high frequency lasted for 31 days. On the other hand, the frequency of precursors to HSV increased on day 7 post recrudescence but it was no longer detectable on day 28. Such immunoregulation clearly does not favor an effective (protective) secondary response.

*In toto*, these findings are consistent with the interpretation that Ts responses, detectable only in subjects with a history of recurrent disease, are strictly antigen-dependent and result from secondary exposure to HSV antigen. Implicit in this interpretation is the assumption that Ts cells (and/or their precursors) have increased antigen sensitivity and thereby preferentially respond to the low levels of antigen generated upon reactivation of latent virus.

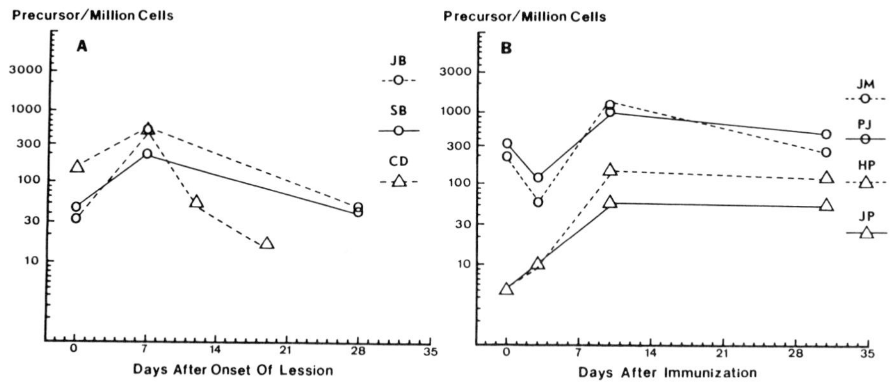


Fig.3 Frequency of precursor cells as determined by limiting dilution assay in patients (initials) with recurrent HSV beginning at recrudescence (A) and in individuals immunized with tetanus toxoid (B).

generated upon reactivation of latent virus. However, since only one half of HSV infected subjects have recurrent disease, the factors that define whether a particular patient will respond to secondary antigen with a negative (Ts) circuitry remain unclear. Presumably this is determined at the time of primary infection, since in a prospective study of 30 patients, the risk of developing recurrent symptoms (first seen within 77.6 ± 21 days p.i.) was correlated with the loss of HSV induced lymphokine responsiveness on day 20-30 p.i. (45).

Antigen presentation and *"relevant immunity"*.

The route of primary virus infection may be a critical factor in determining whether the secondary response will be met with a positive or negative T-cell circuitry by virtue of different antigen presentation by local as compared to circulating APC (for a comprehensive review of antigen presentation see chapter 1). Thus, s.c. infection of mice with HSV-1 was shown to induce T and B cell immunity in the draining lymph nodes. On the other hand, mice infected with HSV-1 by the i.v. route developed Ts cells that specifically suppressed the induction of DTH. These cells were $CD4^+$ at early (7 days) times p.i., possibly representing Ts-inducer cells. They were both $CD4^+$ and $CD8^+$ at 4 weeks p.i. (95). Ts-cell mediated non-responsiveness to HSV antigen was also seen after HSV-1 inoculation in the anterior chamber of the eye (96).

The epithelium contains T lymphocytes [dendritic epithelial cells (dEC)] that, unlike circulating or intestinal T cells, have a severely limited T-cell antigen receptor (TCR) repertoire. It has been suggested that in order to compensate for this restricted repertoire, dEC recognize a common cellular product generated in response to pathogenic stimuli, such as heat shock proteins (hsps) (97,98). Indeed, members of the 70kDa hsp family released from cultured rat cells are involved in antigen processsing and presentation (99,100), and two genes encoding 70kDa hsp were mapped within the major histocompatibility complex (MHC) (101). A human $\gamma\delta^+$ T cell clone established from an individual immune to PPD was recently shown to proliferate in response to PPD and to a recombinant mycobacterial 65kDa hsp (102). Hsps are induced by numerous trauma-related conditions including HSV infection (103), and it might be interesting to search for Ts clones specific for the 70kDa hsp after HSV infection by various routes.

The interpretation that antigen presentation during primary infection plays a crucial role in determining the outcome of secondary exposure to antigen is supported by the finding (79-82) that virus-specific immunity induced by HSV infection or hyperimmunization is down-regulated upon secondary exposure to HSV antigen, while down-regulation is not observed in animals rendered immune by infection with a vaccinia recombinant that expresses the HSV gD glycoprotein (Table 1). Presumably, gD presentation within the context of vaccinia virus infection differs from that occurring during HSV replication. Alternatively, HSV proteins are involved in suppression. The latter

**Table 1.** Effect of reinfection on HSV-specific lymphoproliferation*

| Infection | $^{3}$H-TdR incorporation (CPM ± SEM) |
|---|---|
| 1° HSV; 2° HSV | 1,545 ± 421 |
| 1° HSV; 2° VP176 | 10,294 ± 3,020 |
| 1° HSV; 2° Saline | 9,901 ± 4,131 |
| 1° VP176; 2° VP176 | 8,080 ± 600 |
| 1° VP176; 2° HSV | 9,000 ± 400 |
| 1° VP176; 2° Saline | 10,000 ± 1,810 |

*SC from guinea pigs primed (1°) and reinfected (2°) with HSV-2 or VP176 a vaccinia recombinant that expresses gD, were assayed for HSV induced proliferation

interpretation, easily verifiable with vaccinia recombinants that express other HSV proteins, is presently under investigation.

If it is assumed that antigen presentation is a function restricted to macrophages and resident cells of similar lineage (such as the skin APC), are *"relevant"* immune responses generated only when specific viral antigens are integrated within infected cell membranes in a structurally significant fashion? In other words, does integration occur in such a three-dimensional configuration that only *"relevant"* antigenic domains are exposed? Conversely, are Ts cells the result of these same domains being differently exposed or are there specific proteins and/or domains that induce supression?

We have recently addressed some of these questions using two vaccinia virus recombinants in which the HSV-1 gD gene is respectively driven by an early (VP176) or late (VP254) vaccinia virus promoter. Animals immunized with VP176 developed HSV-specific DTH and lymphoproliferative responses and were protected from HSV-2 challenge (Table 2). Long-term protection was mediated by T cells that transfer DTH and lymphoproliferation. On the other

**Table 2.** Effect of immunization on HSV disease.

| | % with disease | | |
|---|---|---|---|
| | VP176 | VP254 | Vaccinia |
| **Mice** | | | |
| Short term (14 days) | 0 | 0 | 100 |
| Long term (50 days) | 0 | 83* | 100 |
| **Guinea pigs** | | | |
| Long term (Primary) | 12 | 87 | 100 |
| Long term (Recurrent) | 12 | 60 | 66** |

* Mice challenged with HSV-2 on days 14 (short term protection) or 50 (long term protection) after immunization with VP176, VP254 or vaccinia virus were followed for paralysis and death. Paralysis in VP254 immune mice was seen in 83%; death in 58%

** Guinea pigs were challenged with HSVB-2 on days 50 and 100 after immunization (long term protection) and followed for primary and recurrent disease. Maximal proportion of animals with recurrent disease in this model is 65-75% (6,46).

recombinant in which gD is driven by a late vaccinia promoter (VP254) induced significantly reduced levels of CMI responses and failed to induce long-term protection from HSV-2 challenge (79-82). The failure of VP254 to induce *"relevant"* (protective) immunity was correlated with a defect in processing of the pgD precursor to its fully glycosylated gD product. This defect was particularly evident in APC, where gD was not made and was not expressed on the cell surface. Indeed, VP254 infected APC did not have antigen presenting capacity for HSV immune LNC, while accessory cell function was displayed by VP176 infected APC in which the mature gD was expressed on the cell surface (80,82). We interpret these findings to indicate that *"relevant immunity"* depends on the proper expression of significant (in this case fully glycosylated) antigenic domains on the APC surface. However, such expression is not required for antibody production as both recombinants induced equal titers of neutralizing antibody (79-82).

It is assumed that after synthesis in infected cells, viral proteins have to be processed to yield peptides that bind to the MHC-encoded glycoproteins and are presented at the cell surface to the T cell receptor. Processing is poorly understood. In the case of the herpesviruses, defective antigen recognition by CTL was associated with decreased surface levels of MHC class I glycoproteins (104). However, recent findings from other viral systems [influenza, murine cytomegalovirus (105,106)] suggest that there is a highly selective, temporally regulated and reversible presention of viral antigen to class I restricted T lymphocytes. This presentation takes place at an intracellular stage of antigen processing after translation and before presentation of the antigen at the cell surface. Further studies are required in order to develop a better understanding of these issues.

Do various HSV proteins and/or unique antigenic determinants within different proteins induce various responses? As discussed earlier, gB contains antigenic determinants that activate Th cells but it is not a major elicitor of neutralizing antibody (78). However, it is unclear whether: (i) proteins differ in their ability to induce Ts cell circuitries, or (ii) certain antigenic determinants within a protein preferentially induce Ts or Th cell responses.

HSV induced immunosuppression and pathogenesis.

How are local stimuli involved in immunosuppression and recurrent disease? Skin irritants and UVB irradiation (sunlight) induce virus reactivation and recurrent disease associated with selective immunoincompetence (4,107).

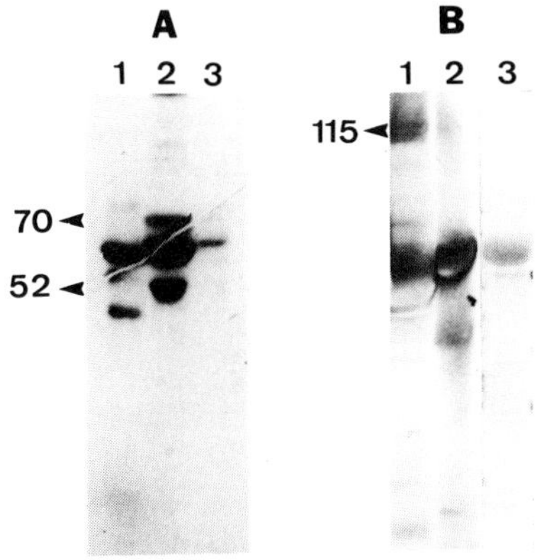


Fig.4 Silver stain of SDS-PAGE of fractions from Sephadex G-100 columns. **A**. Reducing conditions. Lane 1, HSV-SF negative fractions; lane 2, HSV-SF positive fractions; lane 3, bovine serum albumin. **B**. Non-reducing conditions. Lane 1, HSV-SF positive fractions; lane 2, HSV-SF negative fractions; lane 3, bovine serum albumin. Reprinted with permission from reference 109.

Our recent studies indicate that UVB irradiation at the site of HSV-2 infection causes a dose-dependent increase in disease severity expressed as zosteriform lesions, early disease onset and death (at high UV dose). Increased severity was associated with decreased virus-induced CMI responses and the generation of Ts cell circuitries consisting of $CD4^+$ suppressor-inducer T cells, $CD8^+$ suppressor-effector T cells and soluble SFs with both HSV-specific (HSV-SF) and non-specific activity (108,109). The antigen specific factor (HSV-SF) was identified as a 115 kDa protein consisting of two disulfide-bound components with molecular sizes of 70 and 52 kDa (Fig. 4). The development of the UVB induced Ts cell circuitry was directly related to a decrease in the antigen presenting functions of the irradiated skin APC (Fig. 5). When the APC were pulsed with viral antigen prior to irradiation the response was partially restored by addition of the epidermal IL-1 like activity ETAF (110). $PGE_2$ was not involved, as UVB decreased the levels of $PGE_2$ secreted by the epidermal cells and indomethacin did not correct the UVB induced suppression (110).

The exact identity of the APC involved in Ts cell circuitry is not clear. According to some studies the generation of UVB-induced Ts cells is due to a decrease in $T6^+DR^+$ Langerhans cells (111). Others suggest that the cells involved in the generation of the Ts cell circuitry are $T6^-DR^+$ APC, the proportion of which is increased by UVB irradiation (112). $T6^-DR^+$ APC are analogous to a minor population of $OKM1^-OKM5^+$ blood monocytes that have been suggested as potentially involved in immune down-regulation (113).

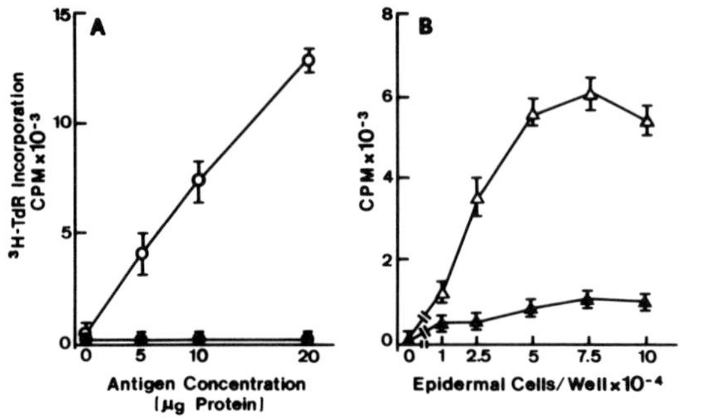


Figure 5. **A.** $^3$H-TdR incorporation (mean cpm ± SEM) in cultures of unfractionated immune LNC ( O ) or Ia$^-$ immune LNC (●) [4 x 10$^5$/well] cultured with various doses (0 to 20 µg protein/ml) of HSV-2 antigen for 4 days. **B.** $^3$H-TdR incorporation (mean cpm ± SEM) in cultures of Ia$^-$ immune LNC [4 x 10$^5$/well] cultured with graded numbers of EC in presence ( Δ ) or absence ( ▲ ) of HSV-2 antigen (20 µg protein/ml). Reprinted with permission from reference 110.

How do SFs reduce proliferation? SF from HSV-stimulated guinea pig immune SC suppressed HSV-induced DNA synthesis when added at early stages (day 0-1) of SC culture, suggesting that it exerts its effect on early event(s). In the presence of SF the onset of proliferation was delayed by 1 day and the maximal levels were significantly decreased (Fig. 6). However as evidenced by identical doubling times in treated and untreated cultures, SF did not modify antigen sensitivity. SF mediated suppression was reversible indicating that suppression is not merely the consequence of toxicity. Taken *in toto*, the data suggest that SF interacts with defined cell surface receptors that represent inhibitory

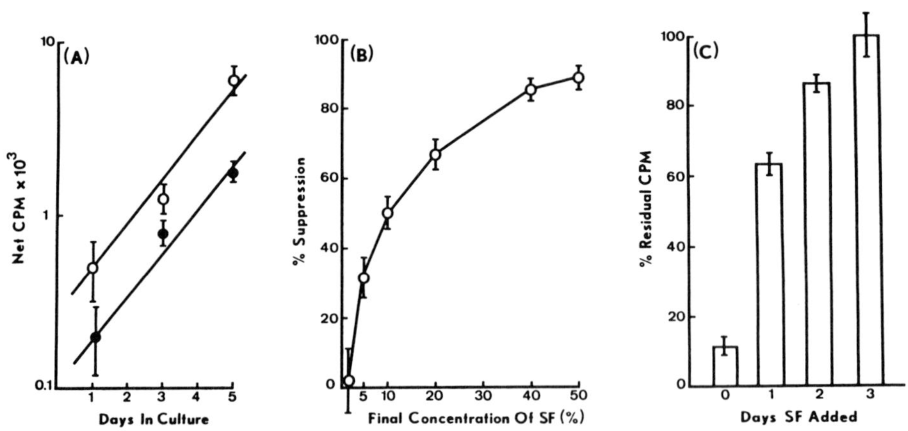


Fig.6 **A.** Time course of SF-mediated suppression. Responder SC were grown in the presence of HSV antigen with (●) or without (O) SF (30%). **B.** Dose effect of SF-mediated suppression. SF was added on day 0; proliferation assayed on day 5. **C.** Effect of delayed SF addition (50%). Proliferation measured on day 5. Data represent mean ± SD. Reprinted with permission from reference 93.

domains within the cell membrane (93). However the molecular identity of the SFs and their exact mechanism of action remain unknown.

The picture that emerges from all these findings is summarized in Fig. 7. Presumably stimulants (such as UVB), acting at the skin level, exert their effect on resident APC decreasing ETAF production and antigen presentation. These APC as well as $PGE_2$ and non-specific (SF) and specific (HSV-SF) suppressive factors are involved in the induction of the Ts cell circuitry. On secondary exposure to reactivated virus, the response is amplified and Ts circuits further activated. HSV-SF may be the product of the $CD4^+$ Ts-inducer cells generated by UVB irradiation, since their suppressive activity is antigen specific. However, $CD4^+$ cells (and presumably HSV-SF) did not mediate suppression directly as evidenced by the failure of SC from UVB irradiated mice to reduce proliferative responses of HSV immune LNC. Instead recognition by another antigen-activated $CD8^+$ Ts cell (suppressor-effector) was required. The suppressive activity of the $CD8^+$ cell appears to be non-specific (108,109). Presumably HSV-SF is involved in the induction of this latter cell subset. The suppressor-effector cells respond by producing nonspecific factors that apparently act as the ultimate suppressors (108,109). Suppression is also mediated by $CD4^+$ $CD8^+$ Ts. Similar conclusions were independently reached in HSV infected mice (111), where HSV-immune SC with only one phenotype marker ($CD4^+$ or $CD8^+$) were found to proliferate in response to HSV antigen. The $CD8^+$ cells had suppressor activity, but suppression also required the contribution of the $CD4^+$ T cells and of $IJ^+$ APC. Suppression was mediated by a SF that contained $IJ^+$ anti-idiotypic protein and was antigen-specific. Non-specific suppressive activity could also

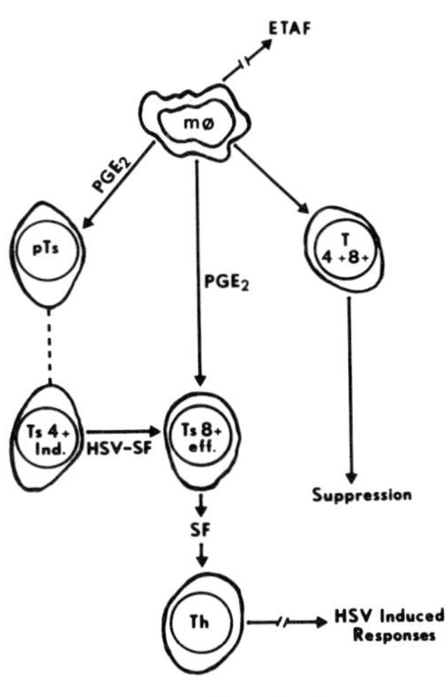

Fig. 7 Proposed circuitry for negative regulation of HSV immunity.

be generated by incubating the $IJ^+$ SF with $CD4^+$ cells from HSV-immune mice.

These interpretations raise two fundametally important questions. First, what is the relevant immune response? Second what role, if any, does antigen presentation actually play in determining the immune function (help versus suppression) that is induced? In this context it might be restated that primary HSV infection or immunization, both of which favor the induction/selection of specific Ts cells, does not lead to a corresponding decrease in HSV-specific neutralizing antibody upon secondary virus challenge, supporting the thesis that in humans, HSV disease is the result of excessive down-regulation of HSV-specific CMI.

**THE HERPESVIRUSES AND AIDS.**

Approximately 1-1.5 million people in the US are infected with HIV. However, only 20-30% of these will develop AIDS within 5 years. Seeking to reconcile the severity of AIDS with the paucity of productively infected cells and the long latency period, one interpretation argues that infection with other viruses converts the latent HIV infection into a productive, cytolytic one and /or causes immunosuppression. Indeed, antigenic stimuli created by viral infections common in the HIV infected individuals can cause T cell activation, thereby converting a latent HIV infection to a cytolytic one. However three criteria must be met by an infectious agent postulated to fulfill such a function: (1) high prevalence of the putative infection in the population at high risk of AIDS, (2) direct HIV activation in co-infected cells, and (3) direct or indirect contribution to immunosuppression.

For a comprehensive discussion of the first two criteria as regards the herpesviruses, the reader is refered to chapters 1, 11 and 12. Briefly, the herpesviruses commonly infect the population at high risk of AIDS and both HSV and HIV replicate in adult human T cells, provided they are activated (83). HSV infection of adult macrophages is abortive (for review see chapters 8 and 9). However, macrophages can be rendered permissive for HSV growth by several days of *in vitro* growth prior to infection (114), growth in conditioned medium or SV40 transformation (115). CMV infection of macrophages, albeit restricted to IE gene expression, presumably accounts for much of the immunosuppression that accompanies CMV infections (121). Presumably these herpesviruses provide the cellular activating conditions that are required in order to support the growth of HIV and vice versa. Direct activation of HIV in co-infected cells has been demonstrated, and it depends on a 5' upstream 73bp HSV recognition sequence [positions -104 to -32 within HIV-LTR (116,117)] that includes the NF-κB and the SP1 binding sites. HSV or CMV infection induces expression of cellular transcription factors that bind to the NF-κB core enhancer sequence thereby mediating HIV activation.

Few studies, have addressed the effects of HSV infection on immunoregulatory functions and the contribution that this may have on the pathogenesis of AIDS. Crucial remaining questions are: How does virus infection alter macrophage functions, particularly antigen processing and presentation, and what is the contribution of HSV-induced Ts cell circuitry to the severe immunosuppression seen in AIDS ? What is the contribution of the other herpesviruses and at what level?

The concept that HSV immunosuppression contributes to AIDS pathogenesis is consistent with recent findings indicating that many immunological abnormalities detected in the AIDS patients are due to $CD4^+$ T cell dysfunction rather than HIV induced cytolysis. Thus, decreased proliferative responses to specific antigens (such as HSV) or mitogens, or the inability of the $CD4^+$ T cells to help B cells in response to mitogens or antigens are observed in the HIV seropositive patient prior to the decrease in $CD4^+$ T cell numbers. At least some of these defects appear to represent immune suppression caused by naturally occuring substances or cells in the course of normal immune reactions such as against HSV. Indeed: (i) $CD4^+$ T cells from AIDS patients are deficient in helper functions, whereas $CD8^+$ cells have an increased or a normal capacity to act as Ts cells, (ii) helper functions can be restored by removal of the Ts cells, and (iii) T cell lines derived from AIDS patients produce various SFs (reviewed in 118).

Consistent with this interpretation we found that PBM from homosexual males at high risk of AIDS, or from ARC patients followed longitudinally fail to produce lymphokines when exposed to HSV-2 antigen (Table 3). Since the

**Table 3.** Lymphokine (LIF) production by HSV stimulated mononuclear cells of HIV infected subjects

| GROUP | HSV-2 | | HSV-2 + IL-1 | | HSV-2 + IL-2 | |
|---|---|---|---|---|---|---|
| | No[a] | MI[b] | No | MI | No | MI |
| AIDS | 0/5 | 96±3.6 | 0/5 | 96±2.6 | 0/5 | 99±0.5 |
| ARC | 2/5 | 81±7.9 | 5/5 | 63±7.6 | 0/5 | 98±1.6 |
| **CONTROLS** | | | | | | |
| homosexual | 2/5 | 80±7.3 | 5/5 | 68±2.7 | 3/5 | 78±7.5 |
| heterosexual | 5/5 | 75±5.1 | ND | | ND | |

[a] Number positive per number tested
[b] Mean migration index ± SEM. Values <80 are positive

lymphokines are products of the $CD4^+$ T cells, the number of which is normal in these patients, the failure does not represent $CD4^+$ cell lysis. Indeed, the function was restored by the addition of IL-1, consistent with the interpretation that at this stage of the disease, the failure to respond to HSV is due to a defect in IL-1 production i.e. at the level of the macrophage. The Ts cascade induced by secondary exposure to HSV could easily explain these findings in that it involves $PGE_2$ synthesis by the macrophage, the induction of Ts cells and SFs that inhibit normal $CD4^+$ responses to antigens other than HSV. Restoration of PHA responsiveness in AIDS patients by $PGE_2$ inhibition with indomethacin, and the demonstration that the cell mediating the suppression is a macrophage (118), support such an interpretation.

Additional levels of potential interaction between HIV and HSV are suggested by the recent finding that HIV gp120 binds to activated $CD4^+$ T cells thereby rendering them susceptible to lysis by HIV specific CTL (see chapter 3 for discussion), and that infection of HSV specific $CD4^+$ (MHC class II restricted) human multifunctional T cell clones with HTLV-1 causes severe

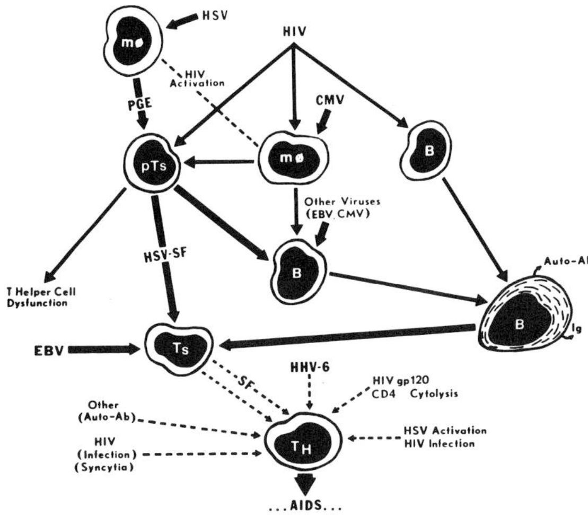


Fig. 8. Schematic representation of the interaction between the herpesviruses and HIV.

Infection

HIV infects T cells, monocytes and maybe B cells; HSV and CMV infect activated monocytes and T cells.

Early events

1. Herpesvirus induction of monokines and cytokines that activate HIV.
2. Ts circuitry induction by HSV and other herpesviruses.
3. Specific and polyclonal B cell stimulation (CMV, EBV).
4. Increased Ig and autoantibody production.
5. Suppressor cells and HIV inhibit Th function.
6. HIV gp120 binding to CD4; lysis.
7. Herpesvirus activation of T cells; HIV infection.
8. T cell co-infection; HIV activation.

Late events

Loss and dysfunction of Th
Clinical AIDS.

impairment of both their cytotoxic and helper activities (119). It would be interesting to determine whether HIV can cause a similar dysfunction of the HSV specific bifunctional $CD4^+$ T cells.

In conclusion, this interpretation argues that the immunosuppressive effects mediated by the herpesviruses and HIV are additive and/or synergistic. As summarized in Fig. 8, HIV, HSV and CMV infect macrophages and skin APC, where direct HIV activation can take place. Additionally, macrophage infection with either virus can cause functional alterations including the failure to produce IL-1, the generation of $PGE_2$ and other monokines and cytokines and the induction of Ts circuitries, or the increased production of IL-1, tumor necrosis factor (TNF$\alpha$) and interferon which can cause HIV activation. SFs and other independently generated cytokines, auto-antibody and EBV and/or CMV mediated polyclonal B cell stimulation accentuate Ts cell induction. EBV also induces Ts cells, although SFs do not seem to be involved (see chapter 2). Human herpesvirus 6 (HHV-6) a predominant T cell tropic virus also co-infects $CD4^+$ T cells with HIV, causing increased HIV gene expression and viral replication and consequently accelerating cell death (120). Activated $CD4^+$ cells (including multifunctional HSV-CTL) preferentially support HIV growth causing further dysfunction. Reactivation of latent HSV, in itself associated with the induction of Ts cell circuitry, provides an additional stimulus for Ts cell generation and the T cell activation required for HIV growth in lymphocytes. Should such combined lysis and dysfunction occur *in vivo*, it could result in immunodefficiency with concomitant activation of latent HIV as well as increased severity of herpetic disease. This interpretation, therefore, visualizes immune dysfunction as a progressive problem related to opportunistic infections during early HIV latency period. It is coupled with $CD4^+$ cell lysis exacerbated by increased HIV growth during the later latency intervals, ultimately resulting in AIDS.

**ACKNOWLEDGEMENT**

The studies done in our laboratory were supported by grant AI22192 from the National Institute of Health. I thank Dr. Albert Donnenberg for help with the mathematical modeling of HSV induced proliferative responses and limiting dilution assays.

**REFERENCES**

1. Shaffer, A.J. Disease of Newborn. W.B.Saunders, Philadelphia, 1965.
2. Nahmias, A.J., Josey, W.E. and Naib, Z.M. JAMA 199:164-168, 1967.
3. Aurelian, L., Wachsman, M., and Burnett, J. Brit. J. Dermatol. 122, in press 1990.
4. Hill, T.J. In: **The Herpesviruses**. (Ed. Roizman, B.) Plenum Press, N.Y. 1985, pp 175-225.
5. Aurelian, L. In: **Biology of human diseases.** (Eds. J.W. Gorrod., O. Albano., S. Papa). Ellis Horwood, Chichester. 1988, pp 253-265.
6. Scriba, M. and Tatzber, F. Infect. Immun. 34:655-661, 1981.
7. McLennan, J.L. and Darby, G. J. Gen. Virol. 51:233-243, 1980.
8. Efstathiou, S., Minson, A.C., Field, H.J., Anderson, J.R., and Wildy, P.

J. Virol. 57:446-455, 1986.
9. Fraser, N.W., Lawrence, W.C., Wroblewska, Z., Gilden, D., and Koprowski, H. Proc. Natl. Acad. Sci. USA 79:6461-6465, 1981.
10. Puga, A., Cantin, E.M., Wohlenberg, C., Openshaw, H., and Notkins, A.L. J. Gen. Virol. 65:437-444, 1984.
11. Kennedy, P.G.E., Al-Saadi, A.A., and Clements, G.B. J. Gen. Virol. 64:1629-1635, 1983.
12. Walz, M.A., Yamamoto, H., and Notkins, A.L. Nature 264:554-556, 1976.
13. Klein, R.J. Arch. Virol. 51:1-13, 1976.
14. Dixon, R., and Schaffer, P.A. J. Virol. 36:189-203, 1980.
15. Gelman, I.H., and Silverstein, S. J. Virol. 61:2286-2296, 1987.
16. O'Hare, P., and Hayward, G.S. J. Virol. 56:723-733, 1985.
17. Gelman, I.H., and Silverstein, S. Proc. Natl. Acad. Sci. USA 82:5265-5269, 1985.
18. Quinlan, M., and Knipe, D.M. Mol. Cell Biol. 5:957-963, 1985.
19. Everett, R. EMBO J. 3:3135-3141, 1984.
20. O'Hare, P., and Hayward, G.S. J. Virol. 53:751-760, 1985.
21. Su, L., and Knipe, D.M. Virology 170:496-504, 1989.
22. Preston, C.M. J. Virol. 32:357-369, 1979.
23. Sacks, W.R., Greene, C.C., Aschman, P.I., and Schaffer, P.A. J. Virol. 55:796-805, 1985.
24. Hayward, G.S. Cancer Cells 4:59-77, 1986.
25. Watson, K., Stevens, J.G., Cook, M.L., and Subak-Sharpe, J.H. J. Gen. Virol. 49:149-159, 1980.
26. Harris, R.A., Everett, R.D., Zhu, X., Silverstien, S., and Preston, C.M. J. Virol. 63:3513-3515, 1989.
27. Stevens, J.G., Wagner, E.K., Devi-Rao, G.B., Cook, M.L., and Feldman, L.T. Science 235:1056-1057, 1987.
28. Cameron, J.M., McDougall, I., Marsden, H.S., Preston, V.G., Ryan, D.M., and Subak-Sharpe, J.H. J. Gen. Virol. 69:2607-2612, 1988.
29. Javier, R.T., Stevens, J.G., Dissette, V.B., and Wagner, E.K. Virology 166:254-257, 1988.
30. Leib, D.A., Coen, D.M., Bogard, C.L., Hicks, K.A., Yager, D.R., Knipe, D.M., Tyler, K.L. and Schaffer, P.A. J. Virol. 63:759-768, 1989.
31. Ho, D.J., and Mocarski, E.S. Proc. Nat. Acad. Sci. USA 86:7596-7600, 1989.
32. Siskind, G.W., and Benacerraf, B. Adv. Immunol. 10:1-50, 1969.
33. Morahan, P.S., Thomson, T.A., Kohl, S., and Murray, B.K. Infect. Immun. 32:180-187, 1981.
34. Morahan, P.S., Breinig, M.C., and McGeorge, M.B. J. Immunol. 119: 2030-2036, 1977.
35. Coleman, R.M., Nahmias, A.J., Williams, S.C., Phillips, D.J., Black, C.M., and Reimer, C.B. J. Infect. Dis. 151:929-936, 1985.
36. Kalino, K.O.K., Marttila, R.J., Granfors, K., and Viljanen, M.K. Infect.Immun. 15:883-889, 1977.
37. Gilman, S.C., Docherty, J.J., and Rawls, W.E. Infect. Immun. 34:880-887, 1981.
38. Norrild, B. In:**The Herpesviruses**. (Ed. Roizman, B) Plenum Press, N.Y. 1985, pp 69-86.

39. Eberle, R., and Courtney, R.J. J. Virol. 35:902-917, 1980.
40. Para, M.F., Bauck, R.B., and Spear, P.G. J. Virol. 41:129-136, 1982.
41. Pereira, L., Dondero, D.V., Gallo, D., Devlin, V., and Woodie, J.D. Infect. Immun. 35:363-367, 1982.
42. Balachandran, N.D., Harnish, W.E., Rawls, W.E., and Bacchetti, S. J. Virol. 44:344-355, 1982.
43. Para, M.F., Parish, M.L., Noble, G., and Spear, P.G. J. Virol. 55:483-488, 1985.
44. Lopez, C., and O'Reilley, R.J. J. Immunol. 118:895-902, 1977.
45. Sheridan, J.F., and Aurelian, L. Diagn. Immunol. 1:245-256, 1983.
46. Donnenberg, A.D., Chaikoff, E., and Aurelian, L. Infect. Immun. 30:99-109, 1980.
47. Darville, J.M., and Blyth, W.A. Arch. Virol. 71:303-310, 1982.
48. Kohl, S., Loo, L.S., and Pickering, L.K. J. Immunol. 127:1273-1275, 1981.
49. Hayashi, Y., Wada, T., and Mori, R. J. Gen. Virol. 64:1007-1012, 1983.
50. Dix, R.D., Pereira, L., and Baringer, J.R. Infect. Immun. 34:192-199, 1981.
51. Eisenberg, R., Cerini, C.P., Heilman, C.J., Joseph, A.D., Dietzchold, B., Golub, E., Long, D., Ponce de Leon, M., and Cohen, G.H. J. Virol. 56:1014-1017, 1985.
52. Eisenberg, R.J., Long, D., Ponce de Leon, M., Matthews,J.T., Spear, P.G., Gibson, M.G., Lasky, L.A., Berman, P., Golub, E. and Cohen, G.H. J. Virol. 53:634-644, 1985.
53. Balachandran, N., Bacchetti, S., and Rawls, W.E. Infect. Immun. 37:1132-1137, 1982.
54. Oakes, J.E., and Lausch, R.N. J. Virol. 51:656-661, 1984.
55. Whitley, R.J., Nahmias, A.J., Visintine, A.M., Fleming, C.L., and Alford, C.A. Pediatrics 66:489- 494, 1980.
56. Kappor, A.K., Nash, A.A., Wildy, P., Phelan, J., Mclean, C.S., and Field, H.J. J. Gen. Virol. 60:225-233, 1982.
57. Nagafuchi, S., Oda, H., Mori, R., and Taniguchi, T. J. Gen. Virol. 44:715-723, 1979.
58. Howes, E.L., Taylor, W., Mitchison, N.A., and Simpson, E. Nature 277:67-68, 1979.
59. Pfizenmaier, K., Jung, H., Starzinski-Powitz, A., Rollinghoff, M., and Wagner, H. J. Immunol. 9:939-944, 1977.
60. Lawman, M.J.P., Rouse, B.T., Courtney, R.J., and Walker, R.D. Infect. Immun. 27:133-139, 1980.
61. Jennings, S.R., Rice, P.L., Pan, S., Knowles, B.B., and Tevethia, S.S. J. Immunol. 132:475-481, 1984.
62. Nash, A.A., Field, H. J., and Quartey-Papafio, R. J. Gen. Virol. 48:351-357, 1980.
63. Nash, A.A., Phelan, J., and Wildy, P. J. Immunol. 126:1260-1261, 1981.
64. Green, J.A., Yeh, T.J., and Overall, J.C. Jr. J. Immunol. 127:1192-1196, 1981.
65. Nash, A.A., Quartey-Papafio, R., and Wildy, P. J. Gen. Virol. 49:309-317, 1980.
66. Pfizenmaier, K., Strazinski-Powitz, A., Rollinghoff, M., Falke, D., and

Wagner, H. Nature 265:630-632, 1977.
67. Larsen, H.S., Feng, M.F., Horohov, D.W., Moore, R.N., and Rouse, B.T. J. Virol. 50:56-59, 1984.
68. Bonneau, R.H., and Jennings, S.R. J. Virol. 63:1480-1484, 1989.
69. Nash, A.A., Jayasunga, A., Phelan, J., Cobbold, S.P., Waldmann, H., and Prospero, T. J. Gen. Virol. 68:825-833, 1987.
70. Yasukawa, M., and Zarling, J.M. J. Immunol. 133:422-427, 1984.
71. Yasukawa, M., and Zarling, J.M. J. Immunol. 134:2679-2682, 1985.
72. Yasukawa, M., and Zarling, J.M. J. Immunol. 133:2736-2742, 1984.
73. Schreir, R.D., Pizer, L.I., and Moorhead, J.W. Infect. Immun. 35:566-571, 1982.
74. Prymowicz, D., Moore, R.N., and Rouse, B.T. J. Immunol. 134:2683-2688, 1985.
75. Jacobs, R.P., Aurelian, L., and Cole, G.H. J. Immunol. 116:1520-1525, 1976.
76. Donenberg, A.D., Bell, R.B., and Aurelian, L. Cell Immunol. 56:526-539, 1980.
77. Chan, W.L., Lukig, M.L., and Liew, F.Y. J. Exp. Med. 162:1304-1318, 1985.
78. Wachsman, M., Aurelian, L., Smith, C.C., Lipinskas, B.R., Perkus, M.E., and Paoletti, E. J. Inf. Dis. 155:1188-1197, 1987.
79. Wachsman, M., Aurelian, L., Hunter, J.C.R., Perkus, M.E., and Paoletti, E. Bioscience Reports 8:323-334, 1988.
80. Wachsman, M., Aurelian, L., Smith, C.C., Perkus, M.E., and Paoletti, E. J. Inf. Dis. 159:625-634, 1989.
81. Wachsman, M., Luo, J.H., Aurelian, L., Perkus, M.E., and Paoletti, E. J. Gen. Virol. 70:2513-2520, 1989.
82. Pelton, B.K., Imrie, R.C., and Denman, A.M. Immunology 32:803-810, 1977.
83. Pelton, B.K., Duncan, I.B., and Denman, A.M. Nature 284:176-177, 1980.
84. Plaeger-Marshall, S.W., and Smith, J.W. J. Infect. Dis. 138:506-511, 1978.
85. Wainberg, M.A., Portnoy, J.D., Clencer, B.,Wubschman, S., Legace-Simard, J., Rabinovitch, N., Remer, Z., and Mendelson, J. J. Infect. Dis. 152:441-448, 1985.
86. Nick, S., Kampe, P., Knoblich, A., Metzger, B., and Falke, D. J. Gen. Virol. 67:1015-1024, 1986.
87. Sheridan, J.F., Donnenberg, A.D., Aurelian, L., and Elpern, D.J. J. Immunol. 129:326-331, 1982.
88. Sheridan, J.F., Beck, M., Aurelian, L., and Radowsky, M. J. Infect. Dis. 152:449-456, 1985.
89. Sheridan, J.F., Beck, M., Smith, C.C., and Aurelian, L. J. Immunol. 138:1234-1239,1987.
90. Shillitoe, E.J., Wilton, J.M.A., and Lehner, T. Infect. Immun. 18:130-137, 1977.
91. Cunningham, A.L., and Merigan, T.C. J. Immunol. 130:2397-2400, 1983.
92. Iwasaka, T., Sheridan, J.F., and Aurelian, L. Infect. Immun. 42:955-967, 1983.

93. Patel, S.S., Wacholtz, M.C., Duby, A.D., Thikle, D.L., and Lipsky, P.E. J. Immunol. 143:1108-1117, 1989.
94. Nash, A.A., and Gell, P.G.H. Cell. Immunol. 75:348-355, 1983.
95. Whittum, J.A., Niederhorn, J.Y., McCulley, J.P., and Streilein, J.W. J. Virol. 51:556-558, 1984.
96. Arsanow, D.M., Kuziel, W.A., Bonyhadi, M., Tigelaar, R.E., Tucker, P.W., and Allison, J.P. Cell 55:837-847, 1988.
97. Arsanow, D.M., Goodman, T., LeFrancois, L., and Allison, J.P. Nature 341:60-62, 1989.
98. Pierce, S.K., Morris, J.F., Grusby, M.J., Kaumaya, P., Van Duskirk, A., Srinivasan, N., Crump, B., and Smolenski, L.A. Immunol. Rev. 106:149-180, 1988.
99. Hightower, L.E., and Guidon, P.T. Jr. Cell Physiol. 138:257-266, 1989.
100. Sargeant, C.A., Dunham, I., Trowsdale, J., and Campbell, R.D. Proc. Natl. Acad. Sci. USA 86:1968-1972, 1989.
101. Haregewoin, A., Soman, G., Hom, R.C., and Finberg, R.W. Nature 340:309-312, 1989.
102. La Thangue, N.B., and Lachtman, D.S. Biosci. Reports. 7:475-482, 1987.
103. Jennings, S.R., Rice, P.L., Kloszewski, E.D., Anderson, R.W., Thompson, D.L., and Tevethia, S.S. J. Virol. 56:757-766, 1985.
104. Townsend, A.R.M., Bastin, J., Gould, K., Brownlee, G., Andrew, M., Coupar, B., Boyle, D., Chan, S., and Smith, G. J. Exp. Med. 168:1211-1224, 1988.
105. Del Val, M., Munch, K., Reddehase, M.J., and Koszinowski, U.H. Cell 58:305-315, 1989.
106. Hill, T.J., and Blyth, W.A. Int. J..Dermatol. 16:274-275, 1977.
107. Yasumoto, S., Hayashi, Y., and Aurelian, L. J. Immunol. 139:2788-2793, 1987.
108. Aurelian, L., Yasumoto, S., and Smith, C.C. J. Virol. 62:2520-2524, 1988.
109. Hayashi, Y., and Aurelian, L. J. Immunol. 136:1087-1092, 1986.
110. Horohov, D.W., Wyckoff, J.H.III., Moore, R.N., and Rouse, B.T. J. Virol. 58:331-338, 1986.
111. Baadsgaard, O., Wulf, H.C., Wantzin, G.L., and Cooper, K.D. J. Invest. Dermatol. 89:113-118, 1987.
112. Cooper, K.D., Neises, G.R., and Katz, S.I. J. Invest. Dermatol. 86:363-370, 1986.
113. Lopez, C., and Dudas, G. Infect. Immun. 23:432-437,1979.
114. Sethi, K.K., and Brandis, H. Arch. Virol. 59:157-172, 1989.
115. Gimble, J.M., Duh, E., and Ostrove, J.M., Gendelman, H.E., Max, E.E., and Rabson A.B. J. Virol. 62:4102-4112, 1988.
116. Nabel, G.J., Rice, S.A., Knipe, D.M. and Baltimore, D. Science 239:1299-1302.
117. Mosca, J.D., Bednarik, D.P., and Raj, N.B.K., Rosen, C.A., Sodroski, J.G., Haseltine, W.A., Hayward, G.S. and Pitha, P.M. Proc. Nat. Acad. Sci. USA 84:7408-7412, 1987.
118. Edelman, A.S., and Zolla-Pazner, S. The FASEB J. 3:22-30, 1989.
119. Inatsuki, A., Yasukawa, M., and Kobayashi, Y. J. Immunol. 143:1327-

1333, 1989.
120. Enzoli, B., Lusso, P., Schechter, F., Josephs, S.F., Rappaport, J., Negro, F., Gallo, R.C., and Wong-Staal, F. EMBO J. 8:3019-3027, 1989.
121. Dudding, L., Haskill, S., Clark, B.D., Auron, P.E., Sporn, S., and Huang, E.S. J. Immunol. 143:3343-3352

# 8

## NATURAL RESISTANCE TO HERPES SIMPLEX VIRUS INFECTIONS: THE MACROPHAGE-INTERFERON AXIS.

INGRID DOMKE-OPITZ AND RAINER ZAWATZKY

**INSTITUTE OF VIRUS RESEARCH, GERMAN CANCER RESEARCH CENTER, INF 280, D 6900 HEIDELBERG, FRG**

**ABSTRACT**

The interactions between Herpes Simplex Virus type 1 and 2 (HSV-1, HSV-2) and the components of the host's natural defense system and immune system have been the subject of numerous clinical studies. In addition, experimental models of HSV infection have been established in a variety of animals. In susceptible hosts, primary systemic infection results in lethal encephalitis, whereas under appropriate conditions local infection causes HSV latency. Evidence has accumulated for a central role of macrophages in controlling resistance to primary HSV infection. In inbred mice, genetically determined natural resistance to HSV correlates with low permissiveness of macrophages leading to abortive infection. Moreover, macrophages can eliminate free extracellular virus and due to their high secretory potential also inhibit HSV replication in neighbouring cells. Agents known to stimulate the secretion of macrophage mediators have been shown to increase their antiviral activity. Interferons type alpha and beta (IFN $\alpha/\beta$) are among the most widely studied macrophage products. Macrophages are able to secrete low amounts of IFN constitutively in response to the macrophage-specific growth factor M-CSF. In addition, there is evidence that in resistant inbred mice, macrophages mainly contribute to the high level of "early" IFN $\alpha/\beta$ measurable *in vivo* after systemic HSV infections. IFN induced by HSV *in vivo* may either act directly by inhibiting virus replication in infected cells, or indirectly by stimulating natural killer cells. Under certain conditions, mouse macrophages can be persistently infected with HSV and may then serve as a vehicle for dissemination of the virus throughout the body. Freshly isolated human monocytes have been shown to restrict HSV replication but become permissive to lytic infection following prolonged *in vitro* culture. Human monocytes have been identified as cells producing IFN $\alpha/\beta$ in response to free infectious virions. HSV infected cells, however, induce IFN in a small population of medium-large sized lymphocytes. Clinical observations suggest that during recurrent HSV infection IFN $\gamma$ is secreted locally by activated T lymphocytes and may synergize with IFN $\alpha/\beta$ in the antiviral activity against HSV. Recently, HSV infection has been shown to transactivate Human Immunodeficiency Virus (HIV) gene expression in persistently infected cells. This observation may be of clinical importance since monocytes can serve as target cells for both viruses.

## INTRODUCTION

There are two types of herpes simplex virus, HSV-1 and HSV-2 (1) and infections with either type are very common in man. Epidemiologic studies in a variety of countries have revealed that more than 90% of the population is seropositive and must therefore have experienced a primary infection with HSV. Characteristically, within two weeks after productive primary infection, a latent stage develops in sensory ganglia and infectious virus is no longer detectable. However, under appropriate conditions, reactivation occurs and infectious virus is again produced. In healthy persons, clinical signs of primary infection are rare. They mainly occur as Herpes labialis and also as Herpes genitalis. Generally, they have a mild course and disappear without serious complications. The same is also true for recurrent infections.

In contrast, in immunosuppressed patients, primary infection with HSV generally results in more severe clinical manifestations and this group also has higher risks of lesions from recurrent infections (for a review, see 2). In newborns infected from their mother during birth, fatal Herpes encephalitis may develop due to virus spread to the central nervous system (3). These observations stress the role of host defense operating during both primary and recurrent infections.

## EXPERIMENTAL MODELS OF HSV INFECTION

In order to study in more detail individual compartments of the host defense system responsible for restriction of virus growth, a number of animal models have been established. Investigators have taken advantage of the fact that HSV is able to cause infection in rodents. The course of infection in these models is influenced by a variety of factors including age and immunocompetence of the host as well as virus strain, dose and route of infection. HSV given by a systemic route may result in the involvement of various organs and therefore, roughly correlates with the generalized infection of newborn children. Infection by a subcutaneous or ocular route shows some parallels to mucocutaneous and corneal infections in humans in that virus replicates at the site of entry and spreads to the sensory ganglia. Table 1 lists the animal models most widely used for the study of various features of experimental HSV infections.

Although animal models in general suffer from the fact that they do not reproduce the clinical course of HSV infection in man in all aspects, they facilitate the analysis of general aspects of pathogenicity and of specific and nonspecific antiviral defense. Among all animal models, HSV infection of mice has been studied most intensively, and it provided invaluable insights into the interaction between HSV and the host defense system. Skin infection into the footpad (4) or the ear of mice (5) is followed by local replication of virus and results in the establishment of latency, whereas systemic infection of HSV susceptible mice via the intracerebral (i.c.) or intraperitoneal (i.p.) route causes dissemination of virus to the central nervous system (reviewed in 6 and 7). Numerous strains of HSV have been tested in mice and profound variations in pathogenicity have been observed following infection of adult mice locally (8) or i.p. but not when given i.c. (9). In contrast, nearly all strains of HSV are highly pathogenic for newborn mice and cause lethal encephalitis following i.p. infection.

Interestingly, the studies by Lopez (9) showed that resistance of adult inbred mice to i.p. infection with a pathogenic strain of HSV-1 is under genetic control. He distinguished mouse strains that were highly susceptible (e.g. A/J), moderately susceptible (e.g. BALB/c, DBA/2) or resistant (e.g. C57BL/6) to the development of Herpes encephalitis. He found that resistance of C57BL/6

Table I
Examples of animal models used in experimental infections with HSV

| Species | Route of Infection Systemic | Local | Aim of Study | References |
|---|---|---|---|---|
| Rabbit | - | cornea | Induction of recurrences of latent virus by surgical means | 18 |
| Rabbit | - | intra-dermal (footpad) | analysis of cell mediated immunity | 19 |
| Rabbit | - | intra-dermal (back) | analysis of cell mediated and humoral immunity | 20 |
| Guinea pig | - | intra-dermal (footpad) | recovery of virus during latency at the primary infection site | 21 |
| Guinea pig | - | vaginal | establishment of latent infection analysis of pathogenicity | 22 |
| Mouse | intra-cerebral | - | analysis of pathogenicity in different mouse strains | 9,23 |
| Mouse | intra-peritoneal | - | genetics of natural resistance | 9,15 |
| Mouse | intra-peritoneal | - | Evaluation of host factors involved in natural resistance | 24-26 |
| Mouse | - | vaginal | development of humoral and cellular immune response | 27 |
| Mouse | - | oral | Protection from oral lesions by vaccines | 28 |
| Mouse | - | intra-dermal (ear) | Demonstration of latency and assessment of recurrent infection | 5,8 |
| Mouse | - | intra-dermal (footpad) | Demonstration of latency | 4,8 |

mice was mediated by cells derived from the bone marrow and inherited as a dominant trait governed by at least two loci not located within the major histocompatibility complex (MHC). The pattern of inheritance of natural resistance to systemic infection with HSV-1 was found to be similar to that observed for genetic resistance to bone marrow allografts that correlated with natural killer cell activity (10). Recently, it was reported that genes located within the MHC are involved in resistance to HSV in a model of local skin infection. MHC congenic mice carrying a C57BL genetic background have been used to study the influence of genes linked to the MHC on limiting viral spread from infected sensory ganglia (11). By measuring infectious virions recovered from homogenates of infected ganglia on day 7 after infection, the author found pronounced differences in the ability of these congenic mice to clear the virus from body fluids. Low titers of HSV and survival of mice were associated with the MHC haplotypes H-2 $^{b}$ and H-2 $^{d}$, but not with H-2 $^{k}$. Clearance of infectious virus one week after infection must also involve elimination of cells productively infected with HSV and can be considered as a specific function of the cellular immune system. Similarly, data have been provided for a central role of epidermal Langerhans cells in limiting viral spread in the host by their ability to efficiently present antigen to T lymphocytes (8). As already claimed in earlier studies (12), it is quite obvious that specific cell mediated immunity constitutes an important factor in resistance of mice to HSV, when local skin infection is studied.

In contrast, cellular immunity does not seem to play a major role in resistance to primary systemic i.p. infection, since athymic C57BL/6 mice were found to be even slightly more resistant to HSV-1 infection than their normal littermates (13) and thymectomy did not increase susceptibility of resistant C58/J mice (14). In a similar model of HSV-2 induced focal necrotic hepatitis, Mogensen (15) observed that genetic resistance of inbred mice was influenced by one dominant X-linked gene and correlated with the ability of macrophages to restrict viral replication thus inhibiting its spread to the liver. In this system athymic nude BALB/c mice also proved to be more resistant to HSV-2 induced hepatitis than the susceptible lttermates (16). Although host defense mechanisms involved in resistance of mice against primary infection with HSV-1 and HSV-2 are similar, there are reports on specific differences in pathogenicity. Most HSV-2 strains required much lower doses than HSV-1 to produce neurological signs and death after intradermal inoculation (5). Additionally, unlike HSV-1 (17), they induced progressive focal necroziting hepatitis in susceptible mice.

## ROLE OF MACROPHAGES IN RESISTANCE TO HSV INFECTIONS

Macrophages are mononuclear cells bearing Fc receptors and capable of phagocytising opsonized foreign particles. By virtue of their secretory activities they contribute to homeostatis under normal conditions, and following stimulation they are able to release a number of hormone-like factors necessary for the initiation and the enhancement of the immune response (for a review see 29). They are widely dispersed throughout the body and originate from hematopoietic stem cells in the bone marrow, from where they enter the circulation as blood monocytes. These cells are believed to represent a reservoir for the recruitment of new macrophages, since following exposure to chemotactic stimuli they are able to actively leave the blood stream and migrate to inflammatory sites in the organism. They are also present as resident cells in most tissues and in serosal cavities. Due to their strategical location at potential sites of entry for pathogens, macrophages are ideally suited as a first line of non-specific defense against invading microorganisms, well before the onset of a specific immune response.

<u>Antiviral activity of macrophages</u>.

Macrophages can inactivate extracellular virus or reduce its replication and release from neighbouring cells, a property designated as "extrinsic" antiviral resistance (30). Hirsch, *et al.* (31) have reported that transfer of macrophages from adult to neonatal mice was able to confer resistance to HSV. Moreover, many agents known to activate macrophage functions such as pyran copolymer or *Corynebacterium parvum* (C.p.) have been found to protect mice efficiently against lethal i.p. challenge with HSV (32-35). On the other hand, impairing macrophage functions *in vivo* by injection of substances toxic for these cells such as silica or bone marrow depletion by $^{89}$Sr strongly increased lethality in otherwise resistant animals (25,36). Extrinsic resistance of macrophages is largely due to their high secretory activity, in particular following stimulation. However, it is important to note that, besides a large variety of natural or synthetic agents, the infecting virus itself can also serve as physiological stimulator of macrophage secretory capacity. For example, proteolytic enzymes or reactive oxygen intermediates (ROI) directly inactivate free extracellular virus particles, and IFNs inhibit viral replication in the surrounding cells.

Secretion of arginase from activated mouse macrophages was also found to prevent productive HSV infection in neighbouring cells (37). In contrast to IFN, the action of arginase was not species specific but interfered with the growth requirement of HSV for arginine. However, Morse and Morahan (38) have reported that C.p. stimulated macrophages also exert a direct antiviral effect on HSV replication in permissive cells, a function apparently not dependent on IFN or other secreted factors but requiring cell to cell contact and interfering directly with host cell protein synthesis. Macrophages of murine and human origin can also be cytotoxic for HSV infected target cells *in vitro* either in the presence or absence of specific antibodies (39,40).

Macrophages as target cells for HSV infection.

Macrophages can fulfill their antiviral functions only if they are themselves nonpermissive to virus infection (41). However, multiple interactions between macrophages and viruses can occur ranging from nonpermissiveness to lytic infection. The outcome of the infection depends on host as well as viral factors (Table 2, reviewed in 41-43). A brief overview of HSV replication will be given prior to discussing in detail possible interactions between macrophages and HSV.

HSV replication in permissive cells.

Absorption of HSV to cellular receptors is the first step to initiate infection. After penetration which is thought to occur by fusion, the capsid is transported to the nucleus and viral DNA is released into the nucleoplasma (44). Infection is accompanied by a rapid shut-off of host cell macromolecular synthesis. Transcription of viral genes proceeds in a cascade manner. The expression

**Table 2**. Factors influencing HSV-macrophage interactions.

| |
|---|
| cellular factors |
| genetic background |
| state of activation |
| state of differentiation |
| *in vitro* ageing |
| proliferation |
| tissue origin |
| viral factors |
| genetics of individual virus strains |

of three classes of HSV genes is coordinately regulated. The synthesis of immediate early (IE), early (E) and late (L) proteins depends on the synthesis of proteins from the preceeding class (45,46). The cascade of HSV protein synthesis is controlled both by transcriptional and post-transcripitonal mechanisms (47). Transcription of HSV DNA is performed by the cellular RNA polymerase II, but viral factors are involved at all stages of infection.

Five IE proteins are known. Three of them (ICP4, ICP27 and ICP0) have been shown to be involved in the regulation of viral gene expression (48-50). Although IE proteins are synthesized independently of other viral functions, their transcription is greatly stimulated by a structural protein of the infecting virion (51-53). The presence of functional IE proteins is necessary for the expression of HSV E genes, some of which like DNA polymerase, thymidine kinase or ribonucleotide reductase, are enzymes required for viral DNA replication and nucleotide metabolism. After the synthesis of sufficient amounts of E proteins, viral DNA is synthesized in the nucleus by a rolling-circle mechanism. The expression of HSV L genes is positively regulated by IE gene products and depends on the presence of E proteins (49,54). In addition, replication of the viral chromosome is required for complete induction of L gene expression (45,46). Many L proteins like the major capsid protein and glycoproteins B and C are structural constituents of the mature virion.

During the synthesis of viral macromolecules, empty viral capsids are assembled and accumulate in the nucleus. Viral DNA is packaged into empty capsids and these capsids become enveloped at the inner lamella of the nuclear membrane. Enveloped particles are transported to the cell surface and are probably released via the Golgi apparatus (55). HSV replication requires approximately 18 hours. During the replication cycle viral glycoproteins are inserted into cytoplasmic and plasma membranes of the host cell. This is of major importance for the induction of antiviral defense mechanisms.

Infection of mouse macrophages with HSV.

Early studies by Johnson have correlated resistance of adult mice to restriction of HSV replication in macrophages (24). This restriction has been described in many studies (reviewed in 41,43). Stevens and Cook were the first to analyse the abortive infection in more detail (56). The interaction of macrophages with HSV, however, does not always result in an abortive infection. Factors influencing the course of HSV infection in macrophages are discussed in the following sections.

Genetic origin of macrophages.

Several studies have correlated resistance to primary i.p. infection with HSV to non-permissiveness of macrophages for virus replication. Thus, resistance of mice to HSV-2 induced focal necrotic hepatitis has been shown to be linked to restriction of HSV-2 replication in macrophages (15). Several studies indicated that macrophages from resistant C57BL/6 mice do not support productive HSV replication (57-63). Brucher *et al.* (62) and Domke-Opitz *et al.* (64) have correlated resistance of these macrophages to IFN production in response to HSV. Cells derived from C57BL/6 mice produced high titers of IFN early after virus infection and resistance of cells could be broken by *in vitro* addition of antibodies to mouse IFN $\alpha/\beta$.

Results from other studies suggested that *in vivo* resistance to HSV infection did not exclusively depend on intrinsic resistance of macrophages to HSV infection. Lopez and Dudas found that macrophages from resistant F1 mice were unable to restrict HSV replication *in vitro* (59). Similar results were reported by Armerding *et al.*, for SJL mice (61).

The mechanism of restriction of HSV growth in macrophages derived from C57BL/6 mice seems to be linked to the IFN induced by the virus. After being produced IFN induces an antiviral state in its target cells. This event usually requires several hours (65). This is in line with the observation that there were no differences in the number of virus producing cells (as judged by infectious center assays) between macrophages derived from resistant C57BL/6 and susceptible DBA/2 mice (62). Viral plaques were observed in peritoneal macrophages of both mouse strains at low multiplicity of infection (MOI). These plaques were much smaller in cells of C57BL/6 origin, suggesting that HSV infected, IFN producing macrophages from these mice protected surrounding cells more efficiently against the second round of replication.

Early after infection viral DNA replication and the expression of viral antigens were comparable in splenic macrophages cultured from resistant C57BL/6 or susceptible DBA/2 mice (64). The study by Sarmiento defined multiple points in the viral replication cycle that were blocked in macrophages derived from different tissues and from different mouse strains (63). Viral macromolecular synthesis was restricted earlier in the replication cycle in C57BL/6 macrophages derived from different sources when compared to macrophages of DBA/2 origin. In all these studies, however, macrophages were grown in the presence of macrophage colony-stimulating factor (M-CSF). This growth factor has been shown to induce minute amounts of IFN which might influence the course of HSV infection in macrophages in addition to the IFN induced in response to HSV (66).

Treatment of macrophages with modulators.

Macrophages display a high degree of heterogeneity which can be modified by various treatments *in vivo* and *in vitro*. Many studies have analysed the effect of eliciting or activating agents on intrinsic resistance of peritoneal macrophages to HSV infection (reviewed in 43). I.p. inoculation of mice with proteose-peptone had no effect on HSV replication in peritoneal macrophages (67). Activation of macrophages by i.p. inoculation of C.p. did not result in increased virus yields when compared to resident peritoneal macrophages

(68). However, in contrast to resident cells, C.p. stimulated cells showed marked cytophatic effects after HSV infection.

Increased permissiveness for HSV-1 and HSV-2 replication has been reported for macrophages elicited with thioglycollate broth (TG; 59, 60, 63, 68). Although an increase in virus titers was found in macrophages derived from genetically susceptible as well as resistant C57BL/6 mice, treatment with TG did not abrogate genetic differences (59,63). In contrast to resident and C.p. elicited cells, TG stimulated macrophages showed moderate levels of viral DNA replication (63,68).

Infection with Bacillus Calmette-Guerin (BCG) has been shown to enhance resistance to HSV *in vivo* as well as in macrophages *in vitro* (59,69). Sethi and Brandis described restriction of HSV growth in peritoneal macrophages derived from mice immune to HSV or *Listeria monocytogenes* (58). These authors also showed that pre-treatment of macrophages with IFN $\gamma$ containing supernatants conferred resistance to HSV replication. It is suggested that in addition to its involvement in extrinsic antiviral resistance activity, macrophage arginase is also involved in intrinsic resistance to HSV replication (57). The antiviral activity of various IFNs against HSV in macrophages has been extensively studied in our group and will be discussed below.

The interesting studies of Bonina *et al.* (67) with rat peritoneal macrophages showed that cells from tumor-bearing rats had lost the capacity to restrict HSV replication (67). Treatment of these rats with immunomodulators stimulated resistance of macrophages against HSV.

*In vitro* ageing and proliferation.

*In vitro* culture of mouse and human macrophages prior to HSV infection has been shown to result in increased viral replication (59,62,70-72). In murine cells increased permissiveness was observed in the presence or absence of M-CSF. Studies analysing the effect of this macrophage-specific growth factor on HSV replication are contradictory. Whereas some authors described a rise in virus titer in M-CSF treated cells, others did not find any effect (58,73). It must be remembered, however, that in addition to its proliferation-promoting activity, M-CSF induces autocrine IFN in macrophages (66). Experiments in our laboratory have clearly shown that bone marrow-derived macrophages grown in the presence of M-CSF are much more resistant to HSV infection than cells promoted under the influence of granulocyte-macrophage colny-stimulating factor (GM-CSF), which does not induce IFN when added to mouse bone marrow cells (74).

HSV replication has been analysed in various macrophage cell lines some of which were obtained by transformation with SV 40. There was a marked heterogeneity among these cell lines in permissiveness for HSV ranging from little to effective replication and to the establishment of a persistent infection (43,58.75,76). *In toto* the results clearly show that proliferation *per se* does not induce permissiveness for HSV. Moreover, one has to be very careful in extrapolating findings obtained with cell lines to primary macrophages.

<u>Tissue origin of macrophages</u>.

HSV replication has been analysed in macrophages derived from the peritoneal cavity, spleen or bone marrow of mice. Bone marrow-derived macrophages grown under the influence of M-CSF have been reported to restrict productive HSV infection but undergo marked cytopathic effects (77,78). Leary *et al*., described HSV DNA replication in bone marrow-derived macrophages but not in resident peritoneal macrophages (77). Conversely, in a study on HSV replication in macrophages derived from various tissues, Sarmiento (63) found that bone marrow-derived macrophages were less permissive than peritoneal macrophages, and splenic macrophages were more permissive than macrophages from the peritoneal cavity. It is not clear whether this apparent contradiction is due to the use of different mouse strains or to the fact that in the latter study macrophages were cultured in the presence of M-CSF prior to HSV infection.

<u>Molecular mechanism of HSV restriction in resident peritoneal macrophages</u>.

In spite of the previously discussed differences in susceptibility for HSV, macrophages in general are much less permissive for HSV replication than other cells. Early stages of HSV infection like absorption, penetration and entry of viral DNA into the nucleus occur in at least some resident peritoneal macrophages (68,77). In earlier studies, Stevens and Cook reported that although HSV DNA and protein synthesis took place, most viral capsids lacked central dense-cores (56). However, they used cells pre-cultured for 2 days in the presence of serum. Recent studies by Morahan *et al.* (68,77,79) clearly showed that HSV DNA did not replicate in resident peritoneal macrophages. Analysis of steady state levels of RNA and protein revealed that all HSV IE, some E, but no L proteins were present in HSV infected cells (79). However, viral gene expression was limited to a small percentage of cells, thus demonstrating the heterogeneity of the resident peritoneal macrophage population.

The differences in HSV gene expression within a given macrophage population remain an enigma. Studies trying to define subpopulations characterized by physical parameters such as size or density with increased permissiveness for HSV have failed so far (43,70). Therefore, new criteria will have to be defined in order to identify macrophage subsets that allow productive viral infection.

Persistent HSV infection of macrophages.

In addition to undergoing abortive or lytic HSV infection, macrophages can be persistently infected with the virus. Studies in our laboratory described persistent HSV infection of macrophages derived from spleen cells of HSV resistant C57BL/6 mice by culture in the presence of M-CSF (64). The infection was characterized by the continuous release of infectious virus particles from a small subpopulation of infected cells. Although IFN was not detectable, treatment of persistently infected cells with antibodies to mouse IFN-$\alpha/\beta$ resulted in increased HSV replication and cytopathic effects. HSV was also reactivated by treatment of macrophages with lipopolysaccharide or tumor necrosis factor (TNF; 80). Splenic macrophages from mouse strains susceptible to HSV *in vivo* were lysed by HSV after 24 hours, indicating that the establishment of persistent infection was due to genetically controlled properties of the host cells.

Tang *et al.* (75) have studied persistent HSV infection in an Ia antigen positive murine macrophage cell line. Consistent with our results with M-CSF promoted primary cells, the infection behaved like a carrier culture consisting of a minority of infected cells and a majority of uninfected cells. Moreover, endogenous IFN was important in the maintenance of HSV persistence.

Genetic factors of the virus.

The molecular genetics of HSV are another factor which influences the interaction of virus and macrophages. Different strains of HSV-1 showed marked variations in their ability to replicate in macrophages although there was no correlation to pathogenicity in mice (70). Ben-Hur *et al.* (73) performed interesting experiments with several HSV-1 strains and recombinants differing in i.p. pathogenicity *in vivo*. Non-pathogenic strains did not replicate in macrophages. Transcription of the IE genes of a non-pathogenic strain was blocked in macrophages, but not in other cells. IE gene expression in macrophages, as well as i.p. virulence, were mapped to a certain DNA sequence in the viral genome.

## HUMAN MONOCYTES AND MACROPHAGES AS TARGET CELLS FOR HSV INFECTION.

Infection of primary monocyte cultures with HSV.

Most studies analysing HSV infection of human mononuclear phagocytes have been performed with blood cells which are less differentiated than tissue macrophages from the lung or the peritoneal cavity (81,82). Several authors reported that HSV infection of freshly isolated human blood monocytes is abortive (71,72,83). During *in vitro* ageing, these cells undergo spontaneous differentiation into macrophages and gain permissiveness for HSV growth (71,72,83,84).

Daniels *et al.* (71) were the first to analyse the mechanism of restriction of HSV growth in freshly isolated human blood monocytes. They reported that viral antigens were expressed in the cells but capsid structures detected by electron microscopy were empty, indicating a block in virus maturation. Linnavuori and Hovi (83) also found viral antigen expression in abortively infected cells. In these studies, cells had either been subjected to adherence and detachment prior to virus infection or the cultures had been infected at an unusually high MOI. Studies in our laboratory analysing the interaction of human blood monocytes with HSV have been performed with cells isolated by counterflow centrifugal elutriation to avoid functional changes due to plastic adherence prior to virus infection. We found that freshly isolated human blood monocytes restricted viral DNA entry into the nucleus (85). Culturing the cells for some hours before inoculation with HSV was sufficient to overcome this early block. There was no expression of HSV IE, E or L genes in fresh monocytes, whereas all classes of genes were expressed in cells cultured for 1 day *in vitro* prior to virus infection (86). However, as previously reported in other studies, virus yield and infectious center assays showed a stepwise increase in susceptibility from day 1 to day 6 *in vitro* culture (71,85). This indicated that at least one additional block occuring after the synthesis of virus proteins was still present in 1 day old monocytes. The factors responsible for this block might be related to the stage of differentiation.

Replication of HSV has also been studied in human tissue derived macrophages (87,88). These cells are considered to represent the *in vivo* counterpart of macrophages derived from peripheral blood monocytes *in vitro* (89). Freshly isolated alveolar or peritoneal macrophages derived from adults restricted HSV replication whereas cells obtained from neonates were permissive (87). As observed for blood monocytes, peritoneal macrophages from adults gained permissiveness for HSV after several days of *in vitro* culture (88). Thus, susceptibility of human macrophages to HSV was linked to *in vitro* ageing whereas the role of differentation remained unclear.

Resistance of human blood monocytes has been correlated to IFN synthesis and inhibition of differentiation (83,84). High titers of IFN were only produced by freshly isolated cells in response to infection with HSV at high MOI. When virus infection was performed in the presence of antibody to IFN, enhanced virus yields were observed only when infection was done at low MOI (84,86). Studies in our laboratory showed that no viral DNA was associated with the nuclear fraction 1 hour after infection of freshly isolated monocytes (85). Viral DNA entry into the nucleus normally precedes the establishment of an antiviral state by IFN, which usually requires several hours. Thus, in contrast to the studies with mouse macrophages, IFN induction is unlikely to be involved in mediating antiviral resistance of human blood monocytes to HSV infection.

HSV replication in myelomonocytic cell lines.

The interaction of the myelogenous leukemic cell line K562 with HSV has been analysed by Rinaldo *et al.* (90). HSV persisted in this cell line for about one year. Whereas 0.5-2.7% of the cells produced infectious virus, expression of viral antigens was observed in a larger proportion of cells.

The promyelocytic cell line HL 60 is capable of differentiating into granulocytes or monocytes depending upon the differentiation stimulus used (91). Pientong *et al.* (92) have found that HSV was replicating in undifferentiated HL-60 cells. Virus replication was accelerated in HL-60 cells caused to differentiate into monocytes by treatment with 1.2-tetradecanoyl-phorbol-13-acetate (TPA). In contrast, replication of HSV was restricted in cells treated with dimethyl-sulfoxide (DMSO), which is known to induce granulocytic differentiation of HL-60 cells. The block in virus replication in DMSO-treated cells was located at the level of translation of HSV E-proteins.

U937 monoblastic cells can be induced to differentiate along the monocytic pathway by various agents (91). Whereas in one study undifferentiated U937 cells were reported to be completely resistant to HSV infection, another study described them to be more susceptible (93,94). Tenney and Morahan found HSV DNA in the nuclei of undifferentiated U937 cells 3 hours after infection but expression of IE genes and viral DNA replication were inhibited (93). After the induction of differentiation by treatment with TPA increased permissiveness for HSV was observed. This was, however, not the case when the cells were induced to differentiate by treatment with other agents. Thus, it seems more likely that the decrease in HSV resistance is due to a nonspecific effect of TPA than to cellular differentiation.

## INVOLVEMENT OF IFN AND NATURAL KILLER CELLS IN ANTIVIRAL DEFENSE AGAINST HSV.

Among all soluble factors generated during *in vivo* and *in vitro* infection with HSV, IFNs are probably the most widely studied. Several clinical reports have shown that IFN is present in body fluids during disseminated HSV infection in newborns (95) and adults (96) and also in vesicle fluids from lesions of recurrent Herpes labialis (97,98). The latter studies revealed a close correlation of HSV titers and levels of IFN α in vesicle fluids, indicating that high titers of IFN reflect excellent virus replication without obvious benefit for the host, although most HSV-1 and HSV-2 clinical isolates were equally sensitive *in vitro* to all classes of human IFN (99). More recently, however, Torseth and Merigan (100) reported on the local production of IFN γ in vesicle fluids of normal individuals with recurrent HSV infection. This IFN was predictive of a long subsequent lesion-free interval and there were decreased levels in immunosuppressed patients during severe recurrent HSV lesions.

In mice, systemic primary infection with HSV clearly showed that the endogenous IFN α/β response is of crucial importance for survival, since simultaneous i.p. injection of HSV and antibodies to IFN α/β rendered the animals susceptible to very low doses of virus (101). In similar experiments we investigated if the genetically determined natural resistance of adult C57BL/6 mice to a highly pathogenic strain of HSV was also due to virus induced production of IFN α/β. Three-4 hours following i.p. infection of adult mice we observed an "early" peak of IFN α/β at the site of infection which was absent in susceptible newborn C57BL/6 mice (102). Accordingly, much higher titers of infectious virus were recovered from the peritoneal fluids of newborn mice one day after infection. This was associated with a peak of "late" IFN, which, however, had no protective effect in these animals. In addition, in adult C57BL/6 mice we observed that resistance to very high virus doses (representing 250 $LD_{50}$) correlated with the amount of "early", HSV-induced IFN production in peritoneal fluids and could be completely abrogated by simultaneous injection of anti-IFN α/β (103). Similar correlations between "early" local IFN production and resistance were also observed for HSV-2, in that susceptible BALB/c mice secreted about 10-fold less "early" IFN into the peritoneal fluid than did resistant C57BL/6 (104). A protective effect of IFN α/β against lethal encephalitis in these animals could be demonstrated by simultaneous injection of IFN together with lethal doses of HSV-1 (103).

In addition to IFN α/β, endogenous levels of IFN γ induced during the first 2 hours after primary infection of mice with HSV-1 were also reported to be crucial for survival (105). Inoculation of anti-IFN γ at various times following i.p.

infection greatly enhanced mortality of the animals and was most effective when given during the first hours after infection, similar to anti-IFN $\alpha/\beta$. The simultaneous presence *in vivo* of IFN $\alpha/\beta$ and IFN $\gamma$ can be of physiological importance, since it is known from *in vitro* studies, that both IFN species act synergistically in inhibiting HSV replication in permissive cells (106).

Much attention has also been paid to the role of natural killer (NK) cells in HSV infections. NK cells are mononuclear cells capable of nonspecifically lysing a number of different targets including certain tumor cells, virus infected and normal cells. They are believed to constitute an important part of the host's natural defense system. Since IFN is known to enhance NK cell mediated lysis (107) and is also induced in the course of HSV infection, studies have been directed to dissect the NK cell response from the IFN response. Attempts were made to correlate natural resistance of inbred mice against primary systemic i.p. infection with HSV-1 or HSV-2 to either "early" local IFN production or induction of NK activity in peritoneal exudate cells which normally are devoid of NK activity (108). HSV induced high titers of "early" IFN in resistant C57BL/6 and C3H/HeJ mice, and these strains also developed high NK activity after one day. SJL mice, however, lacked HSV induced NK activity but were able to mount an "early" IFN response (109). By adoptive transfer of spleen cells from adult C57BL/6 into HSV susceptible suckling mice, Bukowsky and Welsh (110) showed that protection could be mediated by cells depleted of NK activity using antibodies to the specific cell surface marker asialo $GM_1$. In addition, injection of HSV resistant mice with these antibodies did not increase mortality (110,111), unless very high doses were used that also affected the HSV induced "early" IFN response of macrophages (112). These results argue for a predominant role of "early" IFN (induced 2-4 hours after HSV infection) and suggest that NK cell activation measured about 18 hours later is a secondary event that constitutes one of the effector mechanisms by which IFN mediates protection.

Studies performed with human mononuclear cells from peripheral blood (PBML) revealed one important difference, in that these cells already exhibited high levels of constitutive NK activity prior to induction by HSV. A functional heterogeneity of NK cells from PBML has been described in that various subpopulations lyse either K562 cells or HSV infected fibroblasts or both cell types (113). Clinical observations have identified unusually severe HSV disease in newborns or Wiskott-Aldrich patients who have low levels of NK activity against HSV infected fibroblasts but often have normal levels of NK activity against K562 targets (114). From experiments conducted in the presence of either neutralizing antibodies to IFN $\alpha$ or IFN $\beta$, the authors concluded that although IFN $\alpha$ exerted an enhancing effect, constitutive levels of NK activity were independent of the presence of IFN $\alpha$.

**IDENTITY OF THE CELL THAT PRODUCES HSV-INDUCED IFN.**

Studies in mice.

Several lines of evidence suggest that the "early" IFN $\alpha/\beta$ response to HSV in mice is a function of macrophages. First, exposure of mice to lethal doses of X-irradiation revealed that HSV-induced IFN $\alpha/\beta$ production occured in two phases "early" and "late". "Early" levels measured 2-3 hours after infection, were not influenced by X-irradiation in contrast to "late" levels of IFN $\alpha/\beta$, measured 8-10 hours after infection (115). Since macrophages are known to be relatively resistant to X-irradiation, they can be considered as main candidates for "early" IFN production *in vivo*. Second, the amounts of "early" IFN measured in peritoneal fluids from HSV infected C57BL/6, BALB/c and DBA/2 mice correlated well with IFN titers in cultures of infected peritoneal macrophages from these strains, and the producer cells fulfilled the criteria generally attributed to macrophages (62,116,117). Finally, IFN levels in response to HSV-1 and HSV-2 *in vivo* were influenced by the X-chromosome in that female mice produced 2-4 fold more IFN $\alpha/\beta$ than males (118) and this difference was also expressed in peritoneal macrophages *in vitro* (104).

Macrophages not only secrete IFN $\alpha/\beta$ when exposed to HSV but also respond to it. An early function of macrophages after contact with a variety of pathogens is the priming for a respiratory burst followed by the release of ROI (119). This macrophage property is enhanced in a dose-dependent manner by all classes of IFN. Recently, Mogensen *et al.* (120) have shown that peritoneal macrophages from HSV-2 resistant C57BL/6 mice exhibit a higher respiratory burst capacity following HSV infection than macrophages from the more susceptible BALB/c mice, thus reflecting genetically determined differences in HSV induced "early" IFN $\alpha/\beta$ production. They concluded that these differences in early activation of macrophages are mediated by IFN $\alpha/\beta$ in an autocrine fashion.

In an experimental model using highly susceptible neonatal C57BL/6 mice Kohl *et al.* (121) have shown that, following *in vivo* treatment with recombinant interleukin-2 (IL-2), macrophages from these mice can confer resistance to lethal i.p. infection with HSV-1. The protective effect was ascribed to the induction by IL-2 of IFN $\gamma$ which stimulated macrophages from newborn mice to enhanced antibody dependent cytotoxicity towards HSV-1 infected target cells. It is likely that *in vivo* this aspect of the cellular immune response that results in the secretion of immune specific IFN $\gamma$ by activated T lymphocytes is also stimulated by monocytes and macrophages, since under physiological conditions they represent the main source of Interleukin-1 (IL-1) production and IL-1 induces T cells to express IL-2 and its receptor.

Studies in man.

Studies of the cellular origin of HSV induced IFN production in human cells have mainly focused on PBML. The producer cells were shown to vary with the nature of the inducing agent and the immune status of the donor. Monocytes were identified as the cellular source of IFN production in response to infectious HSV by virtue of their capacity to adhere to plastic surfaces (122-124) or by purification using centrifugal elutriation (85). The IFN inducing capacity of HSV in monocytes did not vary significantly among different clinical isolates. In contrast to mouse macrophages, IFN induction in human cells did not require infectious virus and was not dependent on viral replication since UV-irradiation, thermal inactivation or blocking of viral DNA synthesis did not decrease IFN levels in culture supernatants (122).

Cell lines infected with HSV have also been reported to induce IFN $\alpha$ when co-cultured with human PBML (124-126) and Lebon (125) showed that viral glycoprotein D on the surface of infected target cells was the IFN $\alpha$ inducing antigen. There are, however, conflicting reports as to the cellular source of the

IFN response to virus infected cells. Stanwick *et al.* (40) identified human monocytes/macrophages both as mediators of cytotoxic activities against HSV infected cells and as inducers of IFN $\alpha$ production. In contrast, more recent studies presented evidence that HSV infected cells elicit an IFN response in medium-large sized lymphocytes (126) that express HLA class II antigen, are distinct from NK cells (127) and apparently do not belong to the T cell lineage. Interestingly, these cell populations share some characteristics with those reported to secrete IFN $\alpha$ in response to $\beta$-propiolactone inactivated HSV (128). By means of *in situ* hybridization, Gobl *et al.* (126) calculated a frequency of ten IFN producing cells per $10^4$ PBML. This low frequency is in line with earlier observations in other experimental systems (129,130) and suggests that following virus infection only a fraction of cells is expressing the IFN genes efficiently.

Recently, the local presence of IFN $\beta$ in vesicle fluids from patients with HSV skin lesions has been demonstrated (131). From *in vitro* culture systems and immunohistochemical staining of biopsies with anti-IFN $\beta$ the authors concluded that HSV infected keratinocytes probably represent the major source of locally produced IFN $\beta$ during recurrent HSV infections.

IFN-$\gamma$ secreted by primed T cells is a potent stimulator of various immunological functions of monocytes and macrophages. Immunohistochemical staining of biopsies from HSV lesions shortly after onset of symptoms revealed the presence of intracellular and membrane bound IFN $\gamma$ (100). T lymphocytes were found to predominate in these lesions and this observation may explain the early presence of IFN $\gamma$ in vesicle fluids. Earlier studies with PBML *in vitro* have shown that T lymphocytes in cultures of PBML from HSV seropositive donors were able to secrete immune specific IFN $\gamma$ when challenged with HSV antigen, but required the presence of monocytes as accessory cells (132). In contrast to IFN $\alpha$, IFN $\gamma$ was observed only 5 days after challenge and was rarely detectable in PBML from seronegative donors (133). Interestingly, however, Green (134) also detected IFN $\gamma$ production in HSV induced PBML from seronegative donors following depletion of monocytes by adherence. On the other hand, suppression of HSV induced

IFN $\gamma$ production was observed in PBML cultures following addition of increasing numbers of monocytes from seronegative, but not from seropositive donors. He concluded that the normally observed inability of unseparated PBML from seronegative donors to generate immune specific IFN $\gamma$ in response to HSV was due to suppressive effects of monocytes from these persons, whereas monocytes from seropositive donors were functionally quiescent. Taken together, these data reveal an important role of monocytes in regulating the immune specific production of IFN $\gamma$ by T lymphocytes.

**MOLECULAR MECHANISMS OF IFN ACTION AGAINST HSV.**

As described above, IFN induced either in an autocrine fashion during macrophage cultivation or in response to HSV infection, is influencing the interaction between HSV and mouse macrophages. For monocytes of human origin, the situation is less clear. If these cells, however, were pretreated with IFN prior to virus infection, they were protected from HSV replication as well (135). Factors influencing the sensitivity of HSV to IFN include the cell system, the type of IFN and the virus strain (136-139). In general, lytic RNA viruses like Vesicular Stomatitis Virus are more sensitive to IFN than HSV (136). The antiviral activity of recombinant IFN against HSV did not exceed the activity of natural IFN (137). HSV replication in HeLa cells was described as being not very sensitive to IFN whereas splenic mouse macrophages could be protected very efficiently from HSV infection (140-143).

Inhibition of HSV replication by IFN $\alpha$ and IFN $\beta$.

Most authors agree that IFN acts early during the viral replicative cycle. Lipp and Brandner were the first to report a strong impairment of HSV DNA and protein synthesis in cells treated with polyinosinic-polycytidylic acid, an IFN inducer (144). Inhibition of HSV IE protein synthesis in IFN treated cells was described in various experimental systems: HeLa cells or human fibroblasts/ IFN $\alpha$ (141,145), splenic mouse macrophages/mouse IFN $\alpha/\beta$ (143), human blood monocyte derived macrophages/IFN $\beta$ (135) and phytohemagglutinin (PHA) activated human T lymphocytes/IFN $\alpha$ or IFN $\beta$ (146). The subsequent inhibition of HSV E and L gene expression can be explained by a reduction in the levels of HSV IE proteins which are involved in turning on the synthesis of E and L-mRNAs (48-50,143).

The first steps of HSV infection (absorption, penetration, transport to the nucleus, uncoating) were not impaired in IFN-treated cells. In this study, comparable amounts of viral DNA were found in the nuclei of untreated and IFN-treated splenic mouse macrophages early after infection. Expression of IE genes was identified as the major target of IFN-mediated inhibition of HSV replication. Using nuclear run-off transcription analysis with end-specific probes for the ICP4 gene inhibition was shown to occur at the onset of viral transcription (147). Oberman and Panet obtained similar results with IFN $\alpha$ treated HeLa and HEp-2 cells (14). Reduced accumulation of mRNAs coding for HSV IE proteins has also been described by Lipp and Brandner (148) and Sharma *et al.* (149).

There are, however, a few contradictory reports relative to the pattern of HSV inhibition by IFN and suggest that IFN acts late during virus replication. Munoz and Carrasco found no inhibition of viral protein synthesis, but reported that new virus particles grown in IFN treated HeLa cells had reduced infectivity (150). Chatterjee *et al.*, working with human fibroblasts found no impairment of viral DNA synthesis in IFN-treated cells, but the amount of HSV glycoproteins B and D was reduced and there was a decrease in the release of mature virus particles (151,152). In order to explain these inconsistent results, two other studies were performed with different cell lines and virus strains (140,141). HSV IE protein synthesis was always inhibited suggesting the same mechanism of IFN action against HSV in all cells under study.

Inhibition of HSV replication by IFN $\gamma$.

IFN $\gamma$ has been shown to synergise with IFN $\alpha$, IFN $\beta$ and TNF in the antiviral activity against HSV (106,142,153). New results indicate that induction of IFN $\beta$ is a prerequisite for the antiviral activity of TNF (154). The mechanism leading to inhibition of HSV replication in HEp-2 cells treated with IFN $\gamma$ and TNF seems to be comparable to that reported for IFN $\alpha$ and IFN $\beta$. Feduchi *et al.* (155) located the block between virus entry and transcription of IE genes. Studies with splenic mouse macrophages treated exclusively with IFN $\gamma$ revealed an inhibition of synthesis of HSV IE protein ICP4 (156). Contrary to the findings with IFN $\alpha/\beta$ in this cell system, the rate of transcription of IE genes was only slightly delayed in IFN $\gamma$ treated cells. One can speculate that whereas IFN $\alpha/\beta$ acts on HSV transcription, IFN $\gamma$ affects viral translation. Indeed, this would also explain the synergistic effect of IFN $\alpha/\beta$ and IFN $\gamma$ on HSV growth.

In human leucocytes, IFN $\gamma$ exerts only very little antiviral activity against HSV. This has been shown with blood monocyte-derived macrophages and PHA-activated T lymphocytes (135,146). The failure of IFN $\gamma$ to establish an antiviral state in these cells is possibly related to the binding characteristics of IFN $\gamma$ receptors (157,158).

## INTERACTIONS BETWEEN HERPES SIMPLEX AND HUMAN IMMUNODEFICIENCY VIRUS.

The observation that HSV and Human Immunodeficiency Virus (HIV) can both infect $CD4^+$ T-lymphocytes and cells of the monocyte/macrophage lineage has recently created much interest in the study of possible mechanisms of interaction between these two viruses. Indeed, there is evidence from numerous epidemiological studies, that cofactors may be important in the pathogenesis of AIDS. A characteristic feature of HIV infection is a long disease-free stage of latency (159) with low numbers of infected leukocytes (160). The presence of HIV particles has also been demonstrated in tissue macrophages and myeloid precursor cells from the bone marrow of AIDS patients (for a recent review see 161) and it has been proposed that these cells may serve as a reservoir for HIV in the body.

*In vitro* studies have revealed that activation of latent HIV is mediated in *trans* either by cellular transcription factors (162,163) or by transcriptional trans-activator genes from several DNA viruses, including HSV (164-167). In these reports, HSV IE gene products were shown to stimulate, in transfected tissue culture cell lines, the transcription of heterologous reporter genes driven by the HIV long terminal repeats (LTR). Recently, the early appearance of HIV antigens and of reverse transcriptase activity was detected in persistently infected $CD4^+$ lymphoid cells after superinfection with HSV (168). More detailed analysis revealed that trans-activation of HIV did not require full replication competent HSV, but only functional ICP4 (168) and/or ICP0 genes (167,169). Freshly isolated human blood monocytes, however, were found to be resistant to HSV infection and they restrict the entry of viral DNA into the nucleus (85). Thus, it is uncertain if activation of latent HIV in these cells by superinfection with HSV may occur *in vivo*, and therefore we do not know at present if interactions between HIV and HSV are of clinical importance. However, another aspect of HSV/HIV interaction has been recently suggested by epidemiological studies of homosexual men (170,171). The authors suggested that genital ulcers due to HSV-2 infection predispose for the acquisition of HIV infection and thus, may represent a risk factor for AIDS.

## CONCLUSION

There is general agreement that macrophages have an important role in host defense against HSV infections. Although there are numerous differences concerning the course of HSV infection in animal models and man, several interesting correlations in macrophage-HSV interaction exist. Macrophages develop resistance to HSV with age and thus, serve as nonpermissive target cells for viral replication. In addition, they rapidly secrete IFN following contact with HSV, which not only can protect neighbouring cells from infection but can also act as a stimulator of other macrophage functions and of specific and non specific cytotoxicity against infected cells. Considering macrophages as target cells for HSV, however, one aspect has not been investigated so far. Although freshly isolated human blood monocytes have been reported to restrict HSV infection at the early level of viral DNA entry into the nucleus, there are no such studies in the literature on freshly isolated cells of murine origin. Genetic studies in mice have revealed a correlation between the relative resistance to HSV infection and the ability of macrophages to rapidly generate IFN following contact with virus. It is therefore tempting to speculate that in man as well, a genetic defect in this important function of macrophages may contribute to a more severe course of infection.

**ACKNOWLEDGEMENTS**

The authors are indebted to Ms. M. Kulka for skillful preparation of the manuscript.

**REFERENCES**

1. Schneeweis, K.E. and Brandis, M. Zentralbl. Bakteriol. (A) 183:556-558, 1961.
2. Merigan, T.C. In: **The Human Herpesvirus** (Eds. A.J. Nahmias, W.R. Dowdle and R.F. Schinazi), Elsevier North Holland Inc., 1981, pp 309-316.
3. Kohl, S. In: **Immunobiology of HSV infections** (Eds. B.T. Rouse, C. Lopez), CRC press Boca Raton, 1984, pp 121-130.
4. Stevens, J.G. and Cook, M.L. Science 173:843-845, 1971.
5. Hill, T.J., Field, J.J. and Blyth, W.A. J. Gen. Virol. 28:341-353, 1975.
6. Kirchner, H. In: **Monographs in Virology** (Ed. J.L. Melnick), Karger, Basel Vol. 13, 1982, pp 1-104.
7. Lopez, C. In: **Immunobiology of HSV infections** (Eds. B.T. Rouse and C. Lopez), CRC press, Boca Raton, 1984, pp 45-70.
8. Sprecher, E. and Becker, Y. J. Virol. 61:2512-2522, 1987.
9. Lopez, C. Nature 258:152-153, 1975.
10. Lopez, C. In: **The Human Herpesvirus: An Interdisciplinary Perspective** (Eds. A.J. Nahmias, W.R. Dowdle and R.W. Schinazi) Elsevier, 1981, pp 121-131.
11. Simmons, A. J. Exp. Med. 169:1503-1507, 1989.
12. Oakes, J.E. Infect. Immun. 12:166-172, 1975.
13. Zawatzky, R., Hilfenhaus, J. and Kirchner, H. Cell Immunol. 47:424-428, 1979.
14. Schlabach, A.J., Martinez, D., Field, A.K. and Tytell, A.A. Infect. Immun. 26:615-620, 1979.
15. Mogensen, S.C. Infect. Immun. 17:268-273, 1977.
16. Mogensen, S.C. and Andersen, H.K. Infect. Immun. 19:792-798, 1978.
17. Mogensen, S.C. Microbiol. Rev. 43:1-26, 1979.
18. Nesburn, A.B., Green, M.T., Radnoti, M. and Walker, B. Infect. Immun. 15:772-775, 1977.
19. Jacobs, R.P., Aurelian, L. and Cole, G.A. J. Immunol. 116:1520-1525, 1976.
20. Rosenberg, G.L. and Notkins, A.L. J. Immunol. 112:1019-1025, 1974.
21. Scriba, M. Nature 267:529-531, 1977.
22. Anderson, C.A., August, M.J. and Hsiung, G.D. Infect. Immun. 30:159-169, 1980.
23. Schroder, C.H., Engler, H. and Kirchner, H. J. Gen. Virol. 52:159-161, 1981.
24. Johnson, R.T. J. Exp. Med. 120:359-373, 1964.
25. Zisman, B., Hirsch, M.S. and Allison, A.C. J. Immunol. 104:1155-1159, 1970.
26. Zawatzky, R., Hilfenhaus, J., Marcucci, F. and Kirchner, H. J. Gen. Virol.

53:31-38, 1981.
27. Morahan, P.B., Breinig, M.C. and McGeorge, M.B. J. Immunol. 119:2030-2036, 1977.
28. Kitces, E.N., Morahan, P.S., Tew, J.G. and Murray, B.K. Infect. Immun. 16:955-960, 1977.
29. Nathan, C.F. J. Clin. Invest. 79:319-326, 1987.
30. Morahan, P.S. and Morse, S.S. In: **Virus-Lymphocyte Interaction: Implications for Disease** (Ed. M.R. Profitt) Elsevier, 1979, pp 17-35.
31. Hirsch, M.S., Zisman, B. and Allison, A.C. J. Immunol. 104:1160-1165, 1970.
32. Starr, S.E., Visintine, A.M., Tomeh, M.O. and Nahmias, A.J. Proc. Soc. Exp. Biol. Med. 152:57-60, 1976.
33. Glasgow, L.A., Fischbach, J., Bryant, S.M. and Kern, E.R. J. Infect. Dis. 135:763-770, 1977.
34. Morahan, P.S., Kern, E.R. and Glasgow, L. A. Proc. Soc. Exp. Med. 154:615-620, 1977.
35. Kirchner, H., Hirt, H.M. and Munk, K. Infect. Immun. 16:9-11, 1977.
36. McGeorge, M.B. and Morahan, P.S. Infect. Immun. 22:623-626, 1978.
37. Wildy, P., Gell, P.H.G., Rhodes, J. and Newton, A. Infect. Immun. 37:40-45, 1982.
38. Morse, S.S. and Morahan, P.S. Cell Immunol. 58:72-84, 1981.
39. Rager-Zisman, B. and Bloom, B.R. Nature 251:542-543, 1974.
40. Stanwick, T.L., Campbell, D.E. and Nahmias, A.J. Cell Immunol. 53:413-416, 1980.
41. Morahan, P.S., Connor, J.R. and Leary, K.R. Brit. Med. Bull. 41:15-21, 1985.
42. Mogensen, S.C. and Virelizier, J.L. Interferon 8:55-84, 1987.
43. Morahan, P.S. In: **Immunobiology of Herpes Simplex Infection** (Eds. B.T. Rouse and C. Lopez) CRC Press, Boca Raton, USA, 1984, pp 71-89.
44. Little, S.P., Jofre, J.T., Courtney, R.J. and Schaffer, P.A. Virology 115:149-161, 1981.
45. Honess, R.W. and Roizman, B. J. Virol. 14:8-19, 1974.
46. Roizman, B. and Batterson, W. In: **Virology** (Ed. B. Fields), Raven Press, New York, 1985, pp 497-526.
47. Weinheimer, S.P. and McKnight, S.L. J. Mol. Biol. 195:819-833, 1987.
48. Dixon, R.A.F. and Schaffer, P.A. J. Virol. 36:189-203, 1980.
49. Sacks, W.R., Green, C.C., Aschmann, D.P. and Schaffer, P.A. J. Virol. 55:798-805, 1988.
50. Sekalovich, R.E., Leary, K. and Sandri-Goldin, R.M. J. Virol. 62:4510-4522, 1988.
51. Post, L.E., Mackensen, S. and Roizman, B. Cell. 24:55-63, 1981.
52. Campbell, M.E.M., Palfreyman, J.W. and Preston, C.M. J. Mol. Biol. 180:1-19, 1984.
53. O'Hare, P. and Hayward, G.S. J. Virol. 52:522-531, 1984.
54. Watson, R.C. and Clements, J.B. Nature (London) 285:329-330, 1980.
55. Johnson, D.C. and Spear, P.G. Cell 32:987-997, 1983.
56. Stevens, J.G. and Cook, M.L. J. Exp. Med. 133:19-36, 1971.
57. Sethi, K.K. Immunobiol. 165:459-474, 1983.

58. Sethi, K.K. and Brandis, H. Arch. Virol. 59:157-172, 1979.
59. Lopez, C. and Dudas, G. Infect. Immun. 23:432-437, 1979.
60. Frank, U., Schindling, B., Lindermann, J. and Falke, D. Acta Virol. 22:193-202, 1978.
61. Armerding, D., Mayer, P., Scriba, M., Hren, A. and Rossiter, H. Immunobiol. 160:217-227, 1981.
62. Brucher, J., Domke, I., Schroder, C.H. and Kirchner, H. Arch. Virol. 82:83-93, 1984.
63. Sarmiento, M. J. Immunol. 141:2740-2748, 1988.
64. Domke-Opitz, I., Poberschin, P., Mittnacht, S. and Kirchner, H. Virology 159:306-311, 1987.
65. Lindenmann, J., Burke, D. and Isaacs, A. Brit. J. Exp. Pathol. 38:551-562, 1957.
66. Moore, R.M., Larsen, H.J., Horohov, D.W. and Rouse, B.T. Science 223:178-181, 1984.
67. Bonina, L., Ianello, D., Merendino, R., Arena, A. and Mastroeni, P. Infect. Immun. 39:575-579, 1983.
68. Sit, M.F., Tenney, D.J., Rothstein, J.L. and Morahan, P.S. J. Gen. Virol. 69:1999-2010, 1988.
69. Starr, S.E., Visintine, A.M., Tomeh, O.M. and Nahmias, A.J. Proc. Soc. Exp. Biol. Med. 152:57-60, 1976.
70. Gortz, J., Brake, B., Harle-Grupp, V. and Falke, D. Arch. Virol. 79:173-187, 1984.
71. Daniels, C.A., Kleinerman, E.S. and Snyderman, R. Amer. J. Pathol. 91:119-129, 1978.
72. Plaeger-Marshall, S. and Smith, J.W. Proc. Soc. Exp. Biol. Med. 158:263-268, 1978.
73. Ben-Hur, T., Rosen, Wolff, A., Lamade, W., Darai, G. and Becker, Y. Virology 163:397-404, 1988.
74. Kruse, A., Kirchner, H., Zawatzky, R. and Domke-Opitz, I. Scand. J. Immunol., submitted.
75. Tang, J.L., Yamamoto, M., Sakuma, S., Mori, R. and Nagayama, A. Microbiol. Immunol. 32:363-374, 1988.
76. Salo, R. and Ortega, A.P. Antiviral Res. 6:161-169, 1986.
77. Leary, K., Connor, J.R. and Morahan, P.S. J. Gen. Virol. 66:1123-1129, 1985.
78. Hovie, S., Norval, M., Maingay,J. and McBride, W.H. Arch. Virol. 87:229-239, 1986.
79. Morahan, P.S., Mama, S., Anaraki, F. and Leary, K. J. Virol. 63:2300-2307, 1989.
80. Domke-Opitz, I. and Kirchner, H. J. Gen. Virol. submitted.
81. Andreesen, R., Bross, K.J., Osterholz, J. and Emmrich, F. Blood 67:1257-1264, 1986.
82. Morahan, P.S., Volkman, A., Melnikoff, M. and Dempsey, W.L. In: **Macrophages and Cancer** (Eds. G.H. Heppner and A.M. Fulton) CRC Press, Boca Raton, Florida, 1988, pp 1-25.
83. Linnavuori, K. and Hovi, T. J. Gen. Virol. 52:381-385, 1981.
84. Linnavuori, K. and Hovi, T. Virology 130:1-9, 1983.
85. Albers, I., Kirchner, H. and Domke-Opitz, I. Virology 169:466-469,

1989.
86. Albers, I., Kirchner, H. and Domke-Opitz, I. In: **The Biology of the Interferon System 1988** - Proceedings of the Fifth Annual Meeting of the International Society for IFN Research (Eds. Y. Kawade and S. Kobayashi) Kodansha Scientific Ltd., Tokyo, 1989, pp 435-440.
87. Mintz, L., Drew, W.L., Hoo, R. and Finley, T.N. Infect. Immun. 28:417-420, 1980.
88. Albers, I. and Domke-Opitz, I. manuscript in preparation.
89. Hammerstrom, J. Acta Pathol. Microbiol. Scand. Sect. 87:113-120, 1979.
90. Rinaldo, C.R. Jr., Richter, B.S., Black, P.H. and Hirsch, M.S. Infect. Immun. 25:521-525, 1979.
91. Harris, P. and Ralph, P. J. Leukocyte Biol. 37:407-422, 1985.
92. Pientong, C., Weisshart, K., Kuhn, J.E., Knopf, C.W. and Braun, R.W. Virology in press.
93. Tenney, D.J. and Morahan, P.S. J. Immunol. 139:3076-3083, 1987.
94. Rose, R.M., Wasserman, A.S., Wiver, W.Y. and Remold, H.G. Cell. Immunol. 97:397-406, 1986.
95. Catalano, L.W. Jr., Safby, G.M., Museles, M. and Jarzynski, D.J. J. Pediatr. 79:393-400, 1971.
96. Legaspi, R.C., Gatmaitan, B., Bailey, E.J. and Lenor, A.M. Arch. Neurol. 37:76-79, 1980.
97. Overall, J.C.Jr., Spruance, S.L. and Green, J.A. J. Infect. Dis. 143:543-547, 1981.
98. Spruance, S.L., Green, J.A., Chin, G., Yeh, T., Wennerstrom, G. and Overall, J.C. Jr. Infect. Immun. 36:907-910, 1982.
99. Overall, J.C., Yeh, T. and Kern, E.R. J. Infect. Dis. 142:943, 1980.
100. Torseth, J.W. and Merigan, T.C. J. Infect. Dis. 153:979-984, 1986.
101. Gresser, I., Tovey, M.G., Maury, C. and Bandu, M.T. J. Exp. Med. 114:1316-1323, 1976.
102. Zawatzky, R., Engler, H. and Kirchner, H. J. Gen. Virol. 60:25-29, 1982.
103. Zawatzky, R., Gresser, I., DeMaeyer, E. and Kirchner, H. J. Infect. Dis. 146:405-410, 1982.
104. Brandt-Petersen, E., Haahr, S. and Mogensen, S.C. Infect. Immun. 42:740-746, 1983.
105. Stanton, G.J., Jordan, C., Hart, A., Heard, H., Langford, M.P. and Baron, S. Microbiol. Pathogen. 3:179-183, 1987.
106. Czarniecki, C.W., Fennie, C.W., Powers, D.B. and Estell, D.A. J. Virol. 49:490-496, 1984.
107. Gidlund, M., Orn, A., Wigzell, H., Senik, A. and Gresser, I. Nature 273:759-761, 1978.
108. Engler, H., Zawatzky, R., Goldbach, A., Schroder, C.H., Weyand, C., Hammerling, G.J. and Kirchner, H. J. Gen. Virol. 55:25-30, 1981.
109. Engler, H., Zawatzky, R., Kirchner, H. and Armerding, D. Arch. Virol. 74:239-247, 1982.
110. Bukowski, J.F. and Welsh, R.M. J. Immunol. 136:3481-3485, 1986.
111. Chmielarczyk, W., Engler, H., Ernst, R., Opitz, I. and Kirchner, H. J. Gen. Virol. 66:1087-1094, 1985.
112. Habu, S., Akamatsu, K., Tamaoki, N. and Okumura, K. J. Immunol.

133:2743-2747, 1984.
113. Fitzgerald, P.A., Evans, R., Kirkpatrick, D. and Lopez, C. J. Immunol. 130:1663-1667, 1983.
114. Fitzgerald, P.A., Schindler, T.E., Siegal, F.P. and Lopez, C. In: **Natural Killer Activity and Its Regulation** (Eds. T. Hoshino, H.S. Koren and A. Uchida) Exerpta Medica, Amsterdam, 1984, pp 415-421.
115. Zawatzky, R., Dandoy, F., De Maeyer-Guignard, J. and De Maeyer, E. In: **The Biology of the Interferon System** (Eds. E. De Maeyer and H. Schellekens) Elsevier Science Publ., 1983, pp 77-82.
116. Kirchner, H., Engler, H., Schroder, C.H., Zawatzky, R. and Storch, E. J. Gen. Virol. 64:437-441, 1983.
117. Ellerman-Eriksen, S., Liberto, M.C., Ianello, D. and Mogensen, S.C. J. Virol. 67:1025-1033, 1986.
118. Zawatzky, R., Kirchner, H., De Maeyer-Guignard, J. and De Maeyer, E. J. Gen. Virol. 63:325-332, 1982.
119. Babior, B.M. New Engl. J. Med. 298:659-668, 1978.
120. Mogensen, S.C., Ellerman-Eriksen, S. and Sommerlund, M. J. Gen. Virol. in press.
121. Kohl, S., Loo, L.S., Drath, D.B. and Cox, P. J. Infect. Dis. 159:239-247, 1989.
122. Linnavuori, K. and Hovi, T. Antiviral Res. 8:201-208, 1987.
123. Linnavuori, K. J. Med. Virol. 25:61-68, 1988.
124. Capobianchi, M.R., Malavasi, F., DiMarco, P. and Dianzani, F. Arch. Virol. 103:219-229, 1988.
125. Lebon, P. J. Gen. Virol. 66:2781-2786, 1985.
126. Gobl, A.E., Funa, K. and Alm, G.V. J. Immunol. 140:3605-3609, 1988.
127. Fitzgerald-Bocarsky, P., Feldman, M., Mendelsohn, M., Curl, S. and Lopez, C. J. Leukocyte Biol. 43:323-334, 1988.
128. Kirchner, H., Peter, H.H., Hirt, H.M., Zawatzky, R., Dalugge, H. and Bradstreet, P. Immunobiol. 156:65-75, 1979.
129. Zawatzky, R., De Maeyer, E. and De Maeyer-Guignard, J. Proc. Natl. Acad. Sci. USA 82:1136-1140, 1985.
130. Enoch, T., Zinn, K. and Maniatis, T. Mol. Cell. Biol. 6:801-810, 1986.
131. Torseth, J.W., Nickoloff, B.J., Basham, T.Y. and Merigan, T.C. J. Infect. Dis. 155:641-648, 1987.
132. Vallee, M.J., Jordan, G.W., Haaler, S. and Merigan, T.C. J. Immunol. 115:230-238, 1975.
133. Green, J.A., Yeh, T. and Overall, J.C.Jr. J. Immunol. 127:1192-1196, 1981.
134. Green, J.A. J. Med. Virol. 20:381-389, 1986.
135. Domke-Opitz, I., Straub, P. and Kirchner, H. J. Virol. 60:37-42, 1986.
136. Rasmussen, L. and Fairly, L.B. Infect. Immun. 12:104-108, 1975.
137. Rasmussen, L.E., Chen, P.T. and Merigan, T.C. Antimicr. Agents Chemother. 26:599-600, 1984.
138. Fuerst, D.J., Matti, M., Smolin, G. and O'Donnel, J. Curr. Eye Res. 5:331-335, 1986.
139. Leventon-Kriss, S., Morshovitz, M., Smetana, Z., Shewach-Miller, M., Doerner, T. and Gotlieb-Stematsky, T. Med. Microbiol. Immun. 176:151-159, 1987.

140. Oberman, F. and Panet, A. J. Gen. Virol. 69:1167-1177, 1988.
141. Altinkilic, B. and Brandner, G. J. Gen. Virol. 69:3107-3112, 1988.
142. Domke, I., Straub, P., Jacobsen, H., Kirchner, H. and Panet, A. J. Gen. Virol. 66:2231-2236, 1985.
143. Straub, P., Domke, I., Kirchner, H. and Panet, A. Virology 150:411-418, 1986.
144. Lipp, M. and Brandner, G. J. Gen. Virol. 47:97-111, 1980.
145. Gloger, I. and Panet, A. J. Gen. Virol. 65:1107-1111, 1984.
146. Thiele, K. and Kirchner, H. J. Interferon Res. 8:507-515, 1988.
147. Mittnacht, S., Straub, P., Kirchner, H. and Jacobsen, H. Virology 164:201-210, 1988.
148. Lipp, M. and Brandner, G. In: **The Biology of the Interferon System** (Eds. H. Kirchner and H. Schellekens) Elsevier, Amsterdam, 1985, pp 355-360.
149. Sharma, O.K., Goswami, B.B. and Cohrs, R.J. In: **The 2-5 A System** (Eds. B.R.G. Williams and R.H. Silverman) A.R. Liss, New York, 1985, pp 317-324.
150. Munoz, A. and Carrasco, L. J. Gen. Virol. 65:1069-1078, 1984.
151. Chatterjee, S., Lakeman, A.D., Whitley, R.J. and Hunter, E. Virus Res. 1:81-87, 1984.
152. Chatterjee, S., Hunter, E. and Whitley, R. J. Virol. 56:419-425, 1985.
153. Wong, G.H.W. and Goeddel, D.V. Nature (London) 323:819-822, 1986.
154. Jacobsen, H., Mestan, J., Mittnacht, S. and Dieffenbach, C.W. Mol. Cell. Biol. in press.
155. Feduchi, E., Alonso, M.G. and Carrasco, L. J. Virol. 63:1354-1359, 1989.
156. Klotzbucher, A., Mittnacht, S., Kirchner, H. and Jacobsen, H. Virology, submitted.
157. Orchansky, P., Rubinstein, M. and Fischer, D.G. J. Immunol. 136:169-173, 1986.
158. Faltynek, C.R. and Pander, G.L. J. Interferon Res. 6:639-653, 1986.
159. Mebye, M., Bigger, R.J., Ebbesen, P., Neuland, C., Goedert, J.J., Faber, V., Lorenzen, J.P., Skinh, J.P., Gallo, R.C. and Blattner, W.A. Ann. Intern. Med. 104:496-500, 1986.
160. Harper, M.E., Marselle, L.M., Gallo, R.C. and Wong-Staal, F. Proc. Natl. Acad. Sci. USA 83:772-776, 1986.
161. Roy, S. and Wainberg, M.A. J. Leucoc. Biol. 43:91-97, 1988.
162. Nabel, G. and Baltimore, D. Nature (London) 326:711-713, 1987.
163. Jones, K.A., Kadonaga, J.T., Lucino, P.A. and Tijan, R. Science 232:755-759, 1987.
164. Gendelman, H.E., Phelps, W., Feigenbaum, L., Ostrove, J.M., Adachi, A., Howley, P.M., Khoury, G., Ginsberg, H.S. and Martin, M.A. Proc. Natl. Acad. Sci. USA 83:9759-9763, 1986.
165. Mosca, J.D., Bednarik, D.P., Raj, N.B.K., Rosen, C.A., Sodroski, J.G., Haseltine, W.A. and Pitha, P.M. Nature (London) 325:67-70, 1987.
166. Rando, R.F., Pellett, P.E., Luciw, P.A., Bohan, C.A. and Srinivasan, A. Oncogene 1:13-18, 1987.
167. Nabel, G.J., Rice, S.A., Knipe, D.M. and Baltimore, D. Science 239:1299-1302, 1988.

168. Albrecht, M.A., De Luca, N.A., Byrn, R.A., Schaffer, P.A. and Hammer, S.M. J. Virol. 63:1861-1868, 1989.
169. Ostrove, J.M., Leonard, J., Weck, K.E., Rabson, A.B. and Gendelman, H.E. J. Virol. 61:3726-3732, 1987.
170. Holmberg, S.D., Stewart, J.A., Gerber, A.R., Byers, R.H., Lee, F.K., O'Malley, P.M. and Nahmias, A.J. JAMA 259:1048-1050, 1988.
171. Stamm, W.E., Handsfield, H.H., Rompalo, A.M., Ashley, R.L., Roberts, P.L. and Corey, L. JAMA 260:1429-1433, 1988.

# 9

## INNATE RESISTANCE TO HSV-1 INFECTION: MOUSE MACROPHAGE INHIBITION OF VIRAL REPLICATION

Marion Sarmiento

**Department of Veterinary medicine and Surgery, The University of Texas, M.D. Anderson Cancer Center, Houston, Texas 77030**

## ABSTRACT

Mice of C57BL genetic background are known to be highly resistant to acute (lethal) infection with Herpes Simplex Virus type 1 (HSV-1). This ability of C57BL mice to resist HSV-1 lethal infection is due, at least in part, to the efficacy with which individual cells such as macrophages inhibit viral replication intracellularly. Bone marrow, peritoneum, or spleen macrophages isolated from C57BL/6 mice and infected *in vitro*, consistently inhibit viral replication by restricting viral protein or DNA synthesis at an earlier point in the viral growth cycle than macrophages from susceptible DBA/2 mice. The specific point in the replicative cycle at which viral replication is inhibited is dependent on the tissue source from which the macrophages are isolated.

C57BL/6 macrophages (85-90%) inhibit HSV-1 replication very early in the cycle, prior to the onset of immediate early (IE or $\alpha$) protein synthesis. The remaining 10-15% of the cells inhibit HSV-1 replication after the onset of $\gamma_1$ protein synthesis but before viral DNA or late ($\gamma_2$) protein synthesis. DBA/2 bone marrow macrophages inhibit HSV-1 replication between $\gamma_1$ and $\gamma_2$ protein synthesis. A small amount of viral DNA synthesis is detectable in infected DBA/2 bone marrow macrophages, and viral proteins ( $\alpha$, $\beta$ and $\gamma_1$) are expressed faster and in greater amounts in these macrophages than in the corresponding C57BL/6 cells.

Peritoneal macrophages from either C57BL/6 or DBA/2 mice inhibit HSV-1 replication at a point in the cycle between $\gamma_1$ protein synthesis and viral DNA synthesis. No viral DNA synthesis is detectable in either C57BL/6 or DBA/2 peritoneal macrophages. Viral proteins $\alpha$, $\beta$ and $\gamma_1$, however, are expressed faster and in greater amounts in DBA/2 than in C57BL/6 peritoneal macrophages.

Viral replication is inhibited after the onset of DNA synthesis in either C57BL/6 or DBA/2 spleen macrophages. Significantly, more viral DNA synthesis occurs in DBA/2 than in C57BL/6 spleen macrophages. $\gamma_2$ protein synthesis, however, is detectable in DBA/2 spleen macrophages but not in C57BL/6 macrophages.

Clearly, multiple genes function to inhibit HSV-1 replication in macrophages. These yet unidentified genes are differentially expressed in macrophages isolated from various tissue sources and provide the basis for early intracellular events in immune defense against acute (lethal) infection with HSV-1.

## INTRODUCTION

Natural defense against viral infection is mediated through both innate and adaptive immune systems (1). Innate immunity to viral infection is expressed by many tissues in the body of a genetically resistant individual. Thus, most organ tissues can inhibit viral infection by restricting viral replication intracellularly. Certain specialized cells such as macrophages and T lymphocytes can, in addition, synthesize and secrete anti-viral cytokines such as the interferons (IFNs), which also affect the ability of cells to inhibit viral replication. Macrophages and Natural Killer cells can also lyse virus-infected cells.

Adaptive immunity is expressed by B and T lymphocytes through B cell (antibody-producing) and T cell-mediated (antigen-specific) systems. Macrophages also play an important role in adaptive immunity because they function to process and present antigen to helper T cells. The antigen-stimulated helper T cells in turn stimulate the proliferation and differentiation of antigen-specific B cells and T (cytolytic and suppressor) cells.

Expression of effective innate immunity in response to infection with a highly virulent pathogen such as HSV is of foremost importance to the survival of the host. Innate immunity curbs the pathogenic impact of acute, primary HSV infection, and affords the time necessary for adaptive immunity to develop. The macrophage is of central importance in defense against viral infection because it functions, first in innate immunity and later in adaptive immunity.

Clearly, any virus that inhibits macrophage function will suppress both innate and adaptive immunity. Viruses that infect but do not destroy macrophages may be disseminated throughout the body.

Clinical expression of HSV infection in humans typically appears as vesicles and ulcers on the skin, or the oral or genital mucosa (2). These lesions are usually self-limiting, and heal within a week or two. The virus is known to enter local neuronal tissue and to reside in a latent state in regional ganglia (3). Latent virus in neural tissues is the source of recurrent HSV infections observed in some patients. Epidemiological studies have shown that although most persons in modern societies are seropositive for HSV, clinical recurrences of HSV infections are relatively infrequent and mild (2,3). Severe complications from HSV infections, however, are commonly observed in immunosuppressed individuals such as patients receiving tissue grafts (4,5,6), or in AIDS victims (7,8,9). The generalized or systemic HSV infections often observed in AIDS and other immunosuppressed patients has added impetus to research examining the basis of innate immunity in macrophages.

Most studies on innate immunity to HSV infection in macrophages have been performed using mice or isolated mouse tissues. The mouse provides a unique mammalian system for studying the pathogenesis of human viruses such as HSV because inbred strains of mice that are resistant, moderately resistant, or susceptible to acute (lethal) infection with HSV are available. Inbred mouse strains that vary in resistance to HSV infection can be very useful for identifying differences in gene expression that distinguish between resistant and susceptible phenotypes. Studies on innate immunity to HSV infection in mice have investigated: (a) the role of IFN in resistance to acute (lethal) infection, (b) the association of the resistant phenotype with a known genetic background, and (c) the intracellular molecular events that inhibit the viral replicative cycle.

Several sources of data suggest that IFN may contribute, at least in part, to innate immunity to HSV infection. Protective levels of IFN were induced when mice were infected with sufficiently large doses of either HSV-1 or HSV-2 (10,11). Gresser *et al.* (12) demonstrated that anti-IFN sera injected into mice significantly increased susceptiblity to HSV-1. Straub *et al.* (13) found that IFN $\alpha/\beta$ inhibited viral transcription and protein synthesis in DBA/2 spleeen macrophages infected with HSV-1. IFN production in mice has been shown to be governed by several autosomal loci and one X-linked locus (11,14).

Johnson (15) demonstrated that peritoneal macrophages isolated from certain inbred mice produce few viral progeny after infection with HSV-1. Since this initial investigation, intracellular resistance to HSV-1 infection *in vitro* has been observed to correlate highly with resistance to acute (lethal) infection *in vivo* (16,17,18,19,20); macrophages isolated from inbred mice of C57BL genetic backgrounds have been shown to be more resistant to infection *in vitro* than cells from more susceptible inbred mouse strains.

Other studies have examined the ability of mouse macrophages to inhibit HSV replication at the molecular level. Peritoneal macrophages isolated from BRVS Swiss-Webster (21) or from outbred CD-1 (22) mice, and infected with HSV-1 *in vitro*, have been reported to synthesize little if any viral DNA. More recent investigations (23) have demonstrated that macrophages from (resistant) C57BL/6 mice consistently inhibit HSV-1 protein and DNA syntheses at an earlier point in the virus replicative cycle than macrophages from (susceptible) DBA/2 mice. The point of inhibition of virus replication was found to be dependent on the tissue source of the macrophages.

The studies reported in this chapter examine the ability of macrophages from various tissue sources to express innate immunity to HSV infection by inhibiting viral replication intracellularly. Various cell cultivation conditions, designed to enrich for macrophages but which minimize alterations in gene expression, were investigated. Points in the HSV replicative cycle at which virus growth is inhibited by macrophages were identified.

Mice known to be free of specific pathogens (SPF) that may interfere with HSV replication or with $LD_{50}$ values were used for these studies. Female C57BL/6Cr, C57BL10SNJ, B10.BRJ, B10.D2J, BALB/cN, C3HeB/FeJ, DBA/2Cr, and male PL/J mice, 6 to 12 weeks of age, were obtained from Charles River Breeding Laboratories, Inc. (Wilmington, MA), National Cancer Institute (National Institutes of Health, Bethesda, MD), or The Jackson

Laboratory (Bar Harbor, ME), and were maintained under National Institutes of Health guidelines and SPF conditions. The mice were SPF for minute virus, mouse hepatitis virus, Sendai virus, mouse encephalomyelitis (GDVII) virus, reovirus 3, K virus, ectromelia, lymphochoriomeningitis virus, polyoma virus, mouse adenovirus, *Mycoplasma pulmonis*, *Pseudomonas* spp., *Citrobacter* spp., and *Salmonella* spp.

All media and reagents used in tissue culture contained less than 0.1 ng/ml endotoxin as determined by Limulus assay (Associates of Cape Cod, Woodshole, MA). All tissue culture glassware was baked at $210^{\circ}C$ for 18 hr to destroy endotoxin prior to use. The macrophages used in these studies were phagocytic as determined by ingestion of yeast cells (24), but were not cytolytic. Several conditions for isolating and cultivating macrophages from various tissues wee used in these experiments.

Bone marrow cells isolated from C57BL/6CR mice were cultivated under 3 different conditions at $37^{\circ}C$ in RPMI containing 10% FCS, 4% horse serum, and 16% L cell-conditioned medium (10) as follows: (a) The cells from 2 femurs were placed in a 150 $cm^2$ flask containing 30 ml medium and cultivated for 2 days. Non-adherent cells were then transferred to a gelatin-coated 150 $cm^2$ flask (25). Fresh medium (25 ml) was added to each culture every 2 days until the cells adhered and formed a nearly confluent monolayer (approximately 6 days). On the 6th day, the adherent cells were detached with trypsin-versene, and were cultured at $5 \times 10^6$ cells/25 $cm^2$ tissue culture flask in 5 ml medium for an additional 18 hr., (b) nonadherent cells, pre-cultured for 2 days as described above, were transferred to a bacterial-grade, 100 mm diameter petri dish, and cultured an additional 4 days. The cells were transferred to 25 $cm^2$ tissue culture flasks. All cells were adherent 18 hr later. (c) The cells from 2 femurs were cultured for 2 days in a bacterial-grade, 100 mm diameter petri dish. Non-adherent cells were than transferred to a fresh petri dish and cultured for an additional 4 days. The cells were then transferred to 25 $cm^2$ tissue culture flasks, where they became adherent 18 hr later. All adherent 7-day old bone marrow cells cultivated under each of these condtions were identifiable as macrophages by expression of F4/80 antigen (25,26) as determined by fluorescent antibody assay (27).

Spleen mononuclear cells were isolated by sedimentation over Ficoll-Hypaque (28), and cultivated at 37°C in 5 ml of medium at 25 x $10^6$ cells/25 $cm^2$ tissue culture flask that was either untreated or treated with gelatin. Four days later, non-adherent cells were removed and 5 ml fresh medium was added to each culture. Adherent monolayers formed by the 6th day. More than 98% of the cells were determined to be macrophages by expression of F4/80 antigen (25,26).

Peritoneal macrophages were isolated by peritoneal lavage with PBS (29). The macrophage fraction was estimated by esterase stain (30), and the cells were cultivated at 5 x $10^6$ esterase-positive cells/25 $cm^2$ flask for 1 hr at 37°C in tissue culture flasks that were either untreated or coated with gelatin. The non-adherent cells were then removed, and the adherent cells were cultivated for an additional 18 hr in 5 ml medium. More than 95% of the cells were identifiable as macrophages by expression of F4/80 antigen (25,26).

HSV-1 (strain Justin) (31), a neurovirulent isolate, was maintained at low serial passage in VERO cells. Virus titers were quantitated by plaque forming units (PFU) assay (32). Progeny production quantitation assays, infectious center assays, and $LD_{50}$ assays were done in duplicate as described (23).

To analyse for infected cell proteins mouse macrophage monolayers were infected at 20 PFU/cell with HSV-1 (strain Justin) and were radiolabeled at various time intervals with 25 μCi of $^{35}$S-methionine in 1 ml of methionine-deficient medium. The cells were then collected by scraping and sedimentation; cell pellets were solubilized with SDS sample solution for electrophoresis (33), and viscosity was reduced by sonication. Infected cell proteins [ICP (34)] were separated by SDS-PAGE and visualized by autoradiography (33). Virus antigen expression in infected cells was assessed by fluorescent antibody staining using monoclonal antibodies reactive with ICP4 or ICP8 (35). DNA isolation from infected and uninfected cells and hybridization with a $^{32}$P-labeled, cloned Bam HI fragment D from HSV-1 (F) was done as described previously (23).

## RESISTANCE TO ACUTE (LETHAL) INFECTION WITH HSV IN MICE

The mouse is the animal model of choice for studies concerning genetic resistance to HSV infection because inbred strains of mice that are resistant, moderately resistant, or susceptible to acute lethal infection with HSV are readily available. Resistance to acute, lethal infection with HSV in mice has been demonstrated in certain inbred strains of mice (Table 1). Mice with C57BL genetic backgrounds express significantly higher $LD_{50}$ values when infected with a virulent HSV-1 strain than other inbred strains. Resistance to lethal HSV infection appears not to be associated with the H-2 locus because C57BL congenic recombinants with H-2 loci derived from susceptible inbred mouse strains (i.e., B10.BRJ and B10.D2J) are as resistant to HSV infection as C57BL/6 or C57BL/10 mice (11,36; Table 1). Genetic cross and backcross data using inbred strains resistant to or susceptible to HSV infection have revealed that at least two, and possibly more, non-H-2 linked genes govern resistance to lethal infection with HSV (36).

Table 1. $LD_{50}$ in mice infected with HSV-1 (Justin).

| Strain | $LD_{50}$[a] | H-2 haplotype[b] |
|---|---|---|
| C57BL/6Cr | $10^{7.4}$ | b |
| C57BL/10SNJ | $10^{7.6}$ | b |
| B10.BRJ | $10^{7.1}$ | k |
| B10.D2J | $10^{7.6}$ | d |
| C3HeB/FeJ | $10^{3.2}$ | k |
| BALB/cN | $10^{3.7}$ | d |
| DBA/2Cr | $10^{2.1}$ | d |
| PL/J | $10^{2.3}$ | p |

[a]as determined by the method of Reed and Muench (37).
[b]see ref. (38).

Studies examining resistance to HSV infection *in vivo* can only grossly distinguish between the protective effects contributed by innate and adaptive immune systems. Susceptible mice receiving doses of HSV above $LD_{50}$ usually die within the first 3 to 11 days after infection. Mice receiving higer lethal HSV doses tend to die sooner than those receiving lower lethal doses, probably because lethal systemic concentrations of HSV are rapidly attained in these animals and adaptive immunity never has a chance to develop.

Studies with adult and neonate mice have suggested that the macrophage is a key agent in preventing the systemic spread of HSV from the site of infection (15). Macrophages have been implicated as systemic disseminators of diseases caused by Dengue Virus (39), CMV (40,41), and HSV (41). The immune status of the host appears to play a very important role in determining whether macrophages will be able to defend against or contribute to systemic viral infection.

Many investigations over the past 2 decades have examined the ability of HSV to replicate in mouse macrophages *in vitro*. Several major concepts have emerged from these studies: (1) It is important to use mice that are free of pathogens (i.e., Sendai Virus, Mouse Hepatitis Virus) that may potentially interfere with the replication of HSV or affect immunity to infection with this virus, (2) Because endotoxin is known to activate macrophages (42), reagents and vessels used in these studies must be free of endotoxin, and (3) It is important to infect cells in a uniform and synchronous manner. Interference caused by the shape of the vessel or too low input of virus can prevent the uniform infection of all the cells in the culture.

Protective effects against HSV infection contributed by the innate immune system alone can best be studied by examining isolated tissues *in vitro*. The studies presented in this chapter examine the ability of macrophages isolated from resistant (C57BL/6) or susceptible (DBA/2) mice and from various tissue sources, to inhibit HSV replication at specific points in the virus growth cycle. Growth over gelatin was investigated as a means for cultivating and enriching for macrophages, particularly from tissues containing proportionately few macrophages, such as spleen and bone marrow. Results obtained with macrophages cultivated over gelatin or directly over tissue culture plastic were compared. C57BL/6Cr (resistant) and DBA/2Cr (susceptible) mice were used as prototypes in these investigations. Similar results, however, have been observed in macrophages isolated from C57BL/10SNJ (resistant) or PL/J (susceptible) mice (M. Sarmiento, unpublished data).

## INHIBITION OF HSV REPLICATION IN MOUSE MACROPHAGES

C57BL/6 bone marrow macrophages cultivated over gelatin are nonpermissive to HSV replication.

The following experiment demonstrates that tissue culture conditions for bone marrow macrophages can greatly affect permissiveness to HSV replication. Bone marrow cells isolated from C57BL/6Cr mice were cultivated under 3 different conditions: (a) over tissue culture plastic and then over gelatin, (b) over tissue culture plastic and then over bacterial-grade plastic, and (c) over bacterial grade plastic throughout.

Bone marrow macrophages cultivated for 7 days under these various conditions were infected with HSV-1 (Justin) at 20 PFU per cell and radiolabeled with $^{35}$S-methionine from 5 to 9 hr after infection. The cells were harvested and radiolabeled proteins were analyzed by SDS-PAGE followed by autoradiography. The radiolabeled infected cell protein (ICP) profiles shown in Fig. 1 indicate that 7 day-old bone marrow macrophages express significant differences in permissiveness to HSV replication, depending on the conditions of cultivation. Infected protein band profiles obtained from bone marrow macrophages cultivated under conditions (b) and (c) were very similar. Inhibition of cellular protein synthesis was minimal, but many viral polypeptides, including ICPs 5, 11 and 29, were readily detectable. In contrast, macrophages cultivated over gelatin were significantly nonpermissive to HSV replication (Fig. 1a). The protein band profiles illustrated in Fig. 1a reveal that, with the exception of the appearance of an unindentified band near the ICP32 position (indicated by a triangle), the infected protein band profile was indistinguishable from that observed in gelatin-cultivated mock-infected cells. These data indicate that cultivation of C57BL/6 bone marrow cells over gelatin enriches for a population of macrophages that is less permissive to HSV replication than is obtained by cultivation over bacterial-grade petri dishes. Cultivation without gelatin produced results similar to those obtained using bacterial petri dishes (data not shown).

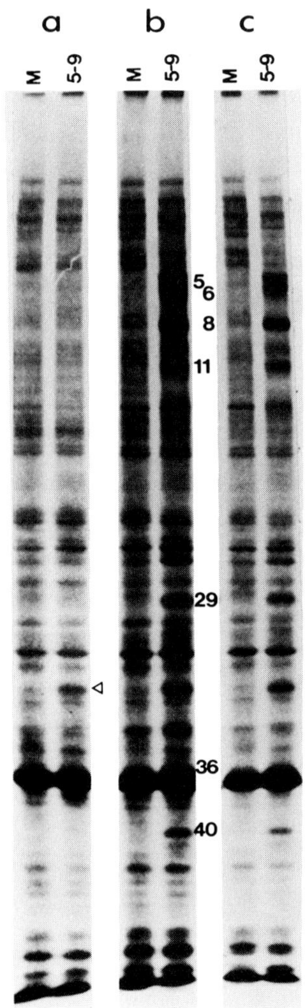


Fig. 1. Autoradiographic images of electrophoretically separated polypeptides from HSV-1 (Justin)-infected C57BL/6 bone marrow macrophages. The macrophages were cultivated over a) gelatin, b) tissue culture plastic, or c) bacterial plastic and tissue culture plastic for 7 days prior to infection. Cells were mock-infected (M) or infected with HSV-1, and then labeled between 5 to 9 hr after infection with $^{35}S$-methionine. The same number of cell equivalents was analyzed in each lane.

DBA/2 bone marrow macrophages cultivated over gelatin are partially permissive to HSV replication.

Bone marrow cells from C57BL/6 and DBA/2 mice wee cultivated over gelatin as described above. Confluent monolayers of 7-day old cells in 25 $cm^2$ flasks were infected with 20 PFU/cell of HSV-1 (Justin) and radiolabeled with $^{35}S$-methionine over various time intervals after infection. Infected HEp-2 cells were included as permissive controls. Fig. 2 demonstrates radiolabeled protein band profiles obtained from these cells.

Inhibition of cellular protein synthesis was detectable in the DBA/2 but not in the C57BL/6 profiles. Viral proteins including $\beta$-ICP8 and $\gamma_1$-ICPs 5,10,11 and 29 were evident in DBA/2 macrophages as early as 2 to 4 hr after infection. Only small amounts of these viral proteins were detectable in C57BL/6 macrophages, and not until 4 to 6 hr after infection. Although DBA/2 macrophages were more permissive to HSV-1 infection than C57BL/6 cells, viral replication appeared to slow down by 6 to 8 hr, and was almost undetectable between 10 to 18 hr after infection. Late ($\gamma_2$) ICP's such as ICP32 were not detectable in infected C57BL/6 or DBA/2 bone marrow macrophages. Fluorescent staining using monoclonal antibodies revealed that the IE and early proteins ICP4 and ICP8 respectively, were expressed in all DBA/2 macrophages by 8 hr after infection. Only 10-15% of infected C57BL/6 bone marrow cells, however, expressed these antigens. HSV antigens were not detectable in the remaining 85-90% of these macrophages (data not shown). Extensive cytopathic effect (CPE) was detectable in infected DBA/2 macrophages by 18 hr after infection. C57BL/6 macrophages showed no CPE even at 7 days after infection.

Cultivation of C57BL/6 and DBA/2 bone marrow macrophages over gelatin enriches for cells that reflect the relative ability of these mouse strains to resist acute (lethal) infection with HSV-1 *in vivo* (see Table 1). Most C57BL/6 bone marrow macrophages (85-90%) cultivated by this method are nonpermissive to HSV replication; the remaining 10-15% of C57BL/6 macrophages, and all DBA/2 macrophages are partially permissive to HSV replication.

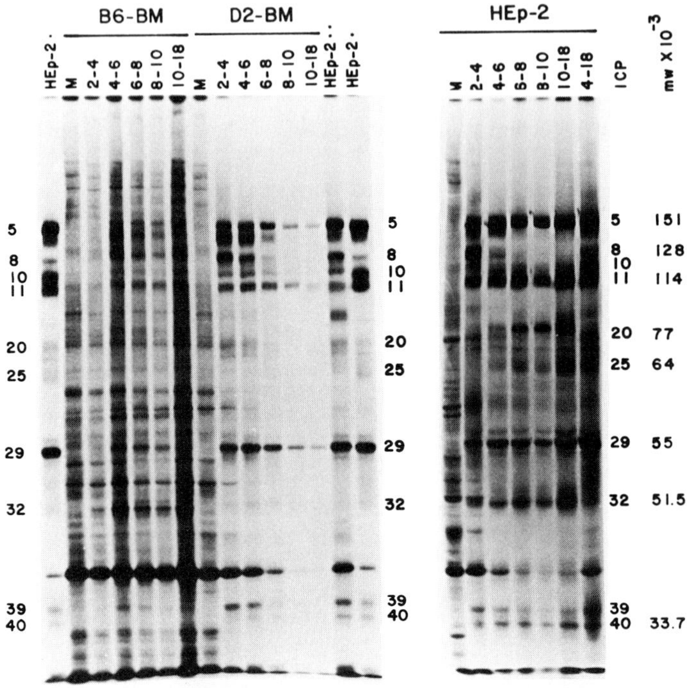


Fig. 2. Autoradiographic images of electrophoretically separated polypeptides from HSV-1 (Justin)-infected C57BL/6 (B6) or DBA/2 (D2) bone marrow (BM) macrophages or HEp-2 cells. Bone marrow macrophages were cultivated over gelatin for 7 days prior to infection. Cells were mock-infected (M) or infected with HSV-1, and then labeled at the intervals (hr) after infection indicated. The same number of cell equivalents was analyzed in each lane. *, HEp-2 cells labeled from 2 to 8 hr after infection; **, HEp-2 cells labeled from 4 to 18 hr after infection. (Reprinted with permission from ref. 19).

DBA/2 peritoneal macrophages are more permissive to HSV replication than C57BL/6 peritoneal macrophages.

Differences in the ability to resist HSV replication *in vitro* that correlate with differences in the ability to resist acute lethal infection with HSV *in vivo* have also been reported in peritoneal and spleen macrophages isolated from C57BL/6 and DBA/2 mice (23). These macrophages express $\alpha$, $\beta$ and $\gamma_1$ ICP after infection with HSV-1; $\gamma_2$ ICPs (e.g., ICP32) are detectable only in DBA/2 spleen macrophages.

Peritoneal macrophages from C57BL/6 and DBA/2 mice were cultivated over gelatin-treated or untreated tissue culture plastic, and the cells were infected and radiolabeled with $^{35}$S-methionine at various time intervals. The ICP profiles in Fig. 3 show that HSV $\alpha$, $\beta$ and $\gamma_1$ ICP were expressed in both macrophages. Similar to the observations with infected bone marrow macrophages, HSV replication appeared to have proceeded at a faster rate and to a later point in the cycle in DBA/2 than in C57BL/6 macrophages. Significantly more $\gamma_1$ ICP5, ICP11, and ICP29 was synthesized at either 4 to 8 hr or 8 to 18 hr in DBA/2 peritoneal macrophages than in C57BL/6 peritoneal macrophages. Viral antigens ICP4 and ICP8 were detectable in more than 98% of the cells as determined by fluorescent antibody assay (data not shown). CPE, however, was not detectable in either C57BL/6 or DBA/2 peritoneal macrophage cultures at 72 hr after infection.

DBA/2 spleen macrophages are more permissive to HSV replication than C57BL/6 spleen macrophages.

Macrophages isolated from spleen, and cultured either in gelatin-treated tissue culture flasks or in untreated flasks, were infected with HSV-1 and radiolabeled at various time intervals after infection with $^{35}$S-methionine. The ICP profiles in Fig. 4 show that HSV $\alpha$, $\beta$ and $\gamma_1$ were expressed in infected C57BL/6 or DBA/2 spleen cells cultivated with or without gelatin. HSV replication progressed further and more rapidly in DBA/2 spleen macrophages than in C57BL/6 spleen macrophages, similar to what was observed with bone marrow and peritoneal macrophages. Significantly more radiolabel was incorporated into $\gamma_1$ ICP5, ICP11, and ICP29 in DBA/2 spleen macrophages

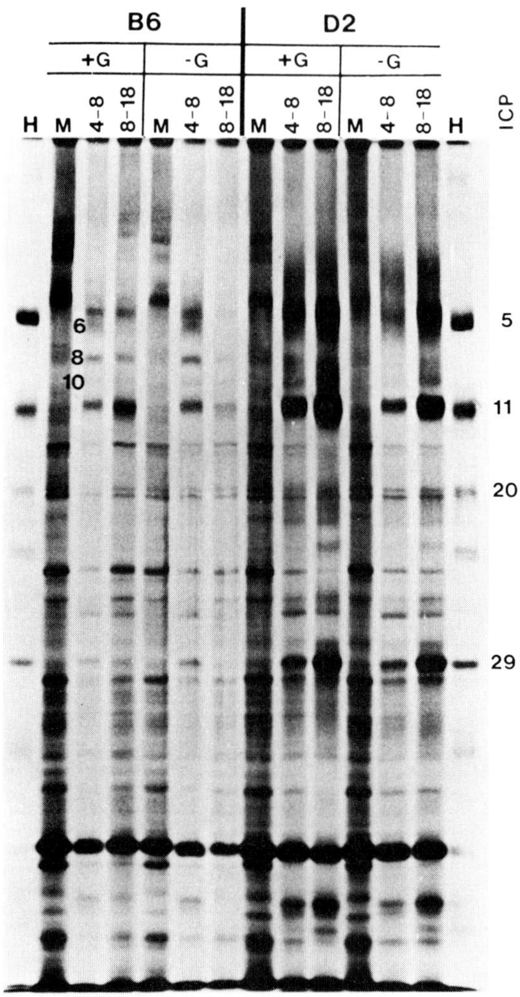


Fig. 3. Autoradiographic images of electrophoretically separated polypeptides from HSV-1 (Justin)-infected C57BL/6 (B6) or DBA/2 (D2) peritoneal macrophages. Macrophages were cultivated over gelatin (+G) or directly over tissue culture plastic (-G) for 7 days prior to infection. Cells were mock-infected (M) or infected with HSV-1, and then labeled between 4 to 8 hr or 8 to 18 hr with $^{35}$S-methionine. The same number of cell equivalents was analyzed in each lane. H, HEp-2 cells labeled from 4 to 18 hr after infecton.

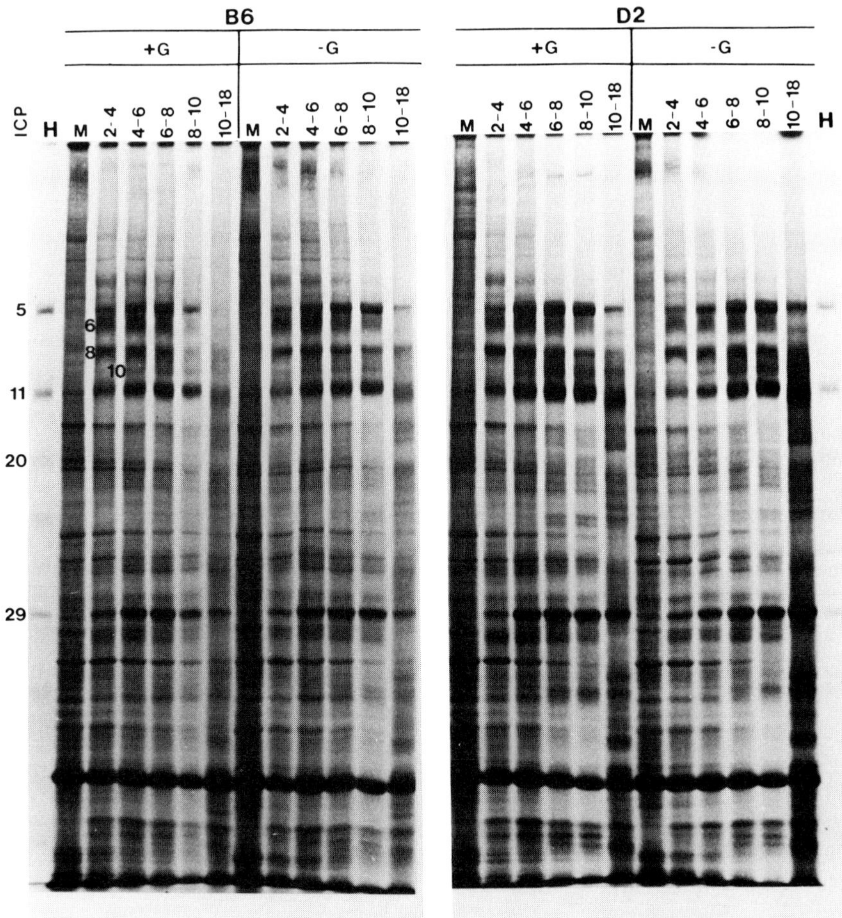


Fig. 4. Autoradiographic images of electrophoretically separated polypeptides from HSV-1 (Justin)-infected C57BL/6 (B6) or DBA/2 (D2) spleen macrophages. Macrophages were cultivated over gelatin (+G) or directly over tissue culture plastic (-G) for 7 days prior to infection. Cells were mock-infected (M) or infected with HSV-1, and then labeled at the indicated intervals (hr) after infection with $^{35}S$-methionine. The same number of cell equivalents was analyzed in each lane. H, HEp-2 cells labeled from 4 to 18 hr after infecton.

than in C57BL/6 spleen macrophages. In addition, $\gamma_2$ ICP25 and ICP32 were detectable in DBA/2 spleen macrophages by 6 to 8 hr after infection. Inhibition of cellular protein synthesis was evident in both C57BL/6 and DBA/2 spleen macrophages by 6 to 8 hr after infection, and extensive CPE was evident in both cultures by 24 hr after infection. Viral antigens ICP4 and ICP8 were detectable by fluorescent antibody assay in more than 98% of either C57BL/6 or DBA/2 spleen macrophages (data not shown). HSV replication was more efficient, however, in HEp-2 cells than in spleen macrophages (compare Fig. 1 with Fig. 4).

HSV progeny and infectious centers production in C56BL/6 and DBA/2 macrophages from bone marrow, spleen and peritoneum.

Very little if any progeny virus was produced in either C57BL/6 or DBA/2 bone marrow or peritoneal macrophages, or in C57BL/6 spleen macrophages, even by 48 hr after infection (Table 2). Slightly higher values (but only up to 10-fold more) were consistently obtained with DBA/2 spleen macrophages than with C57BL/6 spleen macrophages. In sharp contrast to macrophages, HEp-2 cells were found to produce 1.1 x $10^2$ and 8.6 x $10^1$ PFU per cell in two separate experiments. The data in Table 3 indicate that although few infectious

Table 2. Virus production in macrophages from C57BL/6 and DBA/2 mice.

| Tissue Source | C57BL/6 *pfu per cell* | DBA/2 *pfu per cell* |
|---|---|---|
| Bone Marrow | $3.2 \times 10^{-4}$<br>$4.6 \times 10^{-5}$ | $1.5 \times 10^{-5}$<br>$2.5 \times 10^{-4}$ |
| Peritoneum | $4.0 \times 10^{-4}$<br>$7.0 \times 10^{-4}$ | $9.0 \times 10^{-4}$<br>$4.0 \times 10^{-4}$ |
| Spleen | $1.0 \times 10^{-4}$<br>$7.2 \times 10^{-5}$ | $1.1 \times 10^{-3}$<br>$8.2 \times 10^{-4}$ |

Table 3. Infectious center production in mouse macrophage cultures.

| Tissue Source | C57BL/6 *infectious centers per $10^6$ cells* | DBA/2 *infectious centers per $10^6$ cells* |
|---|---|---|
| Bone Marrow | $3.5 \times 10^2$<br>$4.6 \times 10^2$ | $1.3 \times 10^3$<br>$8.3 \times 10^2$ |
| Peritoneum | $1.7 \times 10^3$<br>$3.7 \times 10^3$ | $3.2 \times 10^4$<br>$4.0 \times 10^4$ |
| Spleen | $2.6 \times 10^4$<br>$6.6 \times 10^4$ | $9.7 \times 10^4$<br>$9.9 \times 10^4$ |

centers were produced in the mouse macrophages, spleen macrophages produced more infectious centers by 48 hr after infection than peritoneal macrophages, and peritoneal macrophages produced more infectious centers than bone marrow macrophages. DBA/2 macrophages consistently produced more infectious centers than C57BL/6 macrophages from the same tissue source. HEp-2 cells, however, produced approximately 10-fold more ($9.7 \times 10^5$ and $9.3 \times 10^5$) infectious centers per $10^6$ cells. No increases in either virus progeny or infectious center production were observed in macrophages assayed at 72 and 96 hr after infection.

HSV replication is inhibited prior to viral DNA synthesis in bone marrow and peritoneal macrophages, but not in spleen macrophages.

To determine if HSV DNA synthesis occured in infected macrophages, DNA isolated from cells infected with HSV-1 for 12 hr was bound to nylon membrane and hybridized with a radiolabeled cloned HSV-1 DNA fragment. Infected and uninfected HEp-2 cell DNA was included for positive and negative controls, respectively. The data in Fig. 5 show that very little, if any, viral DNA was detectable in C57BL/6 infected bone marrow macrophages, and in either C57BL/6 or DBA/2 infected peritoneal macrophages. A small amount of viral DNA synthesis was observed in DBA/2 infected bone marrow macrophages. In contrast, significantly greater amounts of HSV DNA were detected in both C57BL/6 and DBA/2 infected spleen macrophages than in peritoneal or bone marrow macrophages. Interestingly, proportionately more viral DNA was synthesized in DBA/2 infected spleen macrophages than in C57BL/6 spleen cells. Even DBA/2 infected spleen macrophages, however, did not synthesize as much viral DNA as was observed in infected HEp-2 cells.

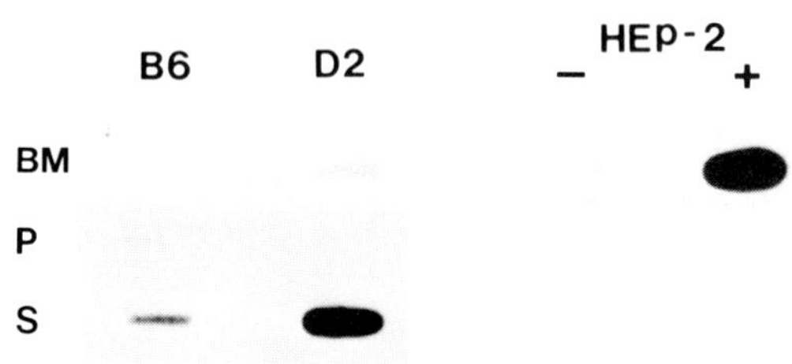


Fig. 5. Slot-blot hybridization of DNA (2 ug/slot), isolated from C57BL/6 (B6) or DBA/2 (D2) bone marrow (BM), peritoneum (P), or spleen (S) macrophages infected with HSV-1 (Justin), and $^{32}$P-labeled cloned HSV-1 (F) Bam H1 restriction fragment (D) probe. Uninfected (-) and infected (+) HEp-2 DNA was included for comparison.

A schematic representation of restrictions of HSV replication observed in mouse macrophages in these studies is shown in Fig. 6. The data may be summarized as follows: (1) HSV replication in DBA/2 bone marrow macrophages is inhibited at a point prior to, or during, DNA synthesis; (2) HSV replication in C57BL/6 and DBA/2 peritoneal macrophages, and in 10-15% of C57BL/6 bone marrow macrophages is inhibited at a point prior to DNA synthesis; (3) the remaining 85-90% of C57BL/6 bone marrow macrophages inhibit HSV replication at a very early point in the cycle, prior to the onset of $\alpha$-ICP protein synthesis; (4) HSV replication proceeds through DNA synthesis in infected spleen macrophages, although more viral DNA appears to be synthesized in DBA/2 than in C57BL/6 macrophages.

## CONCLUSIONS AND COMMENTS

A major mechanism by which mouse macrophages express innate immunity to HSV infection is by inhibiting viral replication intracellularly. C57BL/6 bone marrow macrophages (85-90%) inhibit HSV replication very early in the viral growth cycle, prior to the onset of $\alpha$ protein synthesis; the remaining 10-15% of the cells inhibit HSV replication after $\gamma_1$ ICP synthesis, but prior to the onset of viral DNA synthesis. Cultivation of C57BL/6 bone marrow macrophages over gelatin is required to observe these effects. C57BL/6 and DBA/2 peritoneal macrophages inhibit HSV replication between $\gamma_1$ ICP synthesis, and

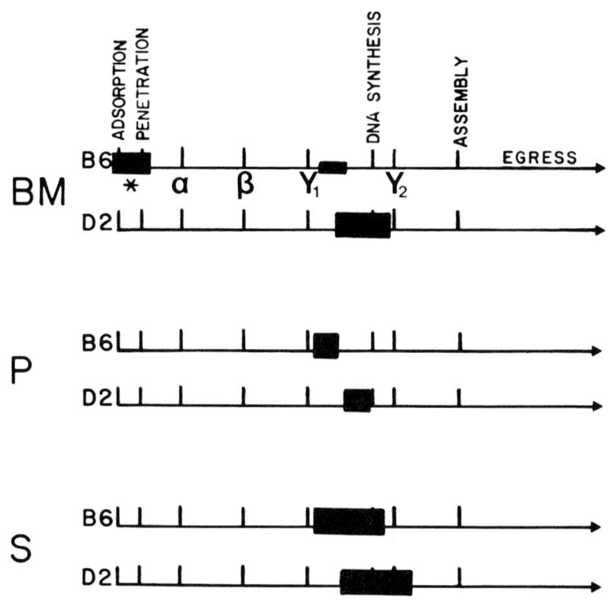


Fig. 6. Schematic representation of HSV replication restrictions evidenced in C57BL/6 (B6) or DBA/2 (D2) mouse macrophages isolated from bone marrow (BM), peritoneum (P), or spleen (S). The primary restriction observed in C57BL/6 bone marrow macrophages is indicated by an *asterisk*.

the onset of viral DNA synthesis; $\alpha$, $\beta$ and $\gamma_1$ ICP are expressed in larger amounts and at faster rates in DBA/2 than in C57BL/6 peritoneal macrophages, indicating that DBA/2 peritoneal macrophages are relatively more permissive. C57BL/6 and DBA/2 spleen macrophages allow HSV replication to proceed through DNA synthesis, although relatively less viral DNA is produced in C57BL/6 than in DBA/2 spleen macrophages. HSV replication in C57BL/6 spleen cells appears to be inhibited before the onset of $\gamma_2$ protein synthesis. DBA/2 spleen cells are more permissive and allow $\gamma_2$ protein synthesis to occur.

The ability of C57BL/6 macrophages to consistently inhibit HSV replication at an earlier point in the viral growth cycle than DBA/2 macrophages from the same tissue source is paralleled by the lower infectious centers yields observed in C57BL/6 relative to DBA/2 macrophages. The data suggest that the earlier in the virus growth cycle that HSV replication is restricted, the more effective is the inhibition of infectious center production. If the restriction occurs prior to or at the onset of DNA synthesis (i.e., bone marrow and peritoneal macrophages), virus yields are very low and almost indistinguishable between C57BL/6 and DBA/2 cells. The relatively higher progeny and infectious center yields observed in spleen cells indicate that if significant amounts of viral DNA are synthesized, infectious progeny and infectious centers are more readily produced. The observation that C57BL/6 spleen macrophages produce significantly more infectious centers than C57BL/6 bone marrow or peritoneal cells, although $\gamma_2$ protein synthesis is not detectable in these cells, suggests that very little $\gamma_2$ protein synthesis may be necessary for proper virus assembly. $\gamma_2$ proteins such as ICP32 (VP19C) are not abundant virion proteins (43). Therefore, the rate-limiting step to infectious progeny production in macrophages after $\alpha$, $\beta$ and $\gamma_1$ protein synthesis appears to be DNA synthesis.

A minimum of 7 HSV genes have been reported to be necessary for origin-dependent plasmid DNA (HSV) replication (44). Genes *pol* and *dbp* encode the DNA polymerase and the major single stranded DNA binding protein ICP8, respectively. Gene UL42 codes for a 65kD protein which binds double stranded DNA. The products of genes UL5, UL8, UL9, and UL52 are expressed in very low amounts and encode proteins of relative molecular weights of 95kD, 75kD, 82kD, and 115kD, respectively.

The relationship between origin-dependent plasmid DNA replication and viral DNA synthesis in infected cells is not known. However, the presence in the HSV genome of genes important for each of these mechanisms is consistent with the concept that all of these contribute either directly or indirectly to viral DNA synthesis in infected cells. Genes *pol* and *dbp* have been shown to be required for viral DNA synthesis in infected cells (45,46,47,48,49,50). The finding that both C57BL/6 and DBA/2 peritoneal macrophages and C57BL/6 bone marrow macrophages (10-15%) inhibit viral replication after $\gamma_1$ but <u>before</u> the onset of DNA synthesis, suggests that these cells may inhibit expression of at least one of the 7 HSV DNA replication genes. Studies using Northern blot and *in situ* hybridization analyses have suggested that transcription of the *pol* gene is repressed in B6C3F$_1$ peritoneal macrophages infected with HSV-1 (KOS) (51). It is not clear from these preliminary studies, however, whether this repression is direct or indirect, or whether other DNA replication genes may also be inhibited. Identification of genes which function to inhibit HSV replication at the time of DNA synthesis will require a careful assessment of the expression of all known DNA replication and stimulatory genes.

In addition to the 7 HSV DNA replication genes, it has been demonstrated that at least 4 genes function to stimulate origin-dependent plasmid DNA replication (44). Three of these stimulatory genes encode IE175 (ICP4), IE110 (ICP0) and IE63. It has been suggested that ICP0 may contribute to HIV activation in AIDS patients (52). Studies with mutant viruses have shown that ICP4 is required for early $\alpha$ gene expression in HSV-infected cells (53,54). It is possible that inhibition of DNA stimulatory gene expression may partially restrict HSV replication, particularly in infected C57BL/6 or DBA/2 spleen macrophages and DBA/2 bone marrow macrophages. On the other hand, expression of these genes in macrophages may provide the stimuli required for HIV as well as HSV growth.

C57BL/6 bone marrow macrophages (85-90%) inhibit viral replication very early in the growth cycle, prior to the onset of $\alpha$ protein synthesis. It is presently not known whether this inhibition occurs at adsorption, penetration, or during uncoating and transport of the capsid to the nucleus. Studies using polyethylene glycol to promote fusion and entry of virus, as has been previously done with mutant virus defective in envelope glycoprotein gB (55) or with non-permissive cell lines (56), may be very informative for determining whether the inhibition in C57BL/6 bone marrow macrophages occurs at penetration; electron microscopy may be useful in elucidating a restriction in uncoating or in capsid transport to the nucleus (57).

Identification of genes that fuction to inhibit HSV replication in macrophages will be important for understanding the genetic basis of innate immunity to HSV. Knowledge of the identities and mechanisms of expression of genes that function in innate resistance to HSV infection will be required for future studies concerned with the development of better clinical approaches to control or prevent the severe HSV pathogenesis that is so often evidenced in AIDS or in immunosuppressed individuals. New drugs or technologies which can alter and control cellular gene expression, so that HSV replication is inhibited intracellularly and viral pathogenesis is curtailed, are needed to prevent HSV-related deaths in these patients.

## ACKNOWLEDGEMENTS

This work was supported by National Institutes of Health Grant AI25358.

## REFERENCES

1. Roitt, I.M., Brostoff, J. and Male, D.K. In: **Immunology** (Ed. D. Bennett), Gower Medical Publ. Ltd., London, 1985, pp 1.1-1.9.
2. Corey, L. and Spear, P.G. N. Engl. J. Med. 314:749-757, 1986.
3. Corey, L. and Spear, P.G. N. Engl. J. Med. 314:686-691, 1986.
4. Naraqi, S., Jacksson, G.G., Jonasson, O. and Yamashiroya, H.M. J. Inf. Dis. 136:531-540, 1977.
5. Pass, R.F., Whitley, R.J., Whelchel, J.D., Diethelm, A.G., Reynolds, D.W. Alford, C.A. J. Inf. Dis. 140:487-492, 1979.
6. Meyers, J.D., Flournoy, N. and Thomas, E.D. J. Inf. Dis. 142:338-346, 1980.
7. Siegal, F.P., Lopez, C., Hammer, G.S., Brown, A.E., Kornfeld, S.J., Gold, J., Hassett, J., Hirschman, S.Z., Cunningham-Rundles, C., Adelsberg, B.R., Parham, D.M., Siegal, M., Cunningham-Rundles, S. and Armstrong, D. N. Engl. J. Med. 305:1439-1444, 1981
8. Quinnan Jr., G.V., Masur, H., Rook, A.H., Armstrong, G., Frederick, W.R., Epstein, J., Manischewitz, J.F., Macher, A.M., Jackson, L., Ames, J., Smith, H.A., Parker, M., Pearson, G.R., Parrillo, J., Mitchell, C. and Straus, S.E., J. Am. Med. Assoc. 252:72-77, 1984.
9. Armstrong, D. Seminars in Oncology 14:40-47, 1987.
10. Zawatsky, R., Gresser, I., DeMaeyer, E. and Kirchner, H. J. Inf. Dis. 146:405-410, 1982.
11. Brandt-Pedersen, E., Haahr, S. and Mogensen, S.C. Infect. Immun. 42:740-746, 1983
12. Gresser, I., Tovey, M.G., Maury, C. and Bandu, M-T. J. Exp. Med. 144:1316-1323, 1976.
13. Straub, P., Domke, I., Kirchner, H., Jacobsen, H. and Panet, A. Virology 150:411-418, 1986.
14. Zawatsky, R., Kirchner, H., DeMaeyer-Guignard, J. and DeMaeyer, E. J. Gen. Virol. 63:325-332, 1982.
15. Johnson, R.T. J. Exp. Med. 120:359-379, 1964.
16. Lopez, C. Nature 258:152-153, 1975.
17. Lopez, C. and Dudas, G. Infect. Immun. 23:432-437, 1979.
18. Collier, L.H., Scott, Q,J, and Pani, A. J. Gen. Virol. 64:1483-1490, 1983.

19. Domke-Opitz, I.P., Poberschin, P., Mittnact, S. and Kirchner, H. Virology 159:306-311, 1987.
20. Shellam, G.R. and Flexman, J.P. J. Virol. 58:152-156, 1986.
21. Stevens, J.G. and Cook, M.L. J. Exp. Med. 133:19-38, 1971.
22. Leary, K.R., Connor, J.R. and Morahan, P.S. J. Gen. Virol. 66:1123-1129, 1985.
23. Sarmiento, M. J. Immunol. 141:2740-2748, 1988.
24. Stewart, C.C. In: **Methods for Studying Mononuclear Phagocytes** (Eds. Adams, D.O., Edelson, P.J. and Koren, H.S.), Academic Press, New York, 1981, pp 5-20.
25. Hume, D.A. and Gordon, S. In: **Mononuclear Phagocytes: Characteristics, Physiology, and Function** (Ed. R. Van Furth), Matinus Nijhoff Publishers, Dordrecht, 1985, pp 9-19.
26. Austyn, J.M. and Gordon, S. Eur. J. Immunol. 11:805-815, 1981.
27. Harlow, E. and Lane, D. In: **Antibodies: A Laboratory Manual**. Cold Spring Harbor Laboratory, Cold Spring Harbor, 1988.
28. Davidson, W.F. and Parish, C.R. J. Immunol. Methods 7:291-300, 1975.
29. Meltzer, M.S. In: **methods for Stydying Mononuclear Phagocytes** (Eds. D.O. Adams, P.J. Edelson and H.S. Koren), Academic Press, New York, 1981, pp 63-67.
30. Koski, I.R., Poplack, D.G. and Blaese, R.M. In: **In Vitro Methods in Cell-Mediated and Tumor Immunigy** (Ed. R.R. Blood), Academic Press, New York, 1976, pp 359-362.
31. Frenkel, N.R., Jacob, R.J., Honess, R.W., Hayward, G.S., Locker, H. and Roizman, B. J. Virol. 16:153-167, 1975.
32. Roizman, B. and Roane, P.R. Jr. J. Immunol. 87:714-727, 1961.
33. Morse, L.S., Pereira, L., Roizman, B. and Shaffer, P.A. J. Virol. 26:389-410, 1978.
34. Roizman, B. and Batterson, W. In: **Fundamental Virology** (Eds. B. Fields and D.M. Knipe), Raven Press, New York, 1986, pp 607-636.
35. Showalter, S.D., Zweig, M. and Hampar, B. Infect. Immun. 34:684-692, 1981.
36. Lopez, C. Immunogenetics 11:87-92, 1980.
37. Reed, L.J. and Muench, H. Am. J. Hyg. 27:493-497, 1938.
38. Roderick, T.H., Stoats, J. and Womack, J.E. In: **Genetic Variants and Strans of the Laboratory Mouse** (Ed. M.C. Green), Gustav Fischer Verlag, New York, 1981, pp 377-396.
39. Halstead, S.B. Science 239:476-481, 1988.
40. Price, P., Winter, J.G. and Shellam, G.R. J. Gen. Virol. 68:2997-3008, 1987.
41. Shellam, G.R. and Flexman, J.P. J. Virol. 58:152-156, 1986.
42. Adams, D.O. and Hamilton, T.A. Ann. Rev. Immunol. 2:283-318, 1984.
43. Gibson, W. and Roizman, B. J. Virol. 10:1044-1052, 1972.
44. Olivo, P.D., Nelson, N.J. and Challberg, M.D. J. Virol. 63:196-204, 1989.
45. Chartrand, P., Stow, N.D., Timbury, M.C. and Wilkie, N.M. J. Virol. 31:265-276, 1979.
46. Coen, D.M., Aschman, D.P., Gelep, P.T., Retondo, M.J., Weller, S.K. and

Schaffer, P.A. J. Virol. 49:236-247, 1984.
47. Conley, A.J., Knipe, D.M., Jones, P.C. and Roizman, B. J. Virol. 37:191-206, 1981.
48. Gibbs, J.S., Chiou, H.C., Hall, J.D., Mount, D.W., Retondo, M.J., Weller, S.K. and Coen, D.M. Proc. Natl. Acad. Sci. USA 82:7969-7973, 1985.
49. Hay, J. and Subak-Sharpe, J.H. J. Gen. Virol. 31:145-148, 1976.
50. Weller, S.K., Spadaro, A., Schaffer, J.E., Murray, A.W., Maxam, A.M. and Schaffer, P.A. Mol. Cell. Biol. 5:930-942, 1985.
51. Morahan, P.S., Mama, S., Anaraki, F. and Leary, K. J. Virol. 63:2300-2307, 1989.
52. Nabel, G.J., Rice, S.A., Knipe, D.M. and Baltimore, D. Science 239:1299-1302.
53. Dixon, R.A.F. and Schaffer, P.A. J. Virol. 36:189-203, 1980.
54. Preston, C.M. J. Virol. 29:275-285, 1979.
55. Sarmiento, M., Haffey, M. and Spear, P.G. J. Virol. 29:1149-1158, 1979.
56. Epstein, A.L., Jacquemont, B. and Machuca, I. Virology 132:315-324, 1984.
57. Batterson, W., Furlong, D. and Roizman, B. J. Virol. 45:397-407, 1983.

# 10

## POTENTIAL INTERACTIONS OF HIV AND THE NEUROTROPIC HERPESVIRUSES IN AIDS ENCEPHALOPATHY

HARRY OPENSHAW, EDOUARD M. CANTIN, and DAVID HINTON

**Department of Neurology, City of Hope National Medical Center, Duarte, California and Department of Pathology, University of Southern California School of Medicine, Los Angeles, California.**

### ABSTRACT

Encephalitis caused by the neurotropic herpesviruses usually occurs from reactivation of a latent infection, and the disease runs an acute clinical course. There is a productive viral infection in neuronal and glial cells, and most often the process is focal. In contrast, HIV encephalitis is a slow virus infection. Viral replication in neuronal cells is not characteristically detected, and the pathology is primarily subcortical and diffuse rather than focal. Certain herpesvirus gene products have been shown to transactivate HIV *in vitro*, and it has been postulated that the herpesviruses may be cofactors for HIV infection. Potential sites of interaction outside the nervous system where both HIV and neurotropic herpesvirus infection occur (productive, abortive, or latent) include monocyte-macrophages, other peripheral blood mononuclear cells, the skin, and ocular tissue (cornea and retina). HIV proviral DNA can be detected in formalin-fixed brain tissue by polymerase chain reaction assays. This technique may prove helpful in screening AIDS brains for herpesvirus and other suspected viral cofactors.

### INTRODUCTION

AIDS dementia complex (1) meets the criteria of a nervous system slow virus infection which Sigurdson established in 1954 on the basis of his study of the lentivirus disease visna in Icelandic sheep (2). Namely, there is a long incubation period of months to years, a protracted course once the nervous system signs or symptoms begin, demonstration of a transmissible agent (the human immunodeficiency virus or HIV), and an invariably fatal outcome. In contrast, the human neurotropic herpesviruses [herpes simplex virus (HSV) type 1 and 2 and varicella zoster virus (VZV)] produce latent infection in primarily nervous system cells and recurrent disease usually on the skin or mucosal surface after reactivation of latent virus in sensory ganglion cells and axonal spread of virus to the skin (3). Clinical HSV brain infection (encephalitis) is not invariably fatal but there is a 70% mortality in untreated patients (4) and both HSV and the usually milder VZV meningoencephalitis run an acute rather than a protracted course. Thus, although both lentiviruses including HIV and the neurotropic herpesviruses can produce latent infection by a block in viral transcription (5,6) the cell type harboring latent virus is different (neurons and possibly other differentiated cells for the neurotropic herpesviruses, T lymphocytes and monocyte-macrophages for HIV) and the time course of the reactivated infection is quite different.

This brief review concerns possible mechanisms and sites of interaction of the neurotropic herpesviruses and HIV with the general speculation that such an interaction may play some role in setting off or exacerbating AIDS dementia complex. What is discussed are recent concepts of herpesvirus latency in the nervous system, experimental demonstration in tissue culture of herpesvirus activation of HIV, potential extra-neural sites of herpesvirus - HIV interactions (drawing in part on recent studies which suggest non-neural HSV latency), and potential nervous system sites of herpesvirus - HIV interactions.

## HERPESVIRUS LATENCY IN THE NERVOUS SYSTEM

Infection of cultured permissive cells with HSV results in viral replication and invariably cell death (7,8). Paradoxically, infection *in vivo* in the natural host and in various animal models leads to the establishment of a latent infection after the primary or acute infection subsides. Latent viral genomes, harbored in peripheral sensory ganglia, can reactivate resulting frequently in recurrent mucocutaneous lesions at or in close proximity to the original site of infection (9,10). A major impediment to understanding latency at a molecular level has been the inability to establish a faithful *in vitro* cell culture latency model. Studies on the molecular basis of latency have therefore utilized ganglia taken from experimental animals or from man at postmortem examination. During the latent phase of infection, the HSV genome is maintained in a nonreplicating state in ganglion cells, and it has been reported that the terminal fragments are rearranged (5,11-13). Latent viral genomes, existing as circular or concatameric entities, can also be detected in brain stem of latently infected mice (12,13). Of interest is the observation that while explantation of latent trigeminal ganglia readily provokes reactivation, this is not the case for latent brain stem (14,15).

The HSV genome comprises at least 70 genes which have been grouped into three major classes: immediate early, early and late, based on their temporal pattern of expression and their coordinate regulation (15-17). The five immediate early genes (ICPO, ICP27, ICP4, ICP22, and ICP47) are expressed first and are distinguished on the basis of their induction by a structural virion component VMW65, also termed VP16 or $\alpha$-transinducing factor ($\alpha$-TIF) (18-20). The immediate early genes are being intensely studied because some of them fulfill essential roles in the regulation of viral gene expression (15,21,22). Moreover, as the first genes expressed they are likely to determine the outcome of infection, namely lytic or latent. The early and late genes in contrast are involved respectively with nucleic acid metabolism and the structural aspect of virion architecture (18).

The transition from an acute productive infection to a latent state is orchestrated by transcriptional silencing of the genome (5,10). However, several groups have recently reported detection of 2-3 transcripts in latently infected mouse and human ganglia, which are transcribed in the opposite orientation to, and thus have 3' termini partially overlapping the 3' end of the ICP0 mRNA (23-25). Initially these latency associated transcripts (LAT) were hypothesized to play a role in the establishment and maintenance of latency, but analysis of LAT deletion mutants suggests their role may be limited to facilitating reactivation (26,27).

To elucidate the action of VMW65 during lytic infection, Preston and colleagues constructed a viable HSV-1 mutant (*in*1814) which by virtue of a 12 base pair insertion in the VMW65 coding sequence has negligible transinducing activity but is normal with respect to virion assembly (28). This mutant has a very interesting phenotype in that at high multiplicities of infection (100 plaque forming units per cell), virus replication is normal; whereas, at low multiplicities the virus shows marked impairment in plaque formation in a cell dependent manner. *In vivo* this mutant (*in*1814) showed a significant reduction in virulence even when high doses were inoculated either intraperitoneally or intracerebrally in mice. This observation is particularly relevant to the establishment of latency. Clearly, transinduction of immediate early genes is important for replication in vivo, probably especially so in nervous tissue where infection is likely to initiate at low multiplicity. It is pertinent to note here that certain ICP0 deletion mutants have a remarkably similar phenotype to the VMW65 (α-TIF) mutant (*in*1814). The ICP0 and VMW65 genes seem to be functionally redundant since both augment levels of critical HSV immediate early gene products that are essential for initiation of replication at low multiplicities of infection. Indeed, it is plausible that the absence or inactivation of VMW65 from HSV nucleocapsids reaching the nerve cell body after retrograde axonal transport from peripheral sites is responsible for the establishment of latency (15,28). The fact that mutant virus (*in*1814) can be retained in cultured cells after inoculation with a low multiplicity of infection and later reactivated supports this contention. However, it is equally plausible that an impairment of ICP0 function could lead to latency since an ICP0 deletion mutant can give rise to a dormant infection in cultured cells (29,30) which can be reactivated by superinfection with human cytomegalovirus (31).

In summary, the currently available experimental data does not support the existence of an HSV gene that functions to induce latency. Rather, it can be hypothesized that latency is the consequence of an inadequate supply of crucial immediate early gene products required to initiate replication in certain cell types, especially at low multiplicities of infection. Parenthetically, it is likely that some of these genes are also involved in reactivation and such role has been proposed for ICP0 (31).

## HERPESVIRUS ACTIVATION OF HIV

The clinical observation that HIV infection is characterized by a long latency period prior to the onset of disease symptoms has focused much attention on the possible role of cofactors in the pathogenesis of AIDS (32). Herpesviruses, particularly HSV, human CMV, and HHV6 are strong contenders for a cofactor role because they are ubiquitous and also especially prevalent in certain high risk populations (33). Moreover, in common with other DNA viruses (e.g., JC virus, BK virus, lymphotropic papovavirus, adenovirus, and bovine papilloma virus) they encode potent transcriptional trans-activators, some of which are functionally promiscuous, activating many different viral and cellular promoters (34,35).

Most studies concerned with activation of HIV by heterologous viruses have utilized transient cotransfection systems in which trans-activation of the HIV-LTR promoter in an HIV-LTRcat reporter construct is manifest as an increase in chloramphenicol acetyl transferase (CAT) activity. Gendelman and colleagues were the first to show transactivation of the HIV-LTR by a genomic HSV-1 fragment encoding the ICP0, ICP4, and ICP27 immediate early genes (36). A fragment of VZV DNA containing the functional homologue of the HSV-1 ICP4 gene was also capable of activating the HIV-LTR promoter (36). Upregulation of CAT activity has also been observed after HSV-1 infection of cells stably transfected with the HIV-LTRcat plasmids (37), and studies with HSV temperature sensitive mutants have implicated VMW65 (the $\alpha$-TIF) in the activation of the HIV-LTR (38).

Results have been contradictory in terms of which of the five HSV immediate early genes play a critical role in trans-activation of the HIV-LTR promoter. Studies using isolated immediate early genes indicated that ICP0 but not ICP4 alone trans-activated the HIV-LTR (37,39); whereas, just the opposite result was reported in another study (38). Clearly, the effects of isolated immediate early genes are cell type specific, and this strongly suggests the involvement of unknown cellular factors in activation of the HIV-LTR. For instance, in SW480 cells, ICP4 alone was active and there was a synergistic effect after addition of the HIV trans-activator *tat*. In contrast, *tat* was required for ICP4 activation in Vero cells but not for HIV augmentation resulting from infection of Vero cells with HSV temperature sensitive mutants at either the permissive (34°C) or the nonpermissive (39°C) temperature (38). This result suggests that a cellular factor constitutively present in SW480 cells but induced by HSV infection even at the nonpermissive temperature in Vero cells, mediates activation of the HIV-LTR by ICP4. This is consistent with the observation that ICP4 trans-activates HSV early genes via an interaction with a cellular protein capable of binding HSV early gene promoter sequences (15,18). It is of considerable interest in this context that HSV infection of HeLa cells results in induction of NF-$\kappa$B, the transcription factor induced during T-cell activation that augments HIV expression by binding to the core enhancer region of the HIV-LTR (40). There is a growing body of evidence which suggests that the NF-$\kappa$B DNA binding protein is involved in gene regulation in a general sense, functioning as a mediator of inducible signal transduction (41,42).

The transient transfection experiments discussed here have been useful in delineating the effects of specific HSV immediate early proteins on expression from the HIV-LTR, but they do not necessarily reflect the activities of the same proteins during co-infection of a cell with HSV and HIV (43). Schaffer and colleagues (44) have used a co-infection system involving HIV and HSV-1 or selected isogenic deletion mutants to examine the role of the three essential HSV trans-activator proteins (ICP0, ICP4, and ICP27) in potentiating HSV replication. The conclusion from this study is that HSV ICP4 is essential while ICP0 and ICP27 are not required. The *cis*-acting target sequence in the HIV-LTR which is specifically required for HIV augmentation by HSV in general and ICP4 in particular has yet to be defined precisely, although the region encompassing the enhancer and SP1 binding sites have been implicated (41,45,46).

Clearly, possible interactions of HSV or other herpesviruses and HIV are multiple, not fully delineated, and potentially quite complex. It is conjectural whether any of these interactions play a role in the pathogenesis of AIDS. Nelson *et al.* (47), by showing co-infection of the same cell with HIV and CMV in the cerebral white matter of a patient with AIDS, provide one instance and a site in which such an interaction could occur. However, recent reports that *tat* in tissue culture media is taken up by fibroblasts and trans-activates HIV through TAR binding (48,49) raise the possibility that *tat* and possibly other, heterologous viral trans-activators may spread extracellularly in AIDS patients. If such is the case then the mere presence of partially expressed, non-replicating herpesviruses in HIV-infected tissue may be of consequence.

## POTENTIAL SITES OF NEUROTROPIC HERPESVIRUS - HIV INTERACTIONS OUTSIDE THE NERVOUS SYSTEM

### HSV Infections of Peripheral Blood Mononuclear Cells

Mononuclear leukocytes are potential extra-neural sites of herpesvirus - HIV interaction. HSV-1 replication has been documented in stimulated human lymphocyte cultures (50), and rarely HSV-1 has been isolated from the buffy coat of patients (51). With a polymerase chain reaction (PCR) assay for an HSV-1 thymidine kinase gene fragment, we have detected HSV-1 in the peripheral blood leukocytes of normal subjects (52). Not yet determined are the infected cell type, the percentage of infected cells, and the state of the infection (productive, abortive, or latent). VZV infection has also been reported in as yet uncharacterized peripheral blood mononuclear cells after varicella and segmental zoster infection (53). In tissue culture lines, long-term persistence of HSV-1 has been reported in the T cell line CEM, the B line NC37, and the myeloid cell line K562 (54,55). In these lines, viral replication occurs in a low percentage of cells and there is partial viral expression without replication in a greater proportion of cells.

The monocyte-macrophage has more relevance to HIV brain infection since this is the carrier cell and the viral reservoir in AIDS dementia complex (1). In tissue culture, HSV-1 produces an abortive infection of adult human monocytes which is characterized by partial expression of viral antigens but little or no virus replication (56). There is some increase in permissiveness in longer-term cultured monocytes, probably indicating differentiation and heterogeneity in the cell population (57). Morahan *et al.* (58) have defined the block in HSV-1 replication in mouse peritoneal macrophages at the level of expression of the early to delayed-early HSV-1 genes. In infected cells there was expression of the immediate early genes ICP4, ICP0, ICP22 and ICP 27 and the early gene thymidine kinase; but the DNA-binding protein ICP8 was barely detectable and not detectable at all were viral DNA polymerase and glycoproteins D and C.

It is uncertain whether the block is at the level of transcription of some of the HSV-1 early genes or if the mRNA or gene product of ICP8 and DNA polymerase is selectively degraded at a higher rate in peritoneal macrophages.

Persistence of HSV DNA at peripheral extra-neural sites

HSV infection and latency in the skin. Recent reports of HIV infection of the bone marrow derived, class II expressing skin Langerhans cells (59) provide another potential site of herpesvirus-HIV interaction. Experimental work in mice indicate that Langerhans cells play an important role in the control of primary HSV infection in skin (60) presumably by presenting viral antigens to T-helper cells (61). Depletion of Langerhans cells by skin abrasion or by steroid treatment enhances skin HSV infection, and an inverse correlation was noted in Langerhans cells density and the virulence of the local HSV infection (comparing the low Langerhans cell density pinna of the mouse ear and the high cell density mouse footpad site of HSV inoculation) (62). If an analogous control is present in man, then HIV-induced depletion or interference of the function of skin Langerhans cells may contribute to the severity of skin herpetic infections in AIDS. Therefore, there exist conditions for bidirectional HIV-HSV potentiation in the skin: augmentation of HIV infection in Langerhans cells by HSV trans-activating factors and augmentation of HSV by HIV interference with the immune function of Langerhans cells.

Opportunities for HSV-HIV interaction would be greater if HSV was persistent or truly latent in the skin. Although ganglionic latency is well-established, views concerning peripheral non-neural latency are controversial. A problem is the differentiation between *bona fide* latency and spontaneous reactivation in ganglia with subclinical peripheral virus shedding. There are species differences in the frequency of spontaneous reactivation: very high frequency with shedding and recrudescent skin lesions in guinea pigs (63), a lower frequency with shedding but usually no clinically apparent recurrent disease in rabbits (64), and very rare reactivation and no peripheral recurrent disease in mice (65). Early reports of HSV latency in the guinea pig footpad were retracted in favor of a persistent HSV infection (66); however, more recent reports in mice are highly suggestive of footpad latency of HSV-2 and HSV-

1 (67-69). In these experiments, over 12 days of explant culture were required for HSV detection, anti-viral drug treatment at a dose that inhibits HSV replication (and therefore would eliminate a persistent infection) had no effect either in explant culture or when given to mice several days prior to sacrifice, and experimental denervation of the footpad by neurectomy prior to HSV inoculation had no effect on HSV detection in footpad explants assayed 2-3 months after viral inoculation.

HSV infection of ocular tissue. In addition to mouse footpad, there are reports now which are compatible with HSV latency in the cornea. HSV isolation from tissue explants was achieved in 11-18% of corneas removed from rabbits and mice experimentally inoculated with HSV (70,71) and in up to 30% of patients undergoing corneal transplantation (72). We have also amplified HSV-1 DNA by the PCR technique from mouse corneal tissue taken at the latent stage of the infection (73). The infected cell type is not known but corneal stromal cells rather than epithelial or endothelial cells are favored (71) and Langerhans cells in the corneal stroma are included among the suspect cell types. It also remains to be determined in any experimental system whether the virus which initiates recurrences on the skin or eye reactivates peripherally or reactives in ganglia and spreads by axonal transport to the skin or eye (or if both mechanisms can occur). In any case, demonstration of HIV in tears and in the cornea of patients with AIDS (74,75) raises the possibility of HIV-HSV interaction in the anterior eye.

The posterior eye, although less well studied at least in terms of HSV latency, is another potential extraneural site of HSV-HIV interaction. HIV has been cultured from AIDS retinal tissue, and viral antigens have been demonstrated by immunohistochemistry in retinal endothelial cells and in neuroretinal cells which have not as yet been fully characterized (76). CMV retinitis is present in as many as a third of AIDS patients in some series (77), and a bidirectional CMV-HIV potentiation which was demonstrated in tissue culture (78) has been speculated to take place in the retina. HSV retinitis in AIDS also occurs but it has been documented in only about 3% of patients (77). There are numerous experimental models of HSV retinitis including animals inoculated with HSV not only by the vitreal route but also by the intracerebral route with HSV spreading to the retina by retrograde axoplasmic transport in the optic nerve (79). There is also an interesting model now being studied by a number of groups of retinitis in the contralateral eye after anterior chamber inoculation (80). The possibility of HSV latency in the retina was raised by the observation of HSV antigens (detected by immunohistochemistry) first appearing in the retina in ocular explants from latently infected mice (81). In support of this, we have recently amplified HSV DNA sequences by the PCR technique from posterior eye tissue of latently infected mice inoculated with HSV by the corneal route, but HSV was detected in more mice when the anterior eye (including the cornea) was assayed (73).

## POTENTIAL SITES OF NEUROTROPIC HERPESVIRUS - HIV INTERACTIONS IN THE NERVOUS SYSTEM

### Animal Models of HSV Encephalitis

The failure so far to establish an adequate animal model of AIDS has forced a reliance on clinicopathological correlations to work out the pathogenesis of particular organ system involvement in AIDS, including AIDS dementia complex. Largely as a consequence of this lack of an animal model, the advances in understanding the biology of HIV infection have lagged behind work on the molecular biology of HIV. HSV, in contrast to the lentiviruses (and also to VZV), does not have a strong species-specific host range, and there are many animal models that have been useful not only in pathogenesis studies of latency and reactivation but also in clinically applied studies of anti-viral drugs and model vaccines.

A problem common to all animal models is to know how well the model mirrors the natural disease. In regard to HSV, certain species are better for one, and others for other aspects of the disease; for example, the mouse generally is preferable for latency studies and the guinea pig for studies of recurrent disease of the skin (63,65). It has been most difficult in any animal to develop a model that parallels HSV encephalitis in man. Two-thirds of the victims of HSV encephalitis have a history of herpes labialis and/or are HSV seropositive at the time of the encephalitis (82). This observation indicates that most cases of encephalitis result from a re-infection with another (exogenous) virus or from reactivation of a latent infection (83), either with reactivation in ganglia and then viral spread to the brain or with reactivation of a latent infection *in situ* in the brain. HSV latency has been demonstrated in the brain of mice inoculated with HSV by the corneal route and also in human subjects dying without neurological signs or symptoms attributable to HSV (14,84,85). As previously noted, explants of brain tissue known to contain HSV DNA sequences seldom yield virus (unlike ganglionic explants) (14). However, under certain experimental situations, HSV reactivation can be achieved *in vivo* in the brain: e.g., hypoglossal neurectomy in latently infected mice with known prior acute HSV infection in the hypoglossal brain stem nucleus (86). In addition to this relative restriction of HSV reactivation in the brain (compared to reactivation in ganglia), it has also been noted in animal models that HSV reactivated in ganglia very seldom spreads centripetally to the brain and causes encephalitis. The only experimental paradigm in which we have observed this centripetal spread involved HSV antibody deficient mice which were given cyclophosphamide as the reactivation stimulus (86). These latently infected mice like those infected by the hypoglossal route, developed a brain stem encephalitis. However, in man HSV encephalitis is usually focal not in the brain stem but in the fronto-temporal area (82). There has been speculation that viral spread to this site may be neural via the trigeminal innervation of the middle cranial fossa (87). The olfactory nerve is another potential route to the temporal lobe, and acute encephalitis by this route has been experimentally produced in rabbits and mice by direct HSV inoculation in the olfactory bulb (89), by intranasal HSV inoculation (89,90), and by corneal HSV inoculation (90) (the last presumably by passage of HSV from the cornea along the nasolacrimal duct to the olfactory mucosa). In one of these olfactory models (89), recurrent temporal lobe HSV encephalitis occurred after apparent viral reactivation induced by cyclophosphamide and dexamethasone in rabbits.

HIV Brain Infection

Clinicopathological Features. Nervous system involvement occurs at some point in most AIDS patients (1,91-93). About one-third suffer either brain lymphoma or opportunistic nervous system infections including toxoplasomis, cryptococcal meningitis, progressive multifocal leukoencephalopathy, and CMV encephalitis. However, the most common neurological problems in AIDS are now thought to be from HIV itself, and although various clinical presentations have been described involving the spinal cord and peripheral nerves, the most commonly recognized syndrome is the so-called AIDS dementia complex as described most extensively by Price and his colleagues (91). Neurological onset is usually subtle with apathy and social withdrawal, difficulty with concentration, memory and cognitive disturbances, and depression or psychomotor retardation. Clinical progression is subacute in most, with a more protracted course in some patients, so that by 2-7 months after onset there is severe neurological disability which characteristically includes motor weakness and ataxia, organic psychosis, and severe dementia with incontinence and mutism.

The neuropathology in AIDS dementia complex corresponds to the clinical characterization as primarily a "subcortical dementia" (92); i.e., diffuse cognitive and motor problems rather than neuroanatomically localized signs or symptoms. This is in sharp distinction to the acute onset and focal nature of HSV encephalitis in which there is most often a temporal lobe presentation with seizures, acute behavioral changes, and delusions or hallucinations progressing in most cases to coma in several days (4,82). The major neuropathological changes in AIDS dementia complex include diffuse white matter pallor, vacuolation and reactive astrocytosis (gliosis), macrophage infiltration and multinucleated cells, and perivascular lymphocytic infiltration. Somewhat less common in most pathological material are microglial nodules and focal areas of necrosis. In general, the extent of the neuropathological involvement correlates with the severity of the dementia, but there are many discrepant cases in which the described pathological abnormalities were milder than anticipated on the basis of the clinical dementia and neurological disability (93).

HIV Infected Cell Types in the Brain. The pathogenesis of AIDS dementia complex is not yet understood. HIV invasion of the brain, or at least the spinal fluid, is known to occur early in the course of HIV infection (1) and clinical signs of a transient aseptic meningitis may be present at this early time. The onset of AIDS dementia complex, however, is usually much later; and it is the monocyte-macrophages harboring HIV in the brain which are thought to be the source of the infection, although invasion of cell-free virus in blood (94) to the brain via endothelial cell or choroid plexus infection is also possible. Another unresolved question is the basis of the diffuse brain cell dysfunction and destruction in this process: how much of this is from direct HIV infection of brain cells and how much is indirect from "bystander" damage (e.g., as a result of toxic monokines or HIV antigens). In regard to neurons, although HIV antigens by immunohistochemistry (95) and HIV RNA by *in situ* hybridization (96) have been described, it is clear now that neuronal infection has not been detected by all groups using these methods and has been recorded only rarely (compared to the monocyte-macrophage, microglial, and multinucleated giant cells) in those few brain blocks where it has been found. Similarly, brain capillary endothelial cell infection has been noted in some but not all *in situ* hybridization reports (96). Rare infected astroglial and oligodendroglial cells have been noted by *in situ* hybridization and by electron microscopy (96, 97); however, the identification of infected oligodendroglial cells has been called into question in the electron microscopy study since the disputed cells bear macrophage or microglial morphological characteristics (98). In contrast, there is no argument that human glial cells can be infected in tissue culture with HIV (99-101), and CMV infection has been shown to trans-activate the HIV-LTR in a glial cell line (102). Not all HIV strains replicate in glial cell lines, and at best the level of replication is low. However, these tissue culture studies raise the possibility that cells other than macrophages-microglia in the brain may be infected in AIDS patients but the level of virus in astroglia (and by implication other brain cell types, including neurons) may be below what can be detected by the techniques of immunohistochemistry and *in situ* hybridization.

Detection of HIV Proviral DNA by PCR in Formalin-Fixed Brain Tissue

Gene amplification by polymerase chain reaction (PCR) is an extremely sensitive technique that may help resolve some of these questions. The PCR technique which was initially used to amplify a globin genomic sequence for the prenatal diagnosis of sickle cell anemia (103) uses two oligonucleotide primers about 20 base pairs in length which flank the sequence to be amplified. One primer is complementary to the coding DNA strand and the other complementary to the non-coding strand. DNA is denatured and annealed to the complementary primers which are then extended by the addition of *Thermus aquaticus* DNA polymerase. Twenty or more cycles of denaturation, primer annealing, and extension results in a 200,000-fold or greater amplification of the targeted sequence. Sequences that are undetectable by standard hybridization techniques can readily be detected after amplification by Southern blot analysis using a synthetic oligonucleotide probe complementary to a region within the amplified segment (the internal probe). Only small amounts of nucleic acids are required for the PCR method and even partially degraded DNA can be amplified since the target sequence is usually small (less than 0.5 kilobases).

Figure 1 is an example of an assay we have done on 9 formalin-fixed, paraffin-embedded brain blocks from 5 AIDS cases (104). The interval between the diagnosis of clinical AIDS and death was 2 months to 3 years (median duration 10 months) and the postmortem interval varied from 15-56 hours. The amplified 128 base pair fragment is defined by the 5' end of the primers which were designed to conserved amino acids sequences in the retroviral reverse transcriptase gene with synthetic oligonucleotide sequence of the primers and the probe based on published HIV sequence data (105). HIV detection was confirmed in the same DNA preparation by additional PCR assays to regions in the HIV-LTR and the *nef* and *env* genes (data not shown). The positive control is the BH-10 plasmid which contains HIV proviral DNA minus the LTR (lane 10), and the negative control is a DNA preparation from peripheral blood mononuclear cells of an uninfected individual (lane 11) or from

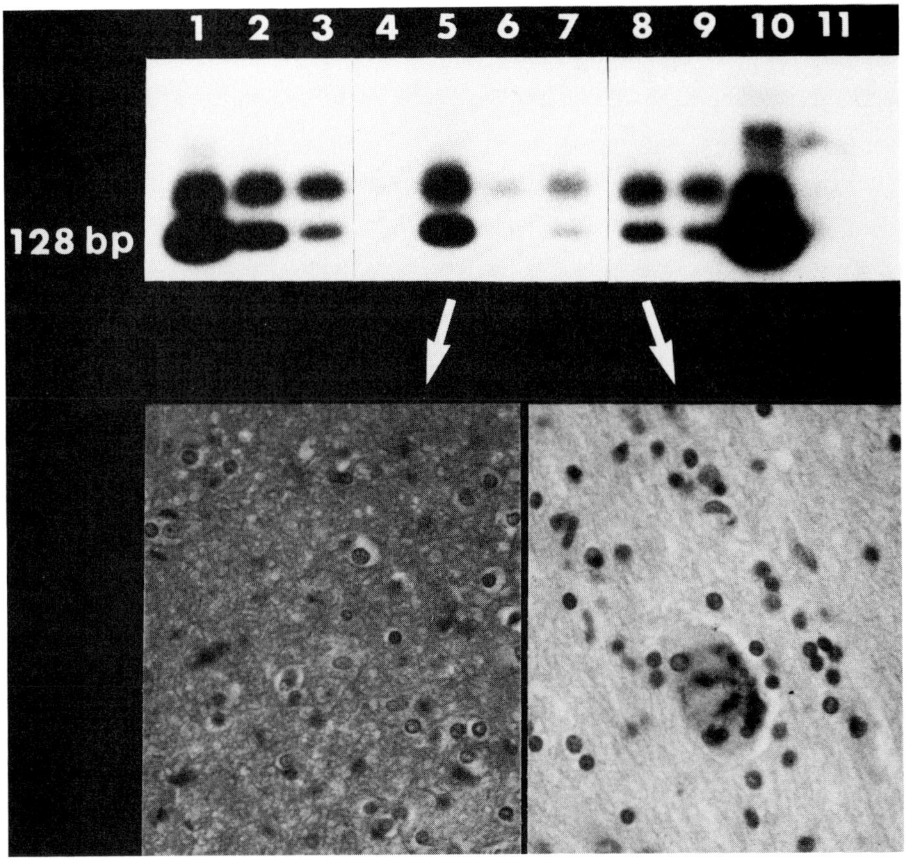


**Fig. 1.** DNA was extracted from single 5 micron paraffin sections of 9 brain blocks from 5 cases of AIDS. Two blocks from different neuroanatomical sites were assayed from one case (lanes 1 and 4), 4 blocks were assayed from a second case (lanes 5, 7, 8, and 9) and 1 block was assayed in the remaining 3 cases. The PCR primers were designed on the basis of HIV nucleotide sequence in the reverse transcriptase gene: primer 1: 3'-**AGGGACACCTTCGTG**-5' and primer 2: 3'-**GTTATGTACCTACTA**-5'. After 50 PCR cycles, an aliquot of the reaction mixture was electrophoresed on an agarose gel, and the gel then transferred to a nylon membrane and hybridized with a $^{32}$P- labelled oligonucleotide corresponding to an internal sequence of the 128 base pair fragment (3'-**ACCTTATAACGACCACTAGG**-5'). Lane 10 is the BH-10 plasmid containing HIV proviral DNA and lane 11 is control, uninfected DNA. The photomicrographs are the adjacent 5 micron sections of the blocks corresponding to lanes 5 and 8. Immunoperoxidase staining for HIV p24 antigen in a multi-nucleated giant cell is shown in the photomicrograph corresponding to lane 8.

non-AIDS paraffin brain blocks (not shown). The method allows a comparison of adjacent 5 micron paraffin sections with one section used for immunohistochemistry and the adjacent section for DNA extraction and PCR assays. In doing this, we have noted PCR amplification corresponding not only to a section with multinucleated giant cells which stain with p24 antigen (lane 8, right-sided photomicrograph) but also corresponding to a section with white matter gliosis only (lane 5, left-side photomicrograph).

Although this result is compatible with a low level HIV infection of brain-derived glial cells, the result can also be explained by HIV in microglia or non-brain cells which were not identified histologically. Also since HIV is present in blood, both cell-free and in peripheral mononuclear cells, amplification of HIV in vascular spaces of the brain block cannot be formally excluded. Clearly, it would be helpful to have additional controls from other organs not known to be infected in AIDS but of similar vascularity to the brain. Also, more experience is needed in brain blocks with this technique, especially in AIDS subjects with early, minimal or no neuropathological abnormalities.

## CONCLUDING COMMENTS

Many aspects of herpesvirus-HIV interaction posed in this review would be more easily approached if there were small laboratory animals permissive for both HIV and HSV. Without this, there must continue to be a reliance on biopsy or autopsy material from AIDS patients and consequently only limited points in time can be evaluated. Moreover, the main histopathological assays used in studies of HIV brain infection - immunocytochemistry, electron microscopy, and *in situ* hybridization - are laborious for extensive sampling and may be of insufficient sensitivity especially for the detection of suspected latency or low level viral persistence in certain cell types. With the exquisitely sensitive PCR technique, tissue screening can be rapidly accomplished for HIV and for herpesviruses or other suspected viral cofactors that may augment HIV infection. Since the PCR technique can be done on DNA extracted from 5 micron paraffin sections, histopathological comparisons with immunohistochemistry or *in situ* hybridization can be done on adjacent sections selected after PCR screening (e.g., screening different brain blocks for viral coinfection). As noted previously, the HIV trans-activating protein in tissue culture media can be taken up by cells (48,49). It is plausible, therefore, to suggest that herpesvirus trans-activating factors may act similarly in brain or other tissue and that coinfection of nearby cells (and not necessarily the same cell) may be sufficient for herpesvirus trans-activation of HIV.

## ACKNOWLEDGEMENT

This work was supported in part by grant RG 2100 from the National Multiple Sclerosis Society and grant CA 33572 from the National Institutes of Health.

## REFERENCES

1. Price, R.W., Brew, B., Sidtis, J., Rosenblum, M., Scheck, A.C., and Cleary P. Science 239: 586-592, 1988.
2. Sigurdsson, B. Brit. Vet. J. 110: 255-270, 1954.
3. Stevens, J.G., and Cook, M.L. Science 173: 843-845, 1971.
4. Whitley, R.J., Soong, S.J., Dolin, R., Galasso, G.J., Ch'ien, L.T., and Alford, C.A. N. Engl. J. Med. 297: 289-294. 1977.
5. Puga, A., Rosenthal, J.D., Openshaw H., and Notkins, A.L. Virology 89:102-111, 1978.
6. Brahic, M.L., Stowring, P., Ventura, P., and Haase, A.T. Nature 292: 240-242, 1981.
7. Strom, T., and Frenkel, N. J. Virol. 61: 2198-2207, 1987.
8. Kwong, A.D., Kruper, J.A., and Frenkel, N. J. Virol. 62: 912-921, 1988.
9. Openshaw, H., Sekizawa, T., Cantin, E., Puga, A., and Notkins, A.L. In: Viral Infections in Oral Medicine (Eds. J. Hooks and G. Jordan), Elsevier North Holland, Inc., New York, N.Y., 1982, pp. 79-88.
10. Cantin, E.M. and Puga, A. In: Concepts in Viral Pathogenesis (Eds. Notkins, A.L. and Oldstone, M.B.A.), Springer-Verlag, New York, 1984, pp. 172-177.
11. Puga, A., Cantin, E.M., Wohlenberg, C., Openshaw, H., Notkins, A.L. J. Gen. Virol. 65: 437-444, 1984.
12. Rock, D.L., Fraser, N.W. J. Virol. 55: 849-852, 1985.
13. Mellerick, D.M., and Fraser, N.W. Virology 158: 265-275, 1987.
14. Cabrera, C., Wohlenberg, C., Openshaw, H., Rey-Mendez, M., Puga, A., and Notkins, A.L. Nature 288: 288-290, 1980.
15. Roizman, B., and Sears, A.E. Am. Rev. Microbiol. 41: 543-71, 1987.
16. Honess, R.W., Roizman, B. J. Virol. 14: 8-19, 1974.
17. Wagner, E.K. In: The Herpesviruses. Vol. 3. Plenum Press, New York, 1985, pp.45-104.
18. Roizman, B., and Batterson, W. In: Virology (Ed. B.N. Fields), Raven Press, New York, 1984, pp. 497-526.

19. Campbell, M.E.M., Palfreyman, J.W., and Preston, C.M. J. Mol. Biol. 180: 1-19, 1984.
20. Bzik, D.J. and Preston, C.M. Nucleic Acids Res. 14: 929-943, 1986.
21. Watson, R.J. and Clements, J.B. Nature 285: 329-330, 1980.
22. Sacks, W.R., Greene, C.C., Aschman, D.P. and Schaffer, P.A. J. Virol. 55: 796-805, 1985.
23. Stevens, J.G., Wagner, E.K., Cook, M.L., and Feldman, L.T. Science 235: 1056-1059, 1987.
24. Puga, A., and Notkins, A.L. J. Virol. 61: 1700-1703, 1987.
25. Deatly, A.M., Spivak, J.G., Lavi, E., and Fraser, N.W. Proc. Natl. Acad. Sci. USA 84: 3204-3208, 1987.
26. Javier, R.T., Stevens, J.G., Dissette, V.B. and Wagner, E.K. Virology 166: 254-257, 1988.
27. Steiner, I., Spivack, J.G., Lirette, R.P., Brown, S.M., MacLean, A.R., Subak-Sharpe, J.H., and Fraser, N.W. EMBO J. 8: 505-511, 1989.
28. Ace, C.I., McKee, T.A., Ryan, J.M., Cameron, J.M. and Preston, C.M. J. Virol. 63: 2260-2269, 1989.
29. Sacks, W.R., and Schaffer, P.A. J. Virol. 61: 829-839, 1987.
30. Everett, R.D. J. Gen. Virol. 70: 1185-1202, 1989.
31. Stow, E.C. and Stow, N.D. J. gen. Virol. 70: 695-704, 1989.
32. Rosenberg, Z.F. and Fauci, A.S. AIDS Res. Hum. Retroviruses 5: 1-4, 1989.
33. Quinn, T.C., Piot, P., McCormick, J.B., Feinsod, F.M., Taelman, H., Kapita, B., Stevens, W., Fauci, A.S. JAMA 257: 2617-2621, 1987.
34. Jones, N.C., Rigby, P.W.J., and Ziff, E.B. Genes Dev. 2: 267-281, 1988.
35. Varmus, H. Genes. Dev. 2: 1055-1062, 1988.
36. Gendelman, H.E., Phelps, W., Feigenbaum, L., Ostrove, J.M., Adachi, A., Howley, P.M., Khoury, G., Ginsberg, H.S., and Martin, M.A. Proc. Natl. Acad. Sci., USA 83: 9759-9763, 1986.
37. Mosca, J.D., Bednarik, D.P., Raj, N.B.K., Rosen, C.A., Sodroski, J.G., Haseltine, W.A., Hayward, G.S., and Pitha, P.M. Proc. Natl. Acad. Sci. USA 84: 7408-7412, 1987.
38. Ostrove, J.M., Leonard, J., Weck, K.E., Rabson, A.B., and Gendelman, H.E. J. Virol. 61: 3726-3732, 1987.
39. Nabel, G.J., Rice, S.A., Knipe, D.M., Baltimore, D. Science 239: 1299-1302, 1988.
40. Gimble, J.M., Duh, E., Ostrove, J.M., Gendelman, H.E., Max, E.E., and Rabson, A.B. J. Virol. 62: 4104-4112, 1988.
41. Lenardo, M.J., Fan, C.M., Maniatis, T., and Baltimore, D. Cell 57: 287-294, 1989.
42. Hiscott, J., Alper, D., Cohen, L., Leblanc, J.F., Sportza, L., Wong, A., and Xanthoudakis, S. J. Virol. 63: 2557-2566, 1989.
43. Everett, R.D. J. Mol. Biol. 203: 87-96, 1988.
44. Albrecht, M.A., DeLuca, N.A., Byrn, R.A., Schaffer, P.A., and Hammer, S.M. J. Virol. 63: 1861-1868, 1989.
45. Harrich, D., Garcia, J., Wu, F., Mitsuyasu, R., Gonzalez, J. and Gaynor, R. J. Virol. 63: 2585-2591, 1989.
46. Garcia, J.A., Wu, F.K., Mitsuyasu, R., and Gaynor, R.B. EMBO J.

6: 3761-3770, 1987.
47. Nelson, J.A., Reynolds-Kohler, C., Oldstone, M.B.A., and Wiley, C.A. Virology 165: 286-290, 1988.
48. Green, M., and Loewenstein, P.M. Cell 55: 1179-1188, 1988.
49. Frankel, A.D., Pabo, C.O. Cell 55: 1189-1193, 1988.
50. Nahmias, A.J., Kibrick, S., and Rosan, R.C. J. Immunol. 93: 69-74, 1964.
51. Craig, C.P. and Nahmias, A.J. J. Infect. Dis. 127: 365-372, 1973.
52. Cantin, E.M., Willey, D.E., Gaidulus, L. Lange, W., Openshaw, H. (abstract) 14th International Herpesvirus Workshop, 1989.
53. Gilden, D.H., Hayward, A.R., Krupp, J. Hunter-Laszlo, M., Huff, J.C. and Vafai, A. Virus Res. 7: 117-129, 1987.
54. Cummings, P.J., Lakomy, R.J. and Rinaldo Jr., C.R. Infect. Immun. 34: 817-827, 1981.
55. Rinaldo Jr., C.R., Richter, B.S., Black, P.H. and Hirsch, M.S. Infect. Immun. 25: 521-525, 1979.
56. Morahan, P.S. In: Immunobiology of Herpes Simplex Virus Infection (Eds. B.T. Rouse and C. Lopez), CRC Press, Inc., Boca Raton, Fla., 1984, pp. 71-89.
57. Daniels, C.A., Kleinerman, E.S. and Snyderman, R. Am. J. Pathol. 91: 119-136, 1978.
58. Morahan, P.S., Mama, S., Anaraki, F. and Leary, K. J. Virol. 63: 2300-2307, 1989.
59. Rappersberger, K., Gartner, S., Schenk, P., Stingl, G., Groh, V., Tschachler, E., Mann, D.L., Wolff, K., Konrad, K. and Popovic, M. Intervirol. 29: 185-194, 1988.
60. Sprecher, E. and Becker, Y. Arch. Virol. 91: 341-349, 1986.
61. Tamaki, K., Stingl, G., Guillino, M., Sachs, D.H., and Katz, S.I. J. Immunol. 123: 784-787, 1979.
62. Sprecher, E. and Becker, Y. J. Virol. 61: 2515-2522, 1987.
63. Donnenberg, A.D., Chaikof, E. and Aurelian, L. Infect. Immun. 30: 99-109, 1980.
64. Nesburn, A.B., Elliot, J.H., and Leibowitz, H.M. Arch. Ophthalmol. 78: 523-529, 1967.
65. Sekizawa, T., Openshaw, H., Wohlenberg, C., and Notkins, A.L., Science 210: 1026-1028, 1980.
66. Scriba, M. Med. Micro. Immunol. 169: 91-96, 1981.
67. Al-Saadi, S.A., Clements, G.B., and Subak-Sharpe, J.H. J. gen. Virol. 64: 1175-1179, 1983.
68. Al-Saadi,S.A., Gross, P., and Wildy P. J. gen. Virol. 69: 433-438, 1988.
69. Clements, G.B., and Subak-Sharpe, J.H. J. gen. Virol. 69: 375-383, 1988.
70. Abghari, S.Z., and Stulting, R.D. Invest. Ophthalmol. Vis. Sci. 29: 239-243, 1988.
71. O'Brien W.J., and Taylor, J.L. Invest. Ophthalmol. Vis. Sci. 30: 357-364, 1989.
72. Tullo, A.B., Easty, D.L., Schimeld, C., Stirling, P.E., and Darville, J.M. Trans. Opthalmol. Soc. U.K. 104: 159-165, 1985.

73. Willey, D.E., Cantin, E., Openshaw, H. (abstract) Invest. Opthalmol. Vis. Sci. 30: 215(5), 1989.
74. Fujikawa, L.S., Salahuddin, S.Z., Ablashi, D., et. al., Opthalmology 93: 1479-1481, 1986.
75. Salahuddin, S.Z., Palestine, A.G., Heck, E., et. al., Am. J. Opthalmol. 101: 149-152, 1986.
76. Pomerantz, R.J., Kuritzkes, D.R., De La Monte, S.M., Rota, T.R., Baker, A.S., Albert, D., Bor, D.H., Feldman, E.L., Schooley, R.T., and Hirsch, M.S. N. Engl. J. Med. 317: 1643-1647, 1987.
77. Pepose, J.S., Holland, G.N., Nester, M.S., Cochran, A.J., and Foos, R.Y. Opthalmology 92: 472-484, 1985.
78. Skolnik, P.R., Kosloff, B.R., and Hirsch, M.S. J. Infect. Dis. 157: 508-514, 1988.
79. Peiffer, R.L., Dekker, C.D., and Siegel, F.L. Invest. Opthalmol. Vis. Sci. 24: 1070-1078, 1983.
80. Whittum, J.A. McCulley, J.P., Niederkorn, J.Y., and Streilin, J.W. Invest. Opthalmol. Vis. Sci. 25: 1065-1073, 1984.
81. Openshaw, H. Infect. Immun. 39: 960-962, 1983.
82. Nahmias, A.J., Whitley, R.J., Visintine, A.N., Takei, Y. and Alford, C.A. J. Infect. Dis. 145: 829-836, 1982.
83. Whitley, R., Lakeman, A.D., Nahmias, A., and Roizman, B. N. Engl. J. Med. 307: 1060-1062, 1982.
84. Rock D., and Fraser, N. Nature 302: 523-525, 1983.
85. Fraser, N.W., Lawrence, W.C. Wroblewska, Z., et al. Proc. Natl. Acad. Sci. USA 78: 6461-6465, 1981.
86. Sekizawa, T., and Openshaw, H. J. Virol. 50: 263-266, 1984.
87. Davis, L.E., and Johnson, R.T. Neurol. 5: 2-5, 1979.
88. Schlitt, M., Lakeman, A.D., Wilson, E.R., To, A., Acoff, R.W., Harsh III, G.R., and Whitley, R.J. J. Infect. Dis. 153: 732-735, 1986.
89. Stroop, W.G., and Schaefer, D.C. J. Infect. Dis. 153: 721-731, 1986.
90. Tomlinson, A.H., and Esiri, M.M. J. Neuro. Sci. 60: 473-484, 1983.
91. Navia, B.A., Jordan, B.D., and Price, R.W. Ann. Neurol. 19: 517-524, 1986.
92. Price, R.W., and Brew, B.J. J. Infect. Dis. 158: 1079-1083, 1988.
93. Navia, B.A., Cho, E.S., Petito, C.K., and Price, R.W. Ann. Neurol. 19: 525-535, 1986.
94. Sonnerborg, A., Ehrnst, A., and Strannegard, O. J. Med. Virol. 27: 258-263, 1989.
95. Pumarola-Sune, T., Navia, B.A., Cordon-Cardo, C., Cho, E.S. and Price, R.W. Ann. Neurol. 21: 490-496, 1987.
96. Wiley, C.A., Schrier, R.D., Nelson, J.A., Lampert, P.W. and Oldstone, M.B.A. Proc. Natl. Acad. Sci. USA 83: 7089-7093, 1986.
97. Gyorkey, F, Melnick, J.L. and Gyorkey, P. J. Infect. Dis. 155: 870-876, 1987.
98. Sharer, L.R., and Prineas, J.W. J. Infect. Dis. 157: 204, 1987.
99. Cheng-Mayer, C., Rutka, J.T., Rosenblum, M.L., McHugh, T., Stites, D.P., and Levy, J.A. Proc. Natl. Acad. Sci. USA 84: 3526-3530, 1987.
100. Dewhurst, S., Sakai, K., Bresser, J., Stevenson, M., Evinger-Hodges, M.J., and Volsky, D.J. J. Virol. 61: 3774-3782, 1987.

101. Chiodi, F., Fuerstenberg, S., Gidlund, M., Asjo, B., and Fenyo, E.M. J. Virol. 61: 1244-1247, 1987.

102. Duclos, H., Elfassi, E., Michelson, S., Arenzana-Seisdedos, F., Munier, A. and Virelizier, J.L. AIDS Res. Hum. Retroviruses 5: 217-224, 1989.

103. Saiki, R., Scharf, S., Faloona, F., Mullis, K.B., Horn, G.T., Erlich, H.A. and Arnheim, N. Science 230: 1350-1354, 1985.

104. Openshaw, H., Cantin, E., and Hinton, D. (abstract) Neurol. 39: 379(s), 1989.

105. Wain-Hobson, S., Sonigo, P., Danos, O., Cole, S., and Alizom, M. Cell 40: 9-17, 1985.

# 11

## HERPESVIRUSES AS CO-FACTORS IN THE IMMUNOPATHOGENESIS OF AIDS

Jeffrey Laurence

**Cornell University Medical College, Division of Hematology-Oncology, New York, NY 10021**

### ABSTRACT

Potential associations between human immunodeficiency viruses (HIV) and the herpesviruses may occur at the host, cell and molecular levels. A review of the biology of the HIV life cycle serves as a foundation to dissect these relationships at points of receptor interactions, shared *cis*-acting transcriptional regulatory elements, and induction of cellular immune responses. The potential for the herpesviruses, particularly HSV-1, -2, HHV-6, CMV, and EBV, to serve as co-factors in the progression of clinical HIV infection is explored, and novel means of disrupting these interactions discussed.

### INTRODUCTION

Retroviruses were among the earliest known viruses, identified some eighty years ago as the filterable agents that cause sarcomas in fowl (1). Yet interest in these microbes has dramatically increased only of late, following their documentation as etiologic agents in human T cell leukemia/lymphoma (2), spastic paraparesis (3), acquired immune deficiency syndrome in human (4,5) and subhuman (6) primates, and identification of mutations and translocations of cellular proto-oncogenes, related to retroviral oncogenes, in human malignancy (7).

Complex and unusual parasites, retroviruses evolved unique associations with their cellular hosts. These include: reverse transcription of genomic RNA into a proviral DNA form; stable integration of viral DNA into the chromosomal structure of somatic and germ line cells; mutation and, occasionally, capture of cellular transcription units; alteration of the growth or function of host cells; and transduction of normal host mechanisms for gene expression under the control of viral elements (8). These regulatory events enable retroviruses to establish a persitent or latent infection from which replication-competent virus may be induced or "rescued" through a variety of pathways.

Latency is defined by establishment of the virus in its host, typically for prolonged periods in the absence of symptomatology, followed by occasional episodes of a lytic replicative cycle accompanied by obvious pathology (9). As such, viral persistence represents an evasion of the host's immune surveillance network. There are thus two essential elements in the establishment of latency: unique components of viral structure, including trans-acting elements and their *cis*-response targets; and immune responses that are ineffectual in recognizing and clearing viral particles and virus-infected cells (10).

Exactly how viruses in general, and retroviruses in particular, maintain this state is an active area of investigation. Its resolution should lead to insights into the treatment and, hopefully, cure of disease related to two important pathogens which become latent in their hosts, herpesviruses and human retroviruses. This chapter will outline the molecular biology of one particular class of human retrovirus, the human immunodeficiency viruses (HIV), etiologic agents of AIDS. Our present understanding of the mechanisms by which a variety of perturbations, particularly herpesvirus co-infections, interact with HIV on a cellular and molecular level will be emphasized.

## IMMUNODEFICIENCY VIRUSES: CLASSIFICATION

HIV-1 and HIV-2 are the etiologic agents of AIDS and its related pathologic processes, outlined in Table 1. HIV is prevalent in North America and most other continents, including the central portion of Africa. HIV-2 is predominantly found in West Africa, although seven cases among West Africans living in the U.S. have been reported (11). They share a common genetic organization, with approximately 50% identity at the nucleotide level (Fig. 1), but little serologic cross-reactivity (11-13). Both cause similar disease, albeit the course of HIV-2 may be attenuated (14, 15). Related lentiviruses, the simian immunodeficiency viruses (SIV), have been isolated from nonhuman primates in Asia and Africa, and are classified according to the species of monkey from which they are primarily obtained (16, 17). All are members of the Old World superfamily (Cercopithecoidea) and include: rhesus (Macaca, SIV/mac);

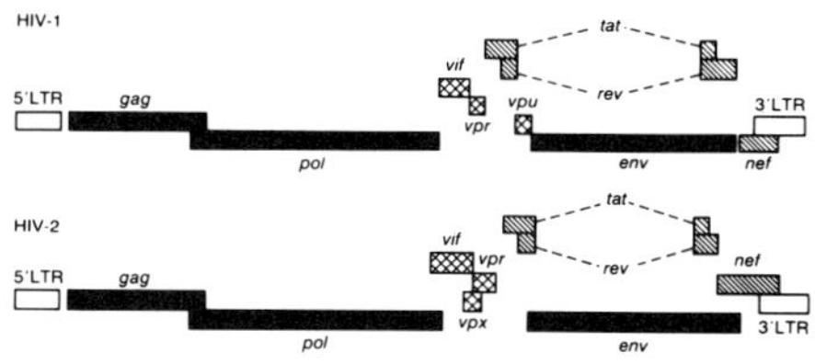


Figure 1. Genetic organization of proviruses of human immunodeficiency viruses types 1 and 2. Transcription units for structural proteins (gag, pol, and env), regulatory proteins (tat, rev, and nef), and genes whose functions have not yet been characterized are illustrated. (Reproduced, with permission, from reference 12.)

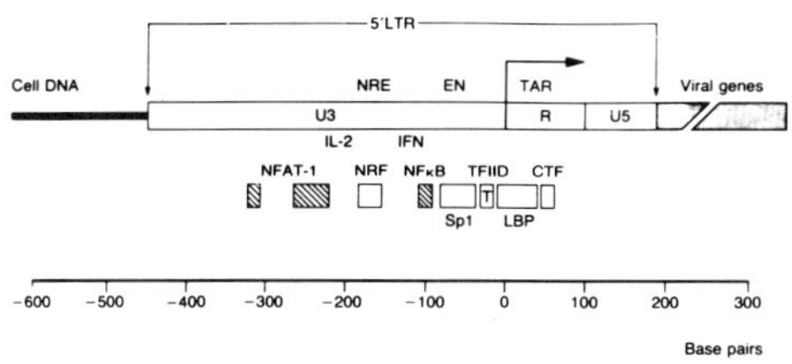


Figure 2. Cis-acting elements in the HIV-1 long terminal repeat (LTR) that regulate viral gene expression. (Reproduced, with permission, from reference 12.)

African green monkeys (Cercopithecus, SIV/agm); mandrills (Papio, SIV/MND); and mangabeys (SIV/SMM). SIV/mac, SIV/SMM and HIV-2 form a closely related subgroup, with SIV/mac and HIV-2 sharing a 70% sequence identity (18) as well as one transcription unit, *vpx*, unique to HIV-2 and SIV/mac (19). This has raised speculation that mangabeys may have been a source for cross-species transmission in the origin of HIV-2 in humans (18) and SIV/mac in captive macaque colonies (20). An early isolate, reported to define a new family of human immunodeficiency viruses, HTLV-IV, proved to be a contaminant of a culture with SIV/mac (21). The genetic diversity among SIV/agm isolates appears to be much greater than for HIV or SIV/mac. Coupled with the high prevalence of natural green monkey infection without symptomatology, it suggests that SIV/agm has been maintained for a prolonged period (20). A further difference illustrated by this isolate is the absence of *vpr*, a transcription unit of uncertain function found in HIV-1, HIV-2 and the SIV/mac group (22), but lacking in SIV/agm (20). Of the SIV subtypes described to date, only those from macaques and mangabeys cause an AIDS-like disease in macaques following experimental inoculation, while among nonhuman primate species, HIV-1 infects only chimpanzees and gibbon apes (23).

Following infection these lentiviruses all cause an immune deficiency. Their major pathway for initiating this infection is via interaction of the glycoprotein outer envelope, gp120, with a receptor, CD4, present on the cell surface of certain T (helper/inducer subset) and B lymphocytes, monocyte-macrophages, and neuroglial cells (24). These viruses resemble other members of the family *Retroviridae* in basic features of genetic organization, structural proteins and enzymes, yet differ in an important aspect: they encode several proteins that regulate viral genes intimately related to a latent/lytic switch characteristic of their life-cycles.

## THE HIV LIFE-CYCLE: EARLY STAGES

### Receptor Biology

Sites for potential interaction with other viral genomes are dictated by the types of cells that are infected. For HIV, the spectrum is wide. Most retroviruses attach to cells via surface membrane proteins recognized by viral coat structures and typically enter through a pH-dependent endocytotic mechanism used to facilitate transport of a variety of extracellular substances, including growth factors. In contrast, HIV appears to penetrate cells via a pH-independent membrane fusion event (25). This is directed by its bipartite

**Table 1.** HIV-Related Disorders

- o AIDS: opportunistic infections, Kaposi's sarcoma, Hodgkins's and non-Hodgkin's (B-cell) lymphomas
- o Lymphadenopathy syndrome (LAS), part of the AIDS-related complex (ARC)
- o Acute mononucleosis-like disease: fever, malaise, pharyngitis, splenomegaly, rash
- o Neuropsychiatric disorders: aseptic meningitis, cerebral atrophy, myelitis, dementia, psychosis
- o Teratogenesis
- o Thrombocytopenia
- o Neoplasms: adenocarcinomas (colon, pancreas), squamous cell carcinomas, malignant melanoma
- o Dermatologic disease: atopic dermatitis, leukoplakia
- o Lymphoid interstitial pneumonitis

envelope glycoprotein, consisting of an exterior portion, gp120, noncovalently linked via amino terminal interactions to a transmembrane domain, gp41, imbedded in the lipid membrane that surrounds the virion (26). The functional envelope ligand appears to be a multimer of this basic gp120-gp41 unit (27). A region of gp120 that specifies binding to its high-affinity receptor, CD4, has been identified by mutational analysis to consist of three noncontiguous regions of the carboxy terminus (28). It is proposed (29) that these regions form a binding pocket capable of multiple interactions with residues in the first domain, at the amino terminal tip, of CD4 on the host cell (30). This domain separates the major histocompatibility complex (MHC)-binding and gp120 - attachment functions of CD4 (30). The multi-unit interaction may account for the high Kd of receptor-ligand association, on the order of $10^{-9}$ M. Following binding, a fusion event must occur. This is encoded in a short hydrophobic alpha-helical span at the amino terminus of gp41. Mutations in this region preclude viral internalization, but it is unclear whether a specific cellular receptor is required for the membrane fusion reaction (29).

Alternate, low-affinity receptors also appear to exist, enabling uptake of HIV into cells which lack detectable surface CD4 or CD4 transcripts. These include not only $CD8^{+}/CD4^{-}$ suppressor/cytotoxic T lymphocytes (31), which may become infected at an early stage of T cell development, when both CD4 and CD8 antigens are co-expressed, but a variety of tissues of ectodermal and endodermal, as well as mesodermal origin. For example, Fc receptors appear to be involved in viral entry into some monocytes (32), complement receptors have been linked to similar events (33), and MHC type II molecules associate with CD4 and may facilitate viral entry (34). The process by which HIV gains entry into such diverse cells as vascular endothelium (35) and fibroblasts (36) is unknown, but may involve non-specific fusion events, or phagocytosis. If Fc receptor-mediated viral entry indeed plays a significant role, possible interactions with herpesviruses such as HSV, which can induce these molecules on endothelial and other cells (37), should be considered as one means by which HIV and herpesviruses interact.

There is also an intriguing report that single amino-acid changes in the HIV envelope can alter viral tropism and receptor binding, abolishing the ability of an isolate to infect a monocyte, while retaining its T cell tropism (38). Modulation of the high-affinity HIV receptor, CD4, may also play a role in initiating and maintaining viral latency. A common feature of persistently infected helper T lymphocyte lines is a loss or dramatic reduction in expression of host CD4, which is directly mediated by the virus (39). These cells may then become resistant to second round "superinfection" by progeny virus in a process of envelope-mediated viral interference (39).

Reverse transcription

Once internalized, HIV is uncoated in the cytoplasm. The virion core consists of a diploid complex of two identical RNA chains of approximately 9.2kb, non-covalently linked at the 5' ends, together with three enzymes, RNA-dependent DNA polymerase, ribonuclease H (RNase H), and integrase. Each viral RNA molecule has sequences unique to the 3' (untranslated 3', U3) and 5' (untranslated 5', U5) ends, as well as repeated (genomic repeat, R) sequences at both termini. It is also base-paired with a lysine transfer RNA that primes DNA synthesis. This synthesis of a linear double-stranded DNA provirus occurs in the cytoplasm, presumably on a replication complex whose component parts have been only tentatively defined (40). The binding of the tRNA-lys primer by the viral DNA polymerase is the first step in the initiation reaction, followed by addition of a deoxynucleotide to the 5' end of the primer (29). Complementary nucleotides are then added to the 3' end of the nascent chain in an elongation reaction. This serves as the site of action for anti-HIV nucleoside analogs such as azidodeoxythymidine (AZT) (41). AZT also has potent anti-EBV, but not CMV or HSV, activity (42).

After the initial stretch of DNA synthesis has occurred at the R-U5 region at the 5' end of one genome, the polymerase switches ("jumps") to the R-U3 region of the second genome (43). This reaction is controlled by R, and requires that the additional RNA template be degraded to reveal the complementary DNA, and the tRNA precursor be removed, steps catalyzed by RNaseH (29). This DNA then anneals to complementary RNA at the 3' end of the second DNA strand. The initial DNA strand now serves as a primer for a second elongation reaction, which continues until a complete copy of the second RNA genome is made, forming a DNA/RNA hybrid. Because this minus strand copy of virion RNA is always made before the plus strand, retroviral DNA synthesis is asymmetrical, in contrast to traditional semiconservative DNA synthesis (8). Following an RNaseH directed cleavage of the RNA genome near the 3' end of the template, and additional RNA-primed DNA synthesis, template RNA is degraded, again by RNaseH, simultaneously with synthesis of the second ("positive") DNA strand, to yield a full-length linear, double-stranded provirus, with duplicated regulatory regions, the long terminal repeats (LTR) structured as U3-R-U5, at both termini. These elements are schematically illustrated in Fig. 2. The significance of these sequences will become apparent as homologous controlling regions are discussed in the context of the herpesviruses. Besides providing transcriptional control elements, the LTR of at least one lentivirus, visna, also is involved in cellular tropism (44). Whether this occurs for HIV is unclear.

Each provirus is approximately 9.5kb. It is thought that virion capsid proteins as well as integrase remain associated with this replication complex throughout reverse transcription (29). The entire process, from virion binding to formation of proviral DNA occurs in the cytoplasm, and requires 2-4 hours (12,45).

Integration

Integration of retroviral DNA into host chromosomes occurs by a specialized and precise recombinational mechanism encoded by the virus, differentiating it from most other viruses whose DNAs are integrated infrequently and haphazardly (8). The signals for transport of the linear duplex DNA form of the provirus with accompanying proteins into the nucleus is unknown, but may occur once the nuclear membrane disperses during mitosis (29). In the nucleus, a fraction of these proviral DNAs form two types of supercoiled circular forms: one with blunt ends joined, yielding a repeated LTR, and one by homologous recombination within the repeated regions, yielding a single

LTR copy. Once in the nucleus, it appears that the immediate precursor for integration is the linear provirus (46). Retroviruses integrate randomly throughout the genome, although about 20% of all integrations involve preferential use of sites ("hot spots") either in host chromosomes (47) or in cell-free integration systems. Viral DNA is seen in the nucleus 4-16 hours after infection, but it is unclear whether integration is a requirement for viral replication in all cell types. While T lymphocytes appear incapable of transcribing unintegrated HIV nuclear proviral DNA (29), visna can replicate in resting cells at least *in vitro* (48).

The mechanics of integration have been defined for murine leukemia virus, and preliminary data indicate that a similar, although perhaps not identical, process occurs with HIV. It requires an enzyme, integrase, which removes the two terminal nucleotides from the proviral LTR, as viral DNA is always joined to host DNA 2 base-pairs from the ends of the LTRs (49). The site of integrase-DNA binding appears to be sequence specific, as conserved regions near the ends of the LTRs are required (49). Integrase also has a weak exonucleolytic activity, and appears to function without a high-energy co-factor, as determined in a cell-free system (50).

Copy number

The copy number of unintegrated HIV DNA has been determined in T lymphoblastoid cell lines and non-transformed T lymphocytes (45). During the first round of infection, completed in 24 h, the copy number per cell rises from 1 to 3. It then rapidly increases, reaching about 20, then declines. Even in chronically infected cultures there is detectable linear DNA, which may have originated from a low level of continual reinfection or intracellular activity of the HIV polymerase (45). These copy numbers are similar to those determined by *in situ* hybridization *in vivo*. Only a fraction of the DNA measured (perhaps 5%) integrates. Although it had been suggested that, by analogy with spleen necrosis virus and Rous-associated virus, large amounts of unintegrated viral DNA might be cytopathic, 20 copies of HIV per cell are not directly toxic (45).

## TRANSCRIPTIONAL CONTROLS

HIV DNA synthesis is rapid and similar to that of standard retroviruses (45), but there are important differences in the temporal control of mRNA species that accumulate. While entry into a host cell initiated conversion of a quiescent, enveloped virion into an enzymatically active, and essentially autonomous, nucleoprotein complex, once integrated the retrovirus becomes highly host-dependent (8). Routine cellular functions, including transcription of the provirus by RNA polymerase II, processing of RNA transcripts, and translation of messenger RNAs, are now regulated by viral signals that determine the efficiency of transcription and splicing, and can bypass termination signals during translation (8). For the creation of new virions, a full length copy of genomic RNA must leave the nucleus and complex to the budding virion, by a process that has not yet been defined (29). Regulation of transcription initiation has been well-studied, however, and depends upon signals found primarily in the U3 region of the LTR.

The Long terminal repeat

The basic distinction between enhancers and promoters in the LTR is operational. An enhancer must be able to stimulate transcription at a distance; this need not be true of a promoter or its components (51). In addition, enhancers are orientation-independent. In general, retroviral regulatory elements are recognized by host transcription factors, determining the microbe's growth potential in cells of divergent type and states of differentiation (8). In HIV, both cellular and viral proteins exert their influence, directly or indirectly, on sequences located near the point of RNA initiation, the "cap" site (+1). RNA polymerase II is dependent upon an initiation complex that forms near this point. Certain sequences are shared by the HIV-1 provirus and cellular genes. These include proteins that recognize the sequences TATAA (-28 to -23); CATAA; and three tandem regulatory factor SP1 binding sequences (between -77 and -45), all of which are responsive, although only two are essential for promoter function (52). Additionally, several *cis*-acting regulatory sequences have been identified in the HIV-1 LTR that are not ubiquitous, and whose function is regulated in part by interaction with cellular regulatory proteins as well as at least one HIV-encoded trans-acting factor, *tat* (52). The trans-acting responsive (*TAR*) sequence for *tat* maps between -17 to +80, with the critical domain +1 to +42, and is regulated in trans by the *tat* (trans-activator) gene (53). Two core transcriptional enhancer elements (GGGACTTTCC; -104 to -95 and -90 to -81), are identical in sequence to the NF-$\kappa$B element of immunoglobulin kappa ($\kappa$) genes (54). In addition, there are regions homologous to the interleukin 2 distal and proximal control elements (-302 to -246) (55); a region homologous to the 5' flanking sequences of $\gamma$-interferon (-125 to -109) (55) and $\beta$-interferon (52,56); a core glucocorticoid responsive element (-257 to -263) (52,53); and sites that recognize the binding proteins for NFAT-1 (nuclear factor of activated T cells-1), and USF (upstream binding factor of adenovirus) (12,29). There is also an apparent binding sequence, CCCCAGCG, for the transcriptional protein AP-2/Jun (52), as well as a negative regulatory element (NRE) (-340 to -185) (59).

As previously stated, HIV-2 is more related to SIV than to HIV-1, and its conserved regions, as well as important distinctions, should offer clues to mechanisms underlying differences in pathogenic potential, latency and interactions with other viruses. In terms of the LTR and transcriptional regulation, the *tat* proteins of HIV-1 and -2 have only short homologous sequences (60), and full activity of the HIV-2 *TAR* requires sequences in the R region not found in the HIV-1 LTR, with the major *TAR*-like element between +1 and +103 (61). The predicted structure for HIV-2 transcripts of *TAR* contains two similar stem-loop structures (61), while the corresponding region for HIV-1 predicts only one (62). Sequences critical for *tat* activity (CUGGGA) are identical in all these loops, however. In addition, the NF-$\kappa$B element is present as a single sequence in HIV-2, at nucleotides -108 to -98 (63), as opposed to the tandem repeat in HIV-1, and a shorter negative regulatory element is present at about +103. These differences are thought to underlie the poor transactivation of the HIV-1 LTR by the HIV-2 *tat* gene product (62).

In most cellular and viral systems, binding sites for initiation factors are located 5' to the cap site (29); HIV-1 has at least two proteins, LBP (leader binding protein) and CTF (CCAAT transcription factor) (64), that bind 3' of the cap site. Deletion of these sequences lowers the rate of RNA synthesis (64).

In addition, proteins that are part of the initiation complex but do not directly bind to DNA may be involved in transcriptional activation. DNA binding may be considered a neutral function which brings the activating surface of the transcriptional activator to the vicinity of a gene (65). Activating regions generally are acidic bearing a significant net negative charge, may form amphipathic alpha-helices (65), and their efficiency is modified by phosphorylation and glycosylation (66). However, an activating region might also interact, perhaps more weakly, with the primary target protein (RNA polymerase II) when both activator and target are off the DNA (65). TFIID (TATA-box transcription factor), which binds to the TATA sequence present in the HIV LTR (Fig. 2), is one potential target for the interaction of such molecules (65). This is an intriguing addition to potential gene regulatory phenomena in HIV, as herpes simplex virus (HSV) protein VP16 appears to function in this manner. VP16 is a virion component that activates transcription of immediate early genes in virally infected cells. It is made only transiently, and can depress ("squelch") transcription from promoters lacking the activator binding sites (67). However, VP16 is not itself a DNA binding protein but rather recognizes molecules bound at either of two DNA sites located 5' of the genes (65,67). The *tat*-equivalent in the human T cell lymphotropic viruses HTLV-I and -II, *tax*, may also function in a similar manner (29).

While much early work centered on *tat*-linked trans-activation, the fact that HIV-2, SIV/agm and HIV-1 possess *tat* and *TAR* elements of equivalent activity and yet appear to have very different pathogenic potentials argues against *tat*-mediated activation of viral gene expression as the sole or major factor in the pathogenesis of these agents. As evident from the complex series of factor and *cis*-acting response elements outlined above, multiple interactions among proteins influence the rate of RNA initiation. The interrelationships among various regulatory components of the HIV life cycle and documented as well as hypothetical points of interaction with herpesviruses, will be discussed in detail.

Cell lineage, position in the cell cycle, metabolic rate, and state of cellular differentiation (69) may determine the relative, and perhaps absolute, concentrations of transcriptional activators. This is true for both ubiquitous cell regulators (68) as well as stimulatory (69) and inhibitory (70) factors limited to HIV and a few other systems. Cytokines, elicited by immune interactions or concomitant infections with other viruses, particularly the herpesviruses, can also affect the levels of such activators and repressors.

Control of the latent/lytic cycle of HIV infection in vitro.

Regulation of chronic HIV infection, with transition to a productive state, may be important in immune pathophysiology. HIV infection of T lymphocytes, monocytes, and possibly other cells includes a chronic or latent phase during which there is little or no viral replication. Transition to productive infection can be initiated *in vitro* by immunologic activation of infected cells (71). These signals induce cellular trans-activating factors, which in turn stimulate transcription from promoter and enhancer elements in the HIV LTR. The virus then synthesizes its own regulatory factors which further affect transcription at the LTR, leading to increased viral replication and cytopathology (72).

HIV mRNA accumulation is also sensitive to a variety of other stimuli. The mechanisms by which influences as disparate as cytomegalovirus (CMV) (73) and HSV (74) infection, monokine and lymphokine treatment, (75, 76) antigenic stimulation (71), and phorbol ester (77) exposure can convert a latent or chronic infection to a productive one appear to be just as varied. For example, NF-κB, an enhancer of HIV-1 and 2, responds to nuclear trans-activating factors induced by antigen and phorbol esters (78). Activation of protein kinases, important enzymes in the triggering of many tissue specific and differentiation functions, may be a key step in the interaction of many of these viral, cellular or chemical factors. In addition, proteins encoded by a variety of herpesviruses as well as other DNA viruses also augment expression from the HIV enhancer, using pathways distinct from NF-κB or the trans-activating gene product of the HIV *tat* gene (69, 79).

Indeed, control of the rate of transcription is one of the primary strategies for selective expression of genetic information. It can arise from specific interactions between a regulatory protein and DNA sequences proximal or distal to the locus being expressed. The following section will describe the molecular biology of various HIV transcription units thought to be involved in the control of viral latency, as a foundation for investigation of how herpesviruses might interact with them.

Transcription Initiation

The HIV genome encodes three genes, *tat* (formerly tatIII), *rev* (formerly art/trs), and *nef* (formerly 3'-orf) whose products regulate the rate of RNA initiation and viral gene expression. cDNA clones corresponding to the same transcription units in SIV/mac (80), and to *rev* in visna (81), have also been obtained, although only the function of the *tat* equivalent has been studied in SIV.

*Nef*, negative regulatory factor, encodes a 27kd membrane protein, possibly anchored to the inner surface of the plasma membrane via myristylation of its amino terminus (82). It can be phosphorylated either through its own protein kinase (PK) activity or by PKC (82). Similarity has been noted between *nef* and sequences in $pp60^{src}$ and c-Ha-ras and, like the latter, it may bind and hydrolyze guanosine triphosphate (82). *Nef* appears to negatively regulate HIV replication, as deletions lead to higher viral titers in tissue culture (83). Two mechanisms have been proposed: interference with cellular signalling (12); or suppression of transcription initiation through upregulation of trans-acting factors which interact with sequences either in the NRE of the LTR (84) or apart from this region (85).

The physiologic role of *nef* is unclear. Mutations that inactivate *nef* or factors that repress its expression or action could be important in inducing HIV from a latent to an actively replicating stage. It appears that the sequence of *nef* is highly variable, and partial or complete attenuation occurs in some strains (85), albeit the possibility that these mutations are induced by the constraints of tissue culture must be entertained.

It has recently been shown that another HIV gene, *rev*, can relieve *nef*-mediated inhibition of transcription from the LTR (86). *Rev*, regulator of virion protein expression, is a bipartite gene which encodes a 19kd nuclear protein through sequences that overlap with, but are in a different reading frame from, those of *tat* (87). As discussed in detail later, *rev* is required for viral replication and for the expression of long HIV transcripts encoding structural (*gag*, *pol* and *env*) precursor polypeptides. However, *rev* may also decrease the rate of RNA initiation through an interaction with NRE sequences (29).

*Tat* is a bipartite gene which encodes a 14kd nuclear protein containing an essential stretch of nine cysteine residues followed by a region of basic amino acids (88). In solution it forms a non-covalently linked dimer capable of binding divalent cations (89). It is essential for HIV replication, with the *cis*-acting response elements (*TAR*) for HIV-1 and 2 described above. The precise mechanism of trans-activation by *tat* has not been defined. Nuclear run-on experiments have suggested a role in RNA initiation (90,91), but the major effects of *tat* appear to occur post-initiation. Using a Xenopus system for microinjection of *tat*, its activity was not altered by transcriptional inhibitors, as it appeared that *tat's* major effect was in potentiating RNA translatability in the nucleus (92). By interacting with *TAR*-containing RNA, *tat* could facilitate transcriptional elongation, increase mRNA stability, and enhance translation. The predominance of each effect might depend upon quantitative differences in *tat* and quantitative as well as qualitative differences in host cellular factors (12).

Splicing, and the importance of rev

Even retroviruses that make a single spliced subgenomic RNA must regulate the ratio of spliced RNA and its precursor, as the latter is used both as genomic and messenger RNA (8). HIV, as other lentiviruses and the oncornaviruses HTLV-I and II, have more complex splicing patterns, encoding viral proteins that influence the ratio of spliced to unspliced transcripts.

The same initiation codon and mRNA is used for expression of capsid and polymerase products: *gag-pol*, from which the core antigens [p55 and its cleavage products, p15, p24 (major) and p18 (myristylated)] and polymerase (p66/p51) are produced. However, suppression and ribosomal frameshifting, specified by consensus sequences present in viral RNA (93), occur with a frequency of about 5% as the ribosome transits the junction between the capsid and *pol* genes, so that *gag-pol* polyproteins are about 5% as abundant as *gag* proteins (29). This translational control, not found among cellular genes, clearly benefits the retrovirus, as structural proteins can be made in large amounts and catalytic proteins in small amounts, to be later incoporated into viral cores via *gag* components (8).

mRNA for the envelope proteins is spliced to remove capsid coding sequences. However, HIV also encodes six other proteins, each of which is specified by a separately spliced mRNA, and two of these (*tat* and *rev*) are encoded by two noncontiguous regions (29). The complex splicing pattern which results provides great flexibility in the regulation of HIV gene expression. One of the major units responsible for this process is *rev*. In the absence of *rev*, doubly-spliced viral RNA species about 2kb in length accumulate (12,45). These are also the first viral transcripts to appear after infection. It takes about 12 h more before steady state expression of the three major transcripts (9.2, 4.3, 2kb) is achieved (45). Indeed, a controlled shift from the early to late transcriptional phases detected *in vitro* may be the equivalent of the transition from latent to acute infection *in vivo* (45).

The exact details of this shift are still being explored. *Cis*-acting repression sequences (CRS), if present on mRNAs for *gag* and *env*, make the long RNA species very unstable (29). They will be translated efficiently only in the presence of a *rev* product (94). RNA made in the absence of *rev* corresponds only to multiply spliced species from which the regulatory proteins (*tat*, *vif*, *nef*, and *rev*) are made. It is unclear how CRS sequences prevent translation. The *rev* protein requires a target sequence known as the *cis*-acting anti-repression sequence (CAR), or REV responsive element (RRE), which is composed of about 400bp in *env* (12,94). Only multiply-spliced RNA transcripts accumulate in viruses with CAR mutations (29). *Rev* does not directly regulate viral RNA splicing, but *rev* protein appears to activate the nuclear export of a constitutively expressed pool of incompletely spliced viral RNAs that are normally excluded from the cytoplasm (95). It is possible that by confining expression of HIV structural proteins to a relatively brief period, *rev* diminishes the capacity of the host's immune system to identify and neutralize these infected cells prior to the release of viral progeny (95).

In addition, as noted above, a *nef-rev* interaction also occurs. Negative regulation of viral protein expression by *rev*-associated *cis*-acting sequences is dependent upon the normally low intracellular concentration of RNA, implying that viral RNA interacts with cellular factors that are limiting (29). The negative feedback control exerted by *nef*-LTR interactions would be a potential buffer against excessive *tat* function early in infection. The decay of *nef* following *rev* expression could ensure sustained viral activation by unhindered *tat*.

In this model (86), during latency the LTR would be regulated by homeostatic mechanisms whose effects might be counterbalanced by *nef* under limited *tat* expression. External stimuli, including protein kinase activators (chemicals, lymphokines, viruses) which are synergistic with *tat* (72) and act through NF-κB as well as other enhancer sequences, would then upregulate HIV. However, the relative importance of *nef* in this scheme is still controversial, as its inhibitory effects on transcriptional activity of the HIV LTR may be of lesser magnitude then previously determined (95).

Monocytes: A special problem?

Most of the work described above has been performed with actively replicating cells or cell lines of the T lymphocyte or promonocytic cell lineages. Monocytes, found in the peripheral blood or tissue-associated, are nondividing cells derived initially from proliferating promonocytes in the bone marrow (96). They would thus not normally be expected to integrate provirus. It has recently been suggested that persistent HIV-1 infection of such cells may involve accumulation of unusual, high molecular weight, extrachromosomal forms (97). These DNA molecules contain multiple copies of the viral genome, probably formed by sequential rounds of LTR-mediated integration of one circular DNA form into pre-existing structures to form "nested self-integrates." It is hypothesized that these genomes replicate as DNA, without going through an RNA intermediate (97). Whether this model, derived from an infected U937 cell line in which virus replication is linked to *c-myc* expression (98), is applicable to nontransformed monocytes must be determined.

Thus far we have outlined the central events in the replicative cycle of HIV: virion receptor binding and entry, integration, and regulated expression. Virus production begins at about 24 h post-infection (45). [Assembly of the mature virus particle is a final and also complex step (29), but is tangential to our review]. We will now explore the relationship of herpesviruses to the regulation of HIV, and their potential involvement in HIV-induced pathology.

**LATENCY: A CLINICAL PERSPECTIVE**

The average interval between HIV infection through sexual intercourse or blood transfusion and development of clinical AIDS is 10 and 8 years, respectively (99,100). [Infants receiving infected blood may have a much more rapid incubation period, however, with a mean interval $<2$ years (100)]. While most individuals develop a serologic response to HIV within 1 week to 9 months after exposure, others may not seroconvert for 3 or more years (101). Indeed, at least three patterns of response to HIV infection have been described (29). The most common is a prolific viremia within the first 6 weeks after exposure, followed by low levels of circulating virus in the presence of continuous antibody production over a prolonged clinical incubation period. Alternatively, infection may occur in the absence of a detectable serologic response or, in rare instances, seropositivity is followed by loss of antiviral antibody despite continual, low level viremia. In all three instances, HIV replication is minimal; amplifying the mechanisms which maintain this depressed level of viral activity is a major goal in AIDS therapeutics.

Quantitation of HIV infection *in vivo*

Early in the course of HIV infection, absolute numbers of $CD4^+$ T lymphocytes may be within normal limits, yet many measures of cellular immunity are markedly perturbed (102). This was initially puzzling as, to the limits of *in situ* hybridization, $<1$ in 10,000 circulating or tissue-associated T and B lymphocytes and monocytes were reported to contain HIV nucleic acid (103). New data indicate, however, that much greater numbers of cells are chronically infected *in vivo*, persisting as stationary cell intermediates from which HIV may be rescued by a variety of seemingly disparate stimuli. Indeed, T lymphocytes (104) and macrophages (105) harbor HIV *in vivo* with several log-fold greater frequencies when molecular hybridizations for both structural and regulatory genes are pursued.

As indicated in the review of the HIV life-cycle, the slow but relentless progression of immunologic dysfunction and clinical symptomatology characteristic of HIV infection is not due to HIV being an intrinsically slow-growing retrovirus, but rather to the complex nature of viral gene regulation, virus-cell, and infected cell-host interactions. HIV regulatory genes facilitate a rapid transition from latent infection to virion release in a manner that may minimize the opportunity for an effective host response (95).

Infection of resting cells may produce one form of HIV latency, maintained in the absence of necessary cellular factors, with the majority of proviral DNA unintegrated and stably maintained for prolonged periods (106). Encounters with appropriate stimuli could then result in cell proliferaton and a permissive infection. Return of these cells into $G_0$ may then establish a second latent form, integrated but quiescent HIV provirus (95). Subsequent exposure to antigen, cytokines, or heterologous viral trans-activators would promote reactivation of the virus (95).

By this scheme, *in vivo* activation signals perform two critical roles: conversion of a latent or chronic infection to a productive one, and generation of a susceptible population of target cells to facilitate viral spread (107). Although it is unclear precisely when the cellular immune network becomes so disabled that it cannot block establishment of the opportunistic infections characteristic of AIDS, it seems likely that the process of continued activation of HIV expression in latently infected cells contributes to this pathology and increases the viral burden by allowing further spread to uninfected cells.

How these signals are exogenously provided or triggered *in vivo* is unclear. The relevance of paradigms established *in vitro*, usually with immortalized cell lines, to clinical disease is based upon epidemiologic data and experimental therapeutics. In terms of cytokines, both interleukin-2 (IL-2) (108) and GM-CSF (granulocyte-monocyte colony stimulating factor) (75,109) can increase the production of HIV *in vitro. In vivo*, IL-2 administration to AIDS patients has been linked to the rapid development of serious bacterial and opportunistic infections (110), whereas no such deleterious effects, either in terms of symptomatology or altered viral burden, have been reported with GM-CSF (111).

Clinical data for herpesvirus infections in relationship to HIV latency are difficult to evaluate, as there is a high incidence of seropositivity for most of these agents among both HIV infected and control populations (Table 2) (112). Attempts to form such correlations are summarized below.

Cytomegalovirus

CMV infections are controlled by the immune system and lead to viral latency in the immunocompetent host. CMV normally persists in salivary and renal tubular epithelial cells, as well as peripheral blood mononuclear cells. By

**Table 2**. Prevalence of Herpesvirus Antibodies in HIV Seronegative Individuals*

| Age category (yr.) | Anti-viral Antibodies[a] | | | |
|---|---|---|---|---|
| | CMV | EBV | HSV | VZV |
| 5-20 | 38 | 67 | 33 | 88 |
| 21-30 | 42 | 82 | 50 | 97 |
| 31-40 | 52 | 82 | 69 | 99 |
| 40-64 | 67 | 100 | 100 | 100 |

* Data taken from reference 112.
[a] % seropositive

*in situ* hybridization for CMV-specific RNA, 0.5-2.0% of $CD4^+$ T lymphocytes and monocytes may harbor CMV (113). In permissive cells, CMV is regulated in a cascade fashion characteristic of all herpesviruses, with proteins expressed in the immediate-early (IE) phase of infection, such as the regulatory phosphoprotein pp89, required for subsequent expression of early and late phase gene products (114). This permissive state appears to be maintained as CMV gene products interfere with presentation of pp89 for recognition by MHC class I restricted cytolytic T cells, without altering expression of immune association molecules themselves (114).

Normally, acute CMV infection leads to a transient cellular immune suppression, with a relative decrease in $CD4^+$ T lymphocytes (115). In the immunocompromised, such as individuals undergoing tissue allografts, CMV-induced immune suppression facilitates development of opportunistic infections, and perhaps malignancies. (116). Indeed, more than 90% of HIV seropositive individuals show evidence of CMV infection (117,118). It is a significant cause of mortality and morbidity, including retinitis, dementia, colitis, glomerulopathy and pneumonia, and is the third most common form of opportunistic infection in AIDS, following *Pneumocystis carinii* pneumonia and *Candida* spp. infection (118). Although the incidence of CMV seropositivity among age-matched, sexually active controls is also high (Table 2), evidence for active CMV infection in AIDS, representing either newly acquired disease or reactivation of latent infection detected by viral isolation from blood, urine or saliva, is extraordinary. It ranges from 39-97%, depending on the clinical stage of HIV infection (119).

There is no significant difference in the prevalence of CMV antibodies among individuals with AIDS in the U.S. or Africa when compared to matched African heterosexual or U.S. homosexual controls (120), nor do antibody titers to CMV differ between homosexual men who did or did not subsequently become infected with HIV over a 12 month surveillance period (121). However, anecdotal reports do suggest that concurrent infection with HIV-1 and CMV produces an atypically severe syndrome (122). In one study of 108 HIV-infected hemophiliacs, the risk of advanced HIV disease in CMV-seropositive individuals was 2.5 times that of a similar CMV seronegative cohort ($p=0.02$) (123), albeit this association was not found in a smaller survey of HIV-infected hemophiliacs (124).

Epstein-Barr virus

EBV persists in a latent form in both epithelial cells and B lymphocytes, but reactivates frequently, perhaps continuously, in the former, resulting in oral shedding of infectious virus (125). The major reservoir for latent virus is still controversial, however; bone marrow eradication and replacement studies suggest a hematopoietic source, most likely the small B lymphocyte (126). EBV shows a narrow target range, with *in vivo* infection restricted to mature resting B lymphocytes bearing the complement receptor 2 (CR 2 or C3d, also known as CD21 and B2), and epithelial cells (127).

EBV infected cells express the B cell activation antigen CD23 (Blast-2) and show increased RNA synthesis by 24 h, with expression of CD23 predictive of subsequent proliferation, at 2-3 days post-infection (128). In the latent state, viral DNA is present as a covalently closed, full-length circular episome (CCC). Circularization of a single genome is crucial to the establishment and maintenance of this state (128). However, in certain long-term B cell lines, EBV may persist as a complete integrated genome or as both integrated and episomal DNA (129).

At least two EBV products are synthesized during persistent infection: EBNAs 1-6 (including EBV nuclear antigen-1, a 75kd DNA binding protein essential to maintain expression of EBV in the immortalized cell, and EBNA-2, a 87kd DNA binding protein which upregulates the CD23 autocrine growth receptor of B cells and may be required for LMP expression and cell immortalization); and LMP (latent membrane protein, a 63kd oncoprotein) (125,130). Evasion of host immune surveillance may be a consequence of the ability of EBV to downregulate critical cell adhesion molecules such as LFA-3 and ICAM-1, leaving inadequate membrane targets for cytotoxic T lymphocytes (131).

In the normal EBV seropositive host, *in situ* hybridization (132) and cell transformation assays (133,134) indicate that 1-2 per $10^6$ B cells are EBV-infected, as opposed to >0.1% during the acute stages of EBV-linked infectious mononucleosis (125). In contrast, symptomatic HIV infected individuals have 6-21 infected cells per $10^6$ circulating B lymphocytes (133,134). If the frequency of EBV infected cells is deduced from the numbers spontaneously producing polyclonal immunoglobulin, however, the latter number is three log-fold greater (133,134), approaching the 0.1% seen in acute EBV infections. This estimate concurs with *in situ* hybridization data on lymph node biopsies from AIDS patients (Table 3) (135). Reflecting this enhanced viral burden, elevated titers of anti-EBV immunoglobulins (136) and increased EBV DNA in oropharyngeal secretions (137) are found in HIV infected individuals. However, as was the case for CMV, there is no difference in prevalence of EBV antibodies among African and U.S. patients with AIDS in comparison to controls (120), nor do antibody titers to EBV differ among those men who either acquired HIV or who remained uninfected (121).

Normally, EBV infection results in a self-limiting lymphoproliferative disease, associated with a relative decrease in $CD4^+$ T cells similar to that seen in CMV mononucleosis. In certain genetic and acquired immunosuppressed states, however, EBV infection may lead to fatal polyclonal B cell proliferations or malignant lymphoma (125). Indeed, AIDS patients are at increased risk for development of aggressive B cell lymphomas of the Burkitt's and non-Burkitt's type, both associated with EBV (138), as well as fatal polyclonal B cell lymphoproliferative disorders (139). Earlier stages of HIV infection are characterized by an unusual lesion, oral hairy leukoplakia, caused by relentless replication of EBV in epithelial cells of the oropharynx (140), as well as generalized lymphadenopathy (LAS). The latter occurs in >60% of individuals, usually within the first six months after HIV infection, and has the histopathologic features of benign follicular hyperplasia (141).

Multiple clonal B cell expansions are present in some 20% of LAS and 60% of lymphoma biopsy samples from HIV infected individuals (141). *C-myc* rearrangements were detectable in 9 of 10 HIV-associated lymphomas, but in none of 22 LAS biopsies (142). HIV genome was not detectable by Southern

**Table 3.** Frequency of herpesvirus DNA and HIV provirus in lymph nodes from HIV-infected individuals*

| Virus | Probe | Positive nodes (%) (n=25) | Frequency of positive cells |
|---|---|---|---|
| HIV | pBH10-R3 | 88 | $10^{-4}$ - $10^{-5}$ |
| EBV | Enzo | 60 | $10^{-3}$ - $10^{-4}$ |
| CMV | Enzo | 16 | $10^{-3}$ - $10^{-4}$ |
| HHV-6 | pZVH1Y | 64 | $10^{-2}$ - $10^{-5}$ |

* Data derived from reference 135. Frequency determined by *in situ* hybridization using biotinylated probes.

analysis in any tumor specimens (142). Tumor karyotypes often reveal at (8;14) translocation, identical to endemic African Burkitt lymphoma (143,144), even though EBV genomes are often absent (145). Less frequently, pleomorphic malignant lymphomas may be seen in AIDS which lack *c-myc* rearrangements yet do contain EBV genome (146). Examination of two cases of AIDS-associated lymphoma revealed presence of EBNA-1 in one and EBNA-1 and -2 in the other, but complete absence of LMP in both, a pattern typical of non-HIV-associated Burkitt lymphoma cells (147).

Herpes simplex viruses

The human neurotropic herpesviruses, HSV-1, HSV-2, and VZV (varicella-zoster), establish a lifelong latent infection as an episome primarily in peripheral sensory ganglia (148). HSV can also infect and remain latent in the same types of cell that are infected by HIV, the monocyte/macrophage (149) and T lymphocyte (150).

In contrast to CMV and EBV, serologic studies suggest that HSV-2 infection is a risk factor for subsequent or concurrent HIV infection (121), but these data give little indication of mechanism. Indeed, HSV-2 may facilitate transmission of HIV, as it causes visible blistering, erosion and bleeding of penile, rectal and vaginal mucosa, and HIV has been isolated from such genital ulcers (151).

Human herpesvirus 6

HHV-6 is a novel human herpesvirus, initially referred to as human B-lymphotropic virus (HBLV) because of its association with certain lymphoproliferative disorders (152). However, despite the fact that it can infect myriad types of hematopoietic and neuroglial cell, it is primarily tropic and cytopathic *in vitro* for $CD4^+$ T lymphocytes (153). In the general population, there is a seroprevalence of 26%, with only 17% of those individuals recalling clinical symptoms characteristic of HHV-6 infection: fever, pharyngitis, skin rash, lymphadenopathy, and abdominal complaints (135). In one survey, <1% progressed to a "chronic fatigue syndrome" which at times has been associated with other members of the herpesvirus family (135).

In contrast, 82% of HIV seropositive individuals with advanced HIV infection [CD4:CD8 T cell ratios below 0.2 (normal $\geq$1.5)], and 44% of such individuals with less progressive disease (CD4:CD8 ratios 0.2 - 0.6), were HHV-6 seropositive (135). *In situ* hybridization of lymph nodes from HIV-associated LAS revealed HHV-6 in 88%, with an estimated frequency of positive cells of about 0.01% (Table 3). This contrasts with the fact that while some 60% of individuals with malignant lymphoma are HHV-6 seropositive, only 2% (3/246) had HHV-6 DNA in their lymphoma cells (154).

While these latter data suggest that HIV infection may activate latent HHV-6 infection, another survey among HIV seropositive blood donors, most of whom were homosexual, showed no difference in HHV-6 antibody titers when compared to controls (155).

## HERPESVIRUSES AND HIV LATENCY: MOLECULAR MECHANISMS

Acute or reactivated herpesvirus infections may contribute to induction of HIV expression, and potentially accelerate the clinical course of HIV disease, by at least three divergent means. A common motif among these pathways is the promiscuous cellular transcriptional regulator, NF-κB, mentioned previously in context of the HIV LTR. NF-κB is a 50kd protein with DNA binding capability, often complexed to a 65kd protein, IκB, which inhibits its activity. Nucleoside triphosphates dramatically stimulate the capacity of NF-κB to bind DNA *in vitro*, while phosphorylation of IκB by PKC may release the active species from its inactive cytosolic form and permit translocation of NF-κB to the nucleus (70). Induction of NF-κB, which does not require new protein synthesis, plays a vital role in T cell activation, cytokine regulation, and transcription of viruses with NF-κB binding sites such as HIV, SIV, and CMV. How it interacts with other transcriptional activators in the HIV-herpesvirus conundrum is an active area of investigation.

1. Soluble factors

Immunologic mediators affect genes that regulate cell proliferation and differentiation. The action of these peptide hormones, or cytokines, at the surface membrane leads to the activation of factors in the nucleus that modulate transcription (156). Sequences resembling NF-$\kappa$B are also associated with immune recognition molecules. These include enhancers for class I MHC antigens, $\beta$2-microglobulin, IL-2, and the IL-2 receptor $\alpha$ chain (CD25 or Tac) (156).

Herpesviruses may provoke HIV expression by induction of certain of these monokines or lymphokines during the course of a normal immune response to infection. Chronically infected cells of the U937 promonocytic (157) and A3.01 $CD4^+$ T lymphoblastoid (158) lineages, as well as non-transformed peripheral monocytes, are susceptible to rescue of HIV following exposure to several such cytokines (Table 4). When human monocytes were cultured in the presence of heat-inactivated HSV-1, HSV-2, HHV-6, VZV (Hall strain), EBV (B95-8), or CMV antigens (1$\mu$g/ml) for 24h, supernatants harvested and incubated with chronically HIV-infected promonocyte or T lymphoblastoid cell lines, virus was induced by the EBV and CMV-supernatants (containing 100 pg/ml and <10 pg/ml, respectively, of endotoxin), but not by the remaining samples (162). HIV-1 can itself induce secretion of cytokines which upregulate latent HIV infection (Table 5).

The exact molecules involved in these *in vitro* and *in vivo* phenomena related to herpesvirus induction of HIV are unknown. TNF (tumor necrosis factor) levels are minimally elevated, if at all, in EBV, CMV, or HSV-1 infections (171), although it is not known whether these herpesviruses might synergize with HIV to further elevate TNF-$\alpha$ levels associated with HIV infection. IL-1, IL-2, and $\gamma$-interferon are possible mediators.

2. Stress responses

Herpesviruses may induce a stress response, with production of factors which subsequently stimulate HIV transcription. For example, heat shock can stimulate both the HIV LTR (172) as well as promoters for the major CMV IE transcription unit (173).

3. Co-infection of a single cell with potential for mutual trans-activation

Like HIV, all herpesviruses encode transcriptional trans-activators of viral gene expression and divide their replication cycle into eary (regulatory) and late

Table 4. Cytokines of the Immune System and HIV Activation

| Cytokine | Reference | Model system | HIV activation | Effect/ mechanism |
|---|---|---|---|---|
| IL-1 | 156 | 70Z pre-B cell, LBRM murine T cell | + | NFkB |
| IL-1 | 161 | early infection of PB monocytes* | - | |
| IL-2, IL-4 | 156 | various cell lines HIV-LTR-CAT activity | - | |
| IL-2, IL-4, IL-6 | 161 | early infection of PB monocytes | - | |
| IL-4 | 159 | early infection of PB monocytes | - | |
| IL-3 | 109 | early infection of PB monocytes | + | increased viral replication |
| IL-3 | 156 | various cell lines HIV-LTR-CAT activity | - | |
| GM-CSF | 156 | various cell lines, HIV-LTR-CAT activity | - | |
| GM-CSF | 75 | chronically infected U937 line | + | increased viral replication |
| | 109 | early infection of PB monocytes | + | |
| gamma-1FN | 109 | early infection of PB monocytes | + | increased viral replication |
| TNF-α | 156 | Jurkat T cells, 70Z pre-B cells U937 promonocytes transfected with HIV-LTR-CAT | + | 8-fold increase in CAT activity linked to NFkB |
| TNF - α | 160 | chronically infected CEM T cell line | + | increased viral replication |

*Peripheral blood monocytes/macrophages.

**Table 5.** Cytokines Upregulated in HIV Infection

| Cytokine | Elevated levels in serum, CSF, or peripheral mononuclear cells (refs) | Upregulated by HIV *in vitro* (refs) |
|---|---|---|
| TNF $\alpha$,ß | + (163,164) | + (223) |
| IL-1 | + (165,170) | + (166) |
| IL-4 (BCGF-I) | + (167) | + (168) |
| IL-6 (BSF-2) | + (170) | + (169) |

(structural) phases of protein synthesis (95). However, because HIV contains only one promoter, the LTR, while the herpesviruses utilize multiple stage-specific promoters, understanding potential reciprocal regulation of gene expression among these agents requires a different perspective.

Two caveats must be considered in interpreting the following data. First, it is important to establish the feasibility of direct co-operation at the single cell level for a herpesvirus and HIV in order to invoke this mechanism. This constraint appeared quite challenging when the frequency of HIV infected circulating or tissue-associated cells appeared to be on the order of 0.01% (103). However, the realization that many more cells harbor latent HIV *in vivo* (104,105), and the fact that at least one HIV trans-activation protein, *tat*, may be directly taken up by cells and function as a soluble mediator (174), make this a reasonable hypothesis. Second, the systems established for examining herpesvirus trans-activation of the HIV LTR typically involve a single reporter gene, bacterial chloramphenicol acetyltransferase (CAT). This model can lead to spurious interpretations. For example, one herpesvirus gene product, derived from the EBV BMLF1 ("M") IE gene, was recently reported to stimulate the HIV promoter (175). However, when growth hormone or $\beta$-galactosidase were used as a reporter gene in lieu of CAT, BMLF1 no longer functioned (176,177). The CAT effect was post-transcriptional, and not directed at promoter gene activation. It calls into question not only the interpretation of the HIV data, but the designation of BMLF1 as a promiscuous transactivator (176). Similar questions need to be raised concerning the ability of molecular clones of several miscellaneous viruses, including the papovaviruses JC, BK and LPV, to stimulate the HIV promoter (178).

Cytomegalovirus

Using double-labeling *in situ* techniques it was shown that CMV and HIV can co-infect the same cell, at least in the central nervous system, and that this is a common event, occurring with a frequency >10% (179). In one *in vitro* study, CMV superinfection of a $CD4^+$ T lymphoblastoid cell line persistently infected with HIV led to a small percentage of cells producing both viruses simultaneously (180). Within 10 days of exposure to CMV, the chronic HIV infection was converted into a highly lytic process. Similarly, another group reported that coinfection of H9 $CD4^+$ T lymphoblastoid cells with CMV and HIV led to a greater number of CMV positive cells per well and prolonged CMV antigen expression (73). It was hypothesized that *tat* enhanced the productive CMV infection, although transfection of CMV-infected T or human embryonic lung cells with *tat* did not alter the amount of CMV DNA that was detected (73). CMV exposure also enhanced HIV expression, with 0.5-2 log-fold increases in HIV titer, in transformed T cells and monocytes (73).

Reciprocal trans-activation mechanisms were explored in these models. The HIV LTR and the CMV IE promoters share binding sites for SP1, AP1 and NF-$\kappa$B, in addition to more common transcriptional activators (Fig. 2) (181). The presence of common *cis*-acting elements suggests common mechanisms of regulation, particularly involving four regions in CMV which are putative NF-$\kappa$B recognition sequences (181). CMV is able to transactivate the HIV LTR when either CMV permissive (MRC-5 human fibroblasts) or non-permissive (LTK murine fibroblasts) cells were transiently transfected with HIV-1-LTRcat (182). The effects of CMV (AD169 strain) were observed as early as 8 h after infection, even in LTK cells, which express only IE and early CMV antigens (182). The latter control was important, as human lymphocytes and monocytes are not fully permissive for CMV replication (183). Enhanced LTR activity was paralleled by an increase in CAT mRNA, and was not seen with the MMTV murine LTR nor with a promoter from hepatitis B virus (182).

The fact that IE and E CMV antigen expression is sufficient for HIV LTR trans-activation has been confirmed (184). Similarly, CMV infection stimulated the LTRs of HIV-1 and -2 in U-373MG, a CMV permissive human astrocytoma cell line (185).

Herpes simplex and Varicella-zoster viruses

Over a dozen papers have appeared describing the ability of the neurotropic herpesviruses to trans-activate HIV-1 and -2. All have used a chimeric construct including the LTR fused to the cat gene in a transient transfection of a variety of types of immortalized cell. Thus, enhanced CAT expression was seen in HeLa cells and human lung fibroblasts infected with HSV-1, HSV-2, and VZV (186); Vero simian fibroblasts with HSV-1 (74,187); and human rhabdomyosarcoma and colon carcinoma cell lines with HSV-1 and VZV (178).

Potential HSV gene candidates have been mapped using transfections with three IE transcription units known to be involved in regulation of HSV expression. ICP0 (188,187) and ICP4 (187) can transactivate the HIV promoter in transient assays in nonlymphoid cells. Recently, however, an acute co-infection system in CEM $CD4^+$ T lymphoblastoid cells was used to examine interactions between HIV and HSV-1 (189). Co-infected cultures demonstrated a marked enhancement of HIV replication. By day 4, $>10^4$ pg/ml of HIV p24 antigen was produced, as opposed to undetectable levels in cells exposed to HIV alone. Once induced, however, HSV had little effect on the rate of HIV replication. HSV-1 mutants with deletions in ICP0, ICP4, and ICP27 were then used to superinfect CEM/HIV cells. ICP0 has been implicated in the reactivation of HSV from the latent state, but is not essential for viral growth, while ICP4 and ICP27 are essential genes. Deletion of ICP0 or ICP27 had no effect on the ability of HSV to induce HIV replication; the ICP4 mutant failed to augment HIV expression (189). It thus appears that functional ICP4, but not HSV replication, is required for potentiation of HIV. It also indicates that VP16, which upregulates IE genes together with cellular elements through *cis*-acting sequences (TAATGARAT) in IE promoters, is insufficient for rescuing HIV, as VP16 is present in the ICP4 mutant (189).

These findings are in sharp contrast to those of two earlier studies (188,190) in which ICP0, but not ICP4 alone, induced the HIV promoter. A fourth group (187) found that ICP4 could upregulate HIV-LTRcat activity in the absence of *tat*, while ICP27 had no effect and ICP0 much less activity. Whether these experimental discrepancies are dependent upon the types of cells used must be determined. Efforts to identify specific *cis*-acting signals responsive to ICP0 in the HIV-LTR were unsuccessful. Activity persisted despite mutation of NF-$\kappa$B sites (188); those sites for ICP4 have not yet been mapped (189).

The mechanism of HSV transcriptional effects on HIV may involve binding of an HSV gene product directly to sequences in the HIV LTR, or an event mediated through cellular transcription factors such as NF-κB. HIV contains an NF-κB element in its LTR (78), which may be similar to the TAATGARAT elements in HSV IE promoters (191). Exonuclease protection and gel retardation assays have shown that HSV-1 mediated activation of the HIV promoter results in induction of a nuclear factor(s) which binds to NF-κB and Sp1 sequences, and that both sites may be required for full activation (192).

Finally, the possibility of a reciprocal effect of HIV on HSV gene regulation was explored. Plasmids containing promoter regions for HSV-1 $\alpha$ and HSV-2 $\beta$ proteins fused to cat were cotransfected with a *tat* plasmid into HeLa and human lung fibroblast cells (186). No enhancement of CAT activity was seen. In order to definitively address this issue, however, transfections with whole HIV provirus are required.

Human Herpesvirus-6

Two-color immunofluorescence studies have demonstrated that HHV-6 and HIV-1 can co-infect an individual $CD4^+$ T lymphocyte (193). Using standard HIV-1-LTRcat constructs transfected into two different $CD4^+$ T lymphoblastoid cell lines, HHV-6 infection was also shown to augment HIV promoter activity (194). The mechanism is unknown.

Epstein-Barr virus

Both EBV and HIV can co-infect human B lymphocytes (195). For many reasons, potential interrelationships between EBV and HIV are the most intriguing of any of the herpesvirus/AIDS interactions. Documentation of the effects of EBV gene products on HIV has not been straightforward, however. As noted above, it was initially reported that an EBV IE gene product, BamHI MLF1, stimulates expression of HIV-LTRcat, and that HIV promoter sequences necessary for trans-activation by EBV do not include the *tat*-responsive sequence, *TAR* (175). Another EBV IE gene, BamHI/ZLF1, as well as the late gene EBNA-1, had no effect in this system (175,196). However, BMLF1 increased HIV-CAT protein activity more than HIV-CAT mRNA, and it was shown in a later experiment that the effect was reporter gene dependent (176). This indicated that the BMLF-1 effect on HIV-LTRcat was biologically irrelevant, and called into question all studies of the effects of BMLF1 on heterologous promoters where CAT was used as an indicator (176).

Recently a third IE gene, BRLF ("R") (197), has been shown to transactivate the HIV-LTR in a reporter gene independent manner (177). Its mechanism of action is unknown, but is not limited to a specific *cis*-acting element, and is independent of *tat* (177).

The possibility that HIV trans-activators might reciprocally affect EBV was also explored (175,196). The B lymphoblastoid cell line BJAB was transfected with a *tat* plasmid in the presence of five different EBV promoters linked 5' to the cat gene in a pCAT3M vector: EBNA-2 (bp 44,183-45,084); LMP (169,924-169,476); BZLF1 (103,625-103,181); BLLF1 (92,703-92,160); and BMLF1 (84,770-84,333). No activation of any of these promoters was seen, despite the fact that *tax*, the HTLV-I equivalent of *tat*, could transactivate EBNA-2 (196). The latter effect occurred only in B cells, and was dependent on a specific sequence in the EBNA-2 promoter (196). This is curious as HTLV-I is primarily associated with T cell malignancies, albeit an indirect role for HTLV-I in B cell leukemogenesis was suggested by the finding of chronic lymphocytic leukemia cells capable of producing anti-HTLV-I antibodies (198).

Finally, the involvement of EBV in HIV-associated B cell lymphomas and Hodgkin's disease is also being explored on the level of single cell interactions. An obvious hypothesis is that HIV-mediated immune suppression permits reactivation of latent EBV infection, followed by uncontrolled polyclonal proliferation of B lymphocytes and, eventually, a *c-myc* rearrangement. Indeed, there are data for dysregulation in the control of EBV late in the clinical course of HIV infection (199). However, lymphadenopathy typically arises, and persists, <6 months after initial HIV infection, a point at which cellular immune suppression, at least as measured by *in vitro* assays, is minimal. At this stage, EBV-HIV interactions may be pertinent. They are being explored using B cell lines derived from HIV infected donors.

Peripheral B lymphocytes from EBV seropositive individuals are polyclonally activated upon exposure to replication competent HIV (200) or inactivated HIV (201), with peak DNA synthetic responses on day four following exposure. In addition, EBV-infected cord blood B cells can be easily infected with HIV, while EBV negative cord blood cells cannot (196). This occurs despite the fact that levels of the high affinity HIV receptor, CD4, do not differ in the presence or absence of EBV, and may be undetectable by indirect immunofluorescence and cytofluorimetry (196). Following immortalization, however, all human B cell lines can be infected with HIV-1 or -2, regardless of the presence of EBV. Integrated HIV can be detected in these long-term, non-HIV producing B cells (202). These data are in contrast to an earlier study, in which only those B cell lines containing transforming (B95-8) or non-transforming (P3HR-1) EBV could be infected (203), a difference potentially related to divergent tropisms of the HIV stocks used (202). Infection of all such lines is susceptible to blockade by anti-CD4 monoclonal antibodies, even in those cells lacking detectable membrane CD4 (204). These latter cells appear to have CD4 mRNA, however (205).

There are conflicting reports as to whether EBV positive B cell lines superinfected with HIV-1 do (202,203,206) or do not (207) produce more EBV. The phenomenon appears unrelated to the capacity of a particular HIV-1 strain to induce cytolysis or persistent infection in a particular cell (207). In one study, reactivation of EBV by HIV was virus dose-dependent, required viral expression, and resulted in only transient induction of EBV EA and VCA antigens in some 20% of the cell population, even though virtually all cells contained HIV RNA (206). The effect of HIV was synergistic with the PKC activator PMA (206). It also appeared that these cells could be repeatedly infected with HIV, resulting each time in a new cycle of EBV induction (206). Normally, HIV-1 infected T lymphocytes cannot be superinfected because of the down-regulation of surface CD4 shortly after infection (208).

Spontaneous, EBV positive cell lines can be easily established from HIV seropositive individuals (204,209). In one study (204), no evidence for HIV-specific proteins could be detected, while in another (209) HIV could be rescued from all of four such lines established. The relevance of these immortalized lines to HIV gene expression is unclear. Oncornaviruses may reactivate latent EBV, as the gibbon ape/simian sarcoma virus can enhance the outgrowth of EBV transformed B cell lines from EBV seropositive donors (210,211). Following such outgrowth, these karyotypically normal lines also produced tumors in nude mice, a phenomenon unusual in spontaneous transformants (210). Isolation of infectious EBV from supernatants of three spontaneous B cell lines derived from an AIDS patient with Burkitt's lymphoma has also been reported (212). These EBV strains had the unusual ability to induce early antigens in Raji cells (212). Implications for the high rate of B cell lymphomas in HIV infection are intriguing, although the lack of effect of the HIV *tat* trans-activator on promoters for EBV genes involved in immortalization requires explanation. Transfections with whole HIV proviral DNA should be performed, as well as consideration of other potential promoter elements.

## HIV-HERPESVIRUS INTERACTIONS: THERAPEUTIC IMPLICATIONS

Attempts to prevent acquisition or reactivation of herpetic infections might block facilitation of HIV transmission through, for example, establishment of genital ulcers. However, as most individuals infected with HIV already show evidence of a prior herpesvirus infection, primary prevention is usually not practical. Therapeutics can be directed at inhibition of active HIV replication with antiviral nucleoside analogs such as AZT, ddI and ddC (213), control of active herpesvirus replication and EBV-induced lymphoproliferation with drugs such as AZT, DHPG, phosphonoformate and acyclovir (214-216), or blockade of HIV induction.

Among herpesviruses, AZT selectively inhibits EBV replication, albeit it has no effect on replication of the EBV CCC episome (42). EBV-induced B cell proliferation can also be blocked *in vitro* by AZT with $\alpha$ or $\gamma$ interferon, combinations which show striking synergy (196). *In vitro*, $\alpha$-interferon (217), protein kinase C antagonists such as the isoquinolone H-7 (218), staurosporine (219), and the triphenylene estrogen antagonist tamoxifen (220), block stimulation of the HIV LTR and/or upregulation of HIV virions in chronically infected promonocyte and $CD4^+$ T lymphocyte lines. Methods of preventing expression of the NF-$\kappa$B protein, or of inhibiting binding of the activated protein to its enhancer sequence, by way of peptides, chelating agents (221), and modified "anti-sense" oligonucleotides (222), are also being evaluated. More attention should be directed toward exploring ways of prolonging the latent phase of HIV infection, with the goal of chronically treating asymptomatic seropositive individuals.

## REFERENCES


1. Rous, P. J. Exp. Med. 13:397-411, 1911.
2. Poiesz, B.J., Ruscetti, F.W., Gazdar, A.F., Bunn, P.A., Minna, J.D. and Gallo, R.C. Proc. Natl. Acad. Sci. USA 77:7415-7419, 1980.
3. McKhann, G. 2nd., Gibbs, C.J. Jr., Mora, C.a., Rodgers-Johnson, P.E., Liberski, P.P., Gdula, W.J. and Zaninovic, V. J. Infect. Dis. 160:371-379, 1989.
4. Barre-Sinoussi, F., Chermann, J.C., Rey, F., Nugeyre, M.T., Chamaret, S., Gruest, J., Dauguet, C., Axler-Blin, C., Viezinet-Brun, F., Rouzioux, C., Rozenbaum, W. and Montagnier, L. Science 220:868-871, 1983.
5. Gallo, R.C., Salahuddin, S.Z., Popovic, M., Shearer, G.M., Kaplan, M., Haynes, B.F., Palker, T.J., Redfield, R., Oleske, J., Safai, B., et al. Science 224: 500-503, 1984.
6. Gardner, M. and Luciw, P.A. AIDS 2:53, 1988.
7. Varmus, H.E. Annu. Rev. Genet. 18:553-612, 1984.

8. Varmus, H.E. Science 240:1427-1435, 1988.
9. Baichwal, V.R. and Sugden, B. Cell 52:787-789, 1988.
10. Oldstone, M.B.A. Cell 56:517-520, 1989.
11. Ruef, C., et al. Morbid. Mortal. Weekly Rep. 38:572-574, 1989.
12. Peterlin, B.M. and Luciw, P.A. AIDS 2: S29-S40, 1988.
13. Guyader, M., Emerman, M., Sonigo, P., Clavel, F., Montagnier, L. and Alizon, M. Nature 326:662-669, 1987.
14. Kong, L.I., Lee, S.W., Kappes, J.C., Parkin, J.S., Decker, D., Hoxie, J.A., Hahn, B.H. and Shaw, G.M. Science 240:1525-1529, 1988.
15. Evans, L.A., Morreau, J., Odehouri, K., Legg, H., Barboza A., Cheng-Mayer, C. and Levy, J.A. Science 240: 1522-1525, 1988.
16. Benveniste, R.E., Raben, D., Hill, R.W., Knott, W.B., Drummond, J.E., Arthur, L.O., Jahrling, P.B., Morton, W.R., Henderson, L.E. and Heidecker, G. J. Med. Primatol. 18:287-303, 1989.
17. Schneider, J. and Hunsmann, G. AIDS 2:1-9, 1988.
18. Smith, T.F., Scrinivasan, A., Schochetman, G., Marcus, M. and Myers, G. Nature 333:573-575, 1988.
19. Kappes, J.C., Morrow, C.D., Lee, S.W., Jameson, B.A., Kent, S.B., Hood, L.E., Shaw, G.M. and Hahn, B.H. J. Virol. 62:3501-3505, 1988.
20. Li, Y., Naidu, Y., Fultz, P., Daniel, M.D. and Desrosiers, R.C. J. Med. Primatol. 18:261-269, 1989.
21. Kestler, H.W. 3rd., Li, Y., Naidu, Y.M., Butler, C.V., Ochs, M.F., Jaenel, G., King, N.W., Daniel, M.D. and Destosiers, R.C. Nature 331: 619-622, 1988.
22. Ogawa, K., Shibata, R., Kiyomasu, T., Higuchi, I., Kishida, Y., Ishimoto, A. and Adachi, A. J. Virol. 63:4110-4114, 1989.
23. Fultz, P.N. J. Med. Primatol. 18:73-83, 1989.
24. Dalgleish, A.G., Beverley, P.C., Clapham, P.R., Crawford, D.H., Greaves, M.F. and Weiss, R.A. Nature 312:763-767, 1984.
25. Stein, B.S., Gowda, S.D., Lifson, J.D., Penhallow, R.C., Bensch, K.G. and Engelman, E.G. Cell 49:659-668, 1987.
26. Gelderblom, H., Hausmann, E., Ozel, M., Pauli, G. and Koch, M. Virology 156:171-176, 1987.
27. Gelderblom, H.R., Ozel, M., Hausmann, E.H.S., Windel, T., Paul, G. and Koch, M.A. Micron. Micosc. 19:41-60, 1988.
28. Matthews, T.J., Langlois, A.J., Robey, W.G., Chang, N.T., Gallo, R.C., Fischinger, P.J. and Bolognesi, D.P. Proc. Natl. Acad. Sci. USA 83:9709-9713, 1986.
29. Haseltine, W.A. J. AIDS 2:311-334, 1989.
30. Lamarre, D., Ashkenazi, A., Fleury, S., Smith, D.H., Sekaly, R.P and Capon, D.J. Science 245:743-746, 1989.
31. Tsubota, H., Ringler, D.J., Kannagi, M., King, N.W., Solomon, K.R., MacKey, J.J., Walsh, D.G. and Letvin, N.L. J. Immunol. 143:858-863, 1989.
32. Homsy, J., Meyer, M., Tateno, M., Clarkson, C. and Levy, J.A. Science 244:1357-1360, 1989.
33. Soelder, B.M., Reisinger, E.C., Koeftler, D., Bitterlich, G., Wachter, H. and Dierich, M.P. Lancet 2:271-272, 1989.
34. Mann, D.L., Read-Connole, E., Arthur, L.O., Robey, W.G., Wernet, P., Schneider, E.M., Blattner, W.A. and Popovic, M. J. Immunol. 141:1131-1136, 1988.
35. Teitel, J., Shore, A., Read, S.E. and Schiavone, A. J. Infect. Dis. 160:551-552, 1989.

36. Tateno, M., Gonzalez-Scarano, F. and Levy, J.A. Proc. Natl. Acad. Sci. USA 86:4287-4290, 1989
37. Cines, D.B., Lyss, H.D., Bina, M., Corkey, R., Kefalides, H.A. and Friedman, H.M. J. Clin. Invest. 69:123-128, 1982.
38. Cordonnier, A., Montagnier, C. and Emerman, M. Nature 340:571-574, 1989.
39. Stevenson, M., Meier, C., Mann, A.M., Chapman, H. and Wasiak, A. Cell 53:483-496, 1988.
40. Kriegler, M. and Botchan, M. Mol. Cell Biol. 3:325-339, 1983.
41. Furman, P.A., Fyfe, J.A., St Clair, M.H., Weinhold, K., Rideout, J.L., Freeman, G.A., Lehrman, S.N., Bolognesi, D.P., Broder, S., Mitsuya, H. et al. Proc. Natl. Acad-Sci. USA. 83:8333-8337, 1986.
42. Lin, J.C., Zhoug, Z.X., Smith, M.C., Biron, K. and Pagano, J.S. Antimicrob. Agents Chemother. 32:265-267, 1988.
43. Panganiban, A.T. and Fiore, D. Science 241:1064-1069, 1988.
44. Small, J.A., Bieberich, C., Ghotbi, Z., Hess, J., Scangos, G.A. and Clements, J.E. J. Virol. 63:1891-1896, 1989.
45. Kim, S., Byrn, R., Groopman, J. and Baltimore, D. J. Virol. 63:3708-3713, 1989.
46. Fujiwara, T. and Mizuuchi, K. Cell 54:497-504, 1988.
47. Shih, C.C., Stoye, J.P. and Coffin, J.M. Cell 53:531-537, 1988.
48. Haase, A.T., Stowring, L., Harris, J.D., Traynor, B., Ventura, P., Peluso, R. and Brahic, M. Virology 119:399-410, 1982.
49. Roth, M.J., Schwartzberg, P.L. and Goff, S.R. Cell 58:47-54, 1989.
50. Brown, P.O., Bowerman, B., Varmus, H.E. and Bishop, J.M. Proc. Natl. Acad. Sci. USA 86:2525-2529, 1989.
51. Dynan, W.S. Cell 58:1-4, 1989.
52. Twu, J.S. and Robinson, W.S. Proc. Natl. Med. Sci. USA 86:2046-2050, 1989.
53. Dayton, A.I., Sodroski, J.G., Rosen, C.A., Goh, W.C. and Haseltine, W.A. Cell 44:941-947, 1986.
54. Kaufman, J.D., Valendra, G., Roderiguez, G., Buchar, G., Giri, C. and Norcross, W.A. Mol. Cell Biol. 7:3759-3766, 1987.
55. Fujita, T., Shibuya, H., Ohashi, T., Yamanishi, K. and Tanguichi, T. Cell 46:401-405, 1986.
56. Goodbourn, S., Burnstein, H. and Maniatis, T. Cell 45:601-610, 1986.
57. Laurence, J., Sellers, M.B. and Sikder, S.K. Blood 74:291-297, 1989.
58. Miksicek R., Heber A., Schmid W., Danesch U., Posseckert G., Beato M. and Schutz, G. Cell 46:283-290, 1986.
59. Rosen, C.A., Sodroski, J.G. and Haseltine, W.A. Cell 41:813-823, 1985.
60. Guyader, M. Emerman, Sonijo, P., Clavel, F., Montagnier, L. and Alizon, M. Nature 326:662-669, 1987.
61. Arya S.K. and Gallo, R.C. Proc. Natl. Acad. Sci. USA 85:9753-9757, 1988.
62. Muessing, M.A., Smith, D.H. and Capon, D.J. Cell 48:691-701, 1987.
63. Tong-Starksen S, Luciw, P.A., Mittelstadt, P. and Peterlin, B.M. Clin. Res. 36:449A, 1988.
64. Jones, K.A., Luciw, P.A. and Ouchange, N. Genes Develop. 2:1101-1114, 1988.
65. Ptashne, M. Nature 335:683-689, 1988.
66. Mitchell, P.J. and Tjian, R. Science 245:371-378, 1989.
67. Triezenberg, S.J., Kingsburg, R.C. and McKnight, S.C. Genes Dev.

2:718-729, 1988.
68. Reinberg, D., Horikoshi, M. and Roeder, R.G. J. Biol. Chem. 62:3322-3330, 1987.
69. Maio, J.J. and Brown, F.L. J. Virol. 62:1398-1407, 1988.
70. Baeuerle, A. and Baltimore, D. Science 242:540-546, 1988.
71. Zagury, D., Bernard, J., Leonard, R., Cheynier, R., Feldman, M., Sarin, P.S. and Gallo, R.C. Science 231:850-853, 1986.
72. Tong-Starksen, S.E., Luciw, P.A. and Peterlin, B.M. Proc. Natl. Acad. Sci. USA 84:6845-6849, 1987.
73. Skolnick, P.R., Kosloff, B.R. and Hirsch, M.S. J. Infect. Dis. 157:508-514, 1988.
74. Mosca, J.D., Bednarik, D.P., Raj, N.B., Rosen, C.A., Sodroski, J.G., Haseltine, W.A. and Pitha, P.M. Nature 325:67-70, 1987.
75. Folks, T.M., Justement, J., Kinter, A., DiSaretto, C.A. and Fauci, A.S. Science 238:800-802, 1987.
76. Foyanagi, Y., O'Brien, W.A., Zhao, J.Q., Golde, O.W., Gasson, J.C. and Chen, I.S.Y. Science 241:1673-1675, 1988.
77. Folks, T.M., Justement, J., Kinter, A., Schnittman, S., Orenstein, J., Poli, G. and Fauci, A.S. J. Immunol. 140:1117-1122, 1988.
78. Nabel, G.J. and Baltimore, D. Nature 326:711-713, 1988.
79. Nabel, G.J., Rise, S.A., Knipe, D.M. and Baltimore, D. Science 279:1299-1302, 1988.
80. Colombini, S., Arya, S.K., Reitz, M.S., Jogodzinski, L., Beaver, B. and Wong-Staal., F. Proc. Natl. Acad. Sci. USA 86:4813-4817, 1989.
81. Sargan, D.R. and Bennet, I.D. J. Gen. Virol. 70:1995-2006, 1989.
82. Guy, B., Kieny, M.P., Riviere, Y., Le Peuch, C., Dott, K., Girard, M., Montagnier, L. and Lecocq, J.P. Nature 330:266-269, 1987.
83. Luciw, P.A., Chen-Mayer C. and Levy, J.A. Proc. Natl. Acad. Sci. USA 84:1434-1438, 1987.
84. Ahmad, N. and Venkatersan, S. Science 241:1481-1485, 1988.
85. Niederman, T.M.J., Thielan, B.J. and Ratner, L. Proc. Natl. Acad. Sci. USA 86:1128-1132, 1989.
86. Ahmad N., Maitra, R.K. and Venkatesan, S. Proc. Natl. Acad. Sci. USA 86:6111-6115, 1989.
87. Sodroski, J., Coh, W.C., Rosen, C., Dayton, A., Terwilliger, E. and Haseltine, W. Nature 321:412-417, 1986.
88. Sodroski, J., Patarca, R., Rosen, C., Wong-Staal, F. and Haseltine, W. Science 229: 74-77, 1985.
89. Frankel, A.D., Bredt, D.S. and Palo C.O. Science 240:70-73, 1988.
90. Cullen, B. Cell 46:973-982, 1986.
91. Jacobovits, A., Smith, D.A., Jacobovits, E.B. and Capon, J.D. Mol. Cell Biol. 8:2555-2561, 1988.
92. Braddock, M., Chambers, A., Wilson, W., Esnouf, M.P., Adams, S.E., Kingsman, A.J. and Kingsman, S.M. Cell 58:269-279, 1989.
93. Jacks, T., Madhani, H.D., Masiriaz, F.R. and Varmus, H.E. Cell 55:447-458, 1988.
94. Rosen, C.A., Terwilliger, E.F., Dayton, A.E., Sodroski, J.G. and Haseltine, W.A. Proc. Natl. Acad. Sci. USA 85:2071-2075, 1988.
95. Cullen, B.R. and Greene, W.C. Cell 58:423-426, 1989.
96. vanFurth, R. In: **Heterogeneity of mononuclear phagocytes** (Eds. Foster, O., Landy, M.), Academic Press, Inc., London, 1981, pp. 3-10.
97. Pauza, C.D. and Galindo, J. J. Virol. 63:3700-3707, 1989.
98. Pauza, C.D., Galindo, J. and Richman, D. J. Virol. 62:3558-3564, 1988.

99. Lui, K.J., Darrow, W.W. and Rutherford, G.W. III. Science 240:1333-1335, 1988.
100. Medley, G.F., Anderson, R.M., Cox, D.R. and Billard, L. Nature 328:719-721, 1987.
101. Imagawa., D.T., Lee, M.H., Wolinsky, S.M., Sano, K., Morales, F., Kwok, S., Sninsky, J.J., Nishanian, P.G., Giorgi, J., Fahey, J.L. et al. N. Engl. J. Med. 320:1458:1462, 1989.
102. Miedema, F., Petit, A.J., Terpstra, F.G., Schattenkerk, J.K., de Wolf, F., Al B.J., Roos, M., Lange, J.M., Danner, S.A., Goudsmit, J. et al. J. Clin. Invest. 82:1908-1914, 1988.
103. Harper, M.E., Marselle, L.M., Gallo, R.C. and Wong-Staal, F. Proc. Natl. Acad. Sci. USA 83:772-776, 1986.
104. Schnittman, S.M., Psallidopoulos, M.C., Lane, H.C., Thompson, L., Baseler, M., Massari, F., Fox, C.H., Salzman, N.P. and Fauci, A.S. Science 245:305-308, 1989.
105. Meltzer, M.S. and Gendelman, H.E. Immunol. Lett. 19:193-198, 1988.
106. Zagury, D., Bernard, J., Leonard, R., Cheynier, R., Feldman, M., Sarin, P.S. and Gallo, R.C. Science 231:850-853, 1988.
107. Rosenberg, Z.F. and Fauci, A.S. AIDS Res. Human Retrov. 5:1-4, 1989.
108. Markham, P.D., Salahuddin, S.Z., Veren, K., Orndorff, S. and Gallo, R.C. Int. J. Cancer 37:67-72, 1986.
109. Koyanagi, Y., O'Brien, W.A., Zhao, J.Q., Golde, D.W., Gasson, J.C. and Chen, I.S. Science 241:1673-1675, 1988.
110. Murphy, P.M., Lane, H.C., Gallin, J.I. and Fauci, A.S. Ann. Intern. Med. 108:36-41, 1988.
111. Groopman, J.E., Mitsuyasu, R.T., DeLeo, M.J., Oette, H. and Golde, D.W. N. Engl. J. Med. 317:593-598, 1987..
112. Gratama, J.W., Naipal, A.M., Oosterveer, M.A., Stijnen, T., Kluin-Nelemans, H.C., Ginsel, L.A., den Ottolander, G.J., Hekker, A.C., D'Amaro, J., van der Giessen, M. et al. Blood 70:516-523, 1987.
113. Sissons, J.G.P., et al. Immunol. Today 7:57, 1986.
114. Del Val, M., Munch, K., Reddehase, M.J. and Koszinoweski, U.H. Cell 58:305-315, 1989.
115. Schrier, R.D. and Oldstone, M.B.A. J. Virol. 59:127, 1986.
116. Hirsch, M.S., Schooley, R.T., Ho, D.D. and Kaplan, J.C. Rev. Infect. Dis. 6:726-731, 1984.
117. Blaser, M.J. and Cohn, D.L. Rev. Infect. Dis. 8:21-30, 1986.
118. Laurence, J. Infect. Med. 3:262, 1986.
119. Quinnan, G.V. Jr., Masur, H., Rook, A.H., Armstrong, G., Frederick, W.R., Epstein, J., Manischewitz, J.F., Macher, A.M., Jackson, L., Ames, J. et al. J. Amer. Med. Assoc. 252:72-77, 1984.
120. Quinn, T.C., Piot, P., McCormick, J.B., Feinsod, F.M., Taelman, H., Kapita, B., Stevens, W. and Fauci, A.S. J. Amer. Med. Assoc. 257:2617-2621, 1987.
121. Holmberg, S.D., Stewart, J.A., Gerber, A.R., Byers, R.H., Lee, F.K., O'Malley, P.M. and Nahmias, A.J. J. Amer. Med. Assoc. 259:1048-1050, 1988.
122. Bonetti, A., Weber, R., Vogt, M.W., Wunderli, W., Siegenthaler, W. and Liuthy, R. Ann. Internm. Med. 111:293-296, 1989.
123. Webster, A., Lee, C.A., Cook, D.G., Grundy, J.E., Emery, V.C., Kernoff, P.B. and Griffiths, P.D. Lancet 2:63-66, 1989.
124. Barnass, S., O'Toole, C. and Colvin, B. Lancet 2:335, 1989.

125. Thorley-Lawson, D.A. Sem. Hematol. 25:247-260, 1988.
126. Gratama, J.W., Oosterveer, M.A.P., Zwaan, F.E., Lepoutre, J., Klein, G. and Ernberg, I. Proc. Natl. Acad. Sci. USA 85:8693-8696, 1988.
127. Ernberg, I., Falk, K. and Hansson, M. Int. J. Cancer 39:190-197, 1987.
128. Hurley, E.A. and Thorley-Lawson, D.A. J. Exp. Med. 168:2059-2075, 1988.
129. Dambaugh, T., Hennessy, K., Fennewald, S. and Kieff, E. In: **The Epstein-Barr Virus: Recent Advances** (Eds. Epstein, M.A., Achong, B.G.), John Wiley & Sons, New York, 1986, pp. 14-45.
130. Sugden, B. Cell 57:5-7, 1989.
131. Gregory, C.D., Murray, R.J., Edwards, C.F. and Rickinson, A.B. J. Exp. Med. 167:1811-1824, 1988.
132. Andiman, W.A., Eastman, R., Martin, K., Katz, B.Z., Rubinstein, A., Pitt, J., Pahwa, S. and Miller, G. Lancet 2:1390-1393, 1985.
133. Birx, D.L., Redfield, R.R. and Tosato, G. N. Engl. J. Med. 314:874-879, 1986.
134. Yarchoan, R., Redfield, R.R. and Broder, S. J. Clin. Invest. 78:439-447, 1986.
135. Krueger, G.R.F., Koch, B., Ramon, A., Ablashi, D.V., Salahuddin, S.Z., Josephs, S.F., Streicher, H.Z., Gallo, R.C. and Habermann, U. J. Virol. Meth. 21:125-131, 1988.
136. Rahman, M.A., Kingsley, L.A., Breinig, M.K., Ho, M., Armstrong, J.A., Atchison, R.W., Lyter, D.W. and Rinaldo, C.R. Jr. J. Infect. Dis. 159:472-479, 1989.
137. Alsip, G.R., Ench, Y., Sumaya, C.V. and Boswell, R.N. J. Infect. Dis. 157:1072-1076, 1988.
138. Ziegler, J.L., Beckstead, J.A., Volberding, P.A., Abrams, D.I., Levine, A.M., Lukes, R.J., Gill, P.S., Burkes, R.L., Meyer, P.R., Metroka, C.E. et al. N. Engl. J. Med. 311:565-570, 1984.
139. Beissner, R.S., Rappaport, E.S. and Diaz, J.A. Arch. Pathol. Lab. Med. 111:250-253, 1987.
140. Greenspan, J.S., Greenspan, D., Lennette, E.T., Abrams, D.I., Conant, M.A., Petersen, V. and Freese, U.K. N. Engl. J. Med. 313:1564-1571, 1985.
141. Mathur-Wagh, U., Enlow, R.W., Spigland, I., Winchester, R.J., Sacks, H.S., Rorat, E., Yancovitz, S.R., Klein, M.J., Williams, D.C. and Mildvan, D. Lancet 1:1033-1038, 1984.
142. Pelicci, P.G, Knowles, D.M. II., Arlin, Z.A., Wieczorek, R., Luciw, P., Dina, D., Basilico, C. and Dalla-Favera, R. J. Exp. Med. 164:2049-2060, 1986.
143. Knowles, D.N. Lab. Med. 17:674-678, 1986.
144. Alonso, M.L., Richardson, M.E., Metroka, C.E., Mouradian, J.A., Koduru, P.R., Filippa, D.A. and Chaganti, R.S. Blood 69:855-858, 1987.
145. Subar, M., Neri, A., Inghirami, G., Knowles, D.M. and Dalla-Favera, R. Blood 72:667-671, 1988.
146. Knowles, D.M., Inghirami, G., Ubriaco, A. and Dalla-Favera, R. Blood 73:792-799, 1989.
147. Thomas, J.A. and Crawford, D.H. Lancet 1:1075-1076, 1989.
148. Steiner, I., Spivack, J.G., Lirette, R.P., Brown, S.M., MacLean, A.R., Subak-Sharpe, J.H. and Fraser, N.W. EMBO J. 8:505-511, 1989.
149. Kirchner, H. Monogr. Virol. 13:1, 1982.
150. Rinaldo, C.R. Jr., Richter, B.S., Black, P.H., Callery, R., Chess, L. and Hirsch, M.S. J. Immunol. 120:130-136, 1978.

151. Kreiss, J.K., Coombs, R., Plummer, F., Holmes, K.K., Nikora, B., Cameron, W., Ngugi, E., Ndinya Achola, J.O. and Corey, L. J. Infect. Dis. 160:380-384, 1989.
152. Salahuddin, S.Z., Ablashi, D.V., Markham, P.D., Josephs, S.F., Sturzenegger, S., Kaplan, M., Halligan, G., Biberfeld, P., Wong-Staal, F., Kramarsky, B. et al. Science 234:596-601, 1986.
153. Takahashi, K., Sonoda, S., Higashi, K., Kondo, T., Takahashi, H., Takahashi, M. and Yamanishi, K. J. Virol. 63:3161-3163, 1989.
154. Josephs, S.F., Buchbinder, A., Streicher, H.Z., Ablashi, D.V., Salahuddin, S.Z., Guo, H.G., Wong-Staal, F., Cossman, J., Raffeld, M., Sundeen, J. et al. Leukemia 2:132-135, 1988.
155. Brown, N.A., Kovacs, A., Lui, C.R., Hur, C., Zaia, J.A. and Mosley, J.W. Lancet 2:1146, 1988.
156. Osborn, L., Kunkel, S. and Nabel, G.J. Proc. Natl. Acad. Sci. USA 86:2336-2340, 1989.
157. Folks, T.M., Justement, J., Kinter, A., Schnittman, S., Orenstein, J., Poli, G. and Fauci, A.S. J. Immunol. 140:1117-1122, 1988.
158. Folks, T.M., Benn, S., Rabson, A., Theodore, T., Hoggan, M.D., Martin, M., Lightfoote, M. and Sell, K. Proc. Natl. Acad. Sci. USA 82:4539-4543, 1985.
159. Ho, J. and Laurence, J. unpublished data, 1989.
160. Folks, T.M., Clouse, K.A., Justement, J., Rabson, A., Duh, E., Kehrl, J.H. and Fauci, A.S. Proc. Natl. Acad. Sci. USA 86:2365-2368, 1989.
161. Kornbluth, R.S., Oh, P.S., Munis, J.R., Cleveland, .H. and Richman, D.D. J. Exp. Med. 169:1137-1151, 1989.
162. Clouse, K.A., Robbins, P.B., Fernie, B., Ostrove, J.M. and Fauci, A.S. J. Immunol. 143:470-475, 1989.
163. Wright, S.C., Jewett, A., Mitsuyasu, R. and Bonavida, B. J. Immunol. 141:99-104, 1988.
164. Lahdevirta, J., Maury, C.P.J., Teppo, A.M. and Repo, H. Am. J. Med. 85:289-291, 1988.
165. Lepe-Zuniga, J.L., Mansell, P.W.A. and Hersh, E.M. J. Clin. Micro. 25:1695-1700, 1987.
166. Wahl, L., Corcoran, N.L., Authur, L.O., Pyle, S., Harle-Bellan, A. and Farrar, W.L. Proc. Natl. Acad. Sci. USA 86:621-625, 1989.
167. Maggi, E., Macchia, D., Parronchi, P., Mazzetti, M., Ravina, A., Milo, D. and Romagnani, S. Eur. J. Immunol. 17:1685-1690, 1987.
168. Laurence, J., Friedman, S.M., Chartash, E.K., Crow, M.K. and Posnett, D.N. J. Clin. Invest. 83:1843-1848, 1989.
169. Nakajima, K., Martinez-Maza, O., Hirano, T., Breen, E.C., Nishanian, P.G., Salazar-Gonzalez, J.F., Fahey, J.L. and Kishimoto, T. J. Immunol. 142:531-536, 1989.
170. Gallo, P., Frie, K., Rordort, C., Lazdirs, J., Tavolato, B. and Fontane, A. J. Neuroimmunol. 23: 109-116, 1989.
171. Scuderi, P., Sterling, K.E., Lam, K.S., Finley, P.R., Ryan, K.J., Ray, C.G., Petersen, E., Slymen, D.J. and Salmon, S.E. Lancet 2:1364-1365, 1986.
172. Geelen, J.L.M.C., Minnaar, R.P., Boom, R., van der Noordaa, J. and Goudsmit, J. J. Gen. Virol. 69:2913-2917, 1988.
173. Geelen, J.L., Boom, R., Klaver, G.P., Minnaar, R.P., Feltkamp, M.C., van Milligen, F.J., Sol, C.J. and van der Noordaa, J. J. Gen. Virol. 68:2925-2931, 1987.
174. Frankel, A.D. and Pabo, C.O. Cell 55:1189-1193, 1988.

175. Kenney, S., Kamine, J., Markovitz, D., Fenrick, R. and Pagano, J. Proc. Natl. Acad. Sci. USA 85:1652, 1988.
176. Kenney, S., Kamine, J., Holley-Guthrie, E., Mar, E.C., Lin, J.C., Markovitz, D. and Pagano, J. J. Virol. 63:3870-3878, 1989.
177. Kenney, S. personal communication, 1989.
178. Gendelman, H.E., Phelps, W., Feigenbaum, L., Ostrove, J.M., Adachi, A., Howley, P.M., Khoury, G., Ginsberg, H.S. and Martin, M.A. Proc. Natl. Acad. Sci. USA 83:9759-9763, 1986.
179. Nelson, J.A., Reynolds-Kohler, C., Oldstone, M.B.A. and Wiley, C.A. Virology 165:286-290, 1988.
180. Casareale, D., Fiala, M., Chang, C.M., Cone, L.A. and Mocarski, E.S. Int. J. Cancer 44:124-130, 1989.
181. Lenardo, M.J. and Baltimore, D. Cell 58:227-229, 1989.
182. Elfassi, E., Michelson, S., Bachelerie, F., Arenzana-Seisdedos, F. and Virelizier, J.L. Ann. Inst. Pasteur/Virol. 138:461-470, 1987.
183. Rice, G.P.A., Schrier, R.D. and Oldstone, M.B.A. Proc. Natl. Acad. Sci. USA 81:6134-6138, 1984.
184. Davis, M.G., Kenney, S.C., Kamine, J., Pagano, J.S. and Huang, E.S. Proc. Natl. Acad. Sci. USA 84:8642-8646, 1987.
185. Duclos, H., Elfassi, E., Michelson, S., Arenzana-Seisdedos, F., Munier, A. and Virelizier, J.L. AIDS Res. Human Retrov. 5:217-224, 1989.
186. Rando, R.F., Pellett, P.E., Luciw, P.A., Bohan, C.A. and Srinivasan, A. Oncogene 1:13-18, 1987.
187. Ostrove, J.M., Leonard, J., Weck, K.E., Rabson, A.B. and Gendelman, H.E. J. Virol. 61:3726-3732, 1987.
188. Nabel, G.J., Rice, S.A., Knipe, D.M. and Baltimore, D. Science 239:1299-1302, 1988.
189. Albrecht, M.A., DeLuca, N.A., Byrn, R.A., Schaffer, P.A. and Hammer, S.M. J. Virol. 63:1861-1868, 1989.
190. Mosca, J.D., Bednarik, D.P., Raj, N.B., Rosen, C.A., Sodroski, J.G., Haseltine, W.A., Hayward, G.S. and Pitha, P.M. Proc. Natl. Acad. Sci. USA 84:7408-7412, 1987.
191. O'Hare, P. and Hayward, G.S. J. Virol. 61:190-199, 1987.
192. Gimble, J.M., Duh, E., Ostrove, J.M., Gendelman, H.E., Max, E.E. and Rabson, A.B. J. Virol. 62:4104-4112, 1988.
193. Lusso, P., Ensoli, B., Markham, P.D., Ablashi, D.V., Salahuddin, S.Z., Tschachler, E., Wong-Staal, F. and Gallo, R.C. Nature 337:370-373, 1989.
194. Horvat, R.T., Wood, C. and Balachandran, N. J. Virol. 63:970-973, 1989.
195. Montagnier, L., Gruest, J., Chamaret, S., Dauguet, C., Axler, C., Guietard, D., Nugeyre, M.T., Barrie-Sinoussi, F., Chermann, J.C., Brunet, J.B. et al. Science 225:63-66, 1984.
196. Pagano, J.S., Kenney, S., Markovitz, D. and Kamine, J. J. Virol. Methods 21:229-239, 1988.
197. Sample, J., Tanaka, A., Lancz, G. and Nonoyama, M. Virology 139:1-10, 1984.
198. Mann, D.L., DeSantis, P., Mark, G., Pfeifer, A., Newman, M., Gibbs, N., Popovic, M., Sarngadharan, M.G., Gallo, R.C., Clark, J. et al. Science 236:1103-1106, 1987.
199. Ragona, G., Sirianni, M.C., Soddu, S., Vercelli, B., Sebastiani, G., Piccoli, M. and Aiuti, F. Clin. Exp. Immunol. 66:17-24, 1986.
200. Schnittman, S.M., Lane, H.C., Higgins, S.E., Folks, T. and Fauci, A.S.

Science 233:1084-1086, 1986.
201. Pahwa, S., Pahwa, R., Saxinger, C., Gallo, R.C. and Good, R.A. Proc. Natl. Acad. Sci. USA 82:8198-8202, 1985.
202. Monroe, J.E., Calender, A. and Mulder, C. J. Virol. 62:3497-3500, 1988.
203. Salahuddin, S.Z., Ablashi, D.V., Hunter, E.A., Gonda, M.A., Sturzenegger, S., Markham, P.D. and Gallo, R.C. Int. J. Cancer 39:198-202, 1987.
204. Tozzi, V., Britton, S., Ehrnst, A., Lenkei, R. and Strannegard, O. J. Med. Virol. 27:19-24, 1989.
205. Malkovsky, M., Philpott, K., Dalgleish, A.G., Mellor, A.L., Patterson, S., Webster, A.D., Edwards, A.J. and Maddon, P.J. Eur. J. Immunol. 18:1315-1321, 1988.
206. Lai, P.K., Li, X. and Volsky, D.J. Int. J. Cancer 43:1104-1111, 1989.
207. Dahl, K., Martin, K. and Miller, G. J. Virol. 61:1602-1608, 1987.
208. Hoxie, J.A., Alpers, J.D., Rackowski, J.L., Huebner, K., Haggarty, B.S., Cedarbaum, A.J. and Reed, J.C. Science 234:1123-1127, 1986.
209. Casareale, D., Sinangil, F., Hedeskog, M., Ward, W. and Volsky, D.J. AIDS Res. 1:253-270, 1984.
210. Markham, P.D., Ruscetti, F., Salahuddin, S.Z., Gallagher, R.E. and Gallo, R.C. Int. J. Cancer 23:148-156, 1979.
211. Markham, P.D., Ruscetti, F.W., Kalyanaraman, V.S., Ceccherini-Nelli, L., Miller, N.R., Reitz, M.S., Salahuddin, S.Z. and Gallo, R.C. Cancer Res. 41:2738-2744, 1981.
212. Magrath, I., Erikson, J., Whang-Peng, J., Sieverts, H., Armstrong, G., Benjamin, D., Triche, T., Alabaster, O. and Croce, C.M. Science 222:1094-1098, 1983.
213. Schinazi, R.F. Ericksson, B.F.H. and Hughes, S.H. Antimicrob. Agents Chemother. 33:115-117, 1989.
214. Laskin, O.L., Cederberg, D.M., Mills, J., Eron, L.J., Mildvan, D. and Spector, S.A. Am. J. Med. 83:201-207, 1987.
215. Ringden, O., Wilczek, H., Lonnqvist, B., Gahrton, G., Wahren, B. and Lernestedt, J.O. Lancet 1:1503-1504, 1985.
216. Sullivan, J.L., Medveczky, P., Forman, S.J., Baker, S.M., Monroe, J.E. and Mulder, C. N. Engl. J. Med. 311:1163-1167, 1984.
217. Poli, G., Orenstein, J.M., Kinter, A., Folks, T.M. and Fauci, A.S. Science 244:575-577, 1989.
218. Tong-Starksen, S.E., Luciw, P.A. and Peterlin, B.M. J. Immunol. 142:702-707, 1989.
219. Laurence, Sikder, S.K., Jhivari, S., Saunders, A. and Salmon, J.E. 1989.
220. Laurence, J., Cooke, H. and Sikder, S.K. Blood, 1989, in press.
221. Muller, W.E.G., Wenger, R., Reuter, P., Renneisen, K. and Schroder, H.C. Biochim. Biophys. Acta 1008:208-212, 1989.
222. Laurence, J., Sikder, S.K., Miller, P. and Ts'o, P.O.P. Clin. Res. 37:414A, 1989.
223. Merrill, J.E., Koyanagi, Y. and Chen, I.S.Y. J. Virol. 63:4404, 1989.

# 12

## *TRANS*-ACTIVATION OF THE HUMAN IMMUNODEFICIENCYVIRUS (HIV) PROMOTER BY HETEROLOGOUS VIRUS INFECTION

P. M. PITHA and D. P. BEDNARIK

**Oncology Center and Department of Molecular Biology and Genetics, The Johns Hopkins University School of Medicine, Baltimore, Maryland 21205 USA**

## ABSTRACT

In infected cells, HIV replication is regulated at least by three virus encoded proteins. The product of *tat* gene is a positive transcriptional regulator, the product of *rev* gene regulates the transport of viral mRNA encoding the structural proteins from the nucleus to the cytoplasm while the products of the *nef* gene may have a negative regulatory effect. However, the transcriptional activity of the promoter (LTR) can also be stimulated by various extracellular stimuli such as mitogens, cytokines and differentiation factors, or by infection with several heterologous viruses including the herpesviruses that often accompany HIV infection in AIDS patients. The molecular mechanism by which the herpesviruses enhance the transcriptional activity of HIV-LTR is not clear. However, it was shown that activation is mediated by the products of herpesvirus immediate early (IE) genes and probably proceeds through stimulation of cellular transcriptional factors such as NF-$\kappa$B. The activity of HIV-LTR can be down-regulated by methylation, which is accompanied by the reduced capacity of this region to bind the trans-activating nuclear protein. The stimulation of methylated HIV-LTR expression by HSV-1 infection seems to involve hypomethylation, while *tat*-mediated trans-activation occurs without an apparent change in LTR methylation.

## INTRODUCTION

One of the particularities of HIV infection is a long time period between infection and onset of disease-acquired immunodeficiency syndrome (AIDS). This period, which is sometimes called latency, is characterized by the presence of antibodies to HIV (p24), and an extremely low number of infected cells (1-2). The mechanisms that trigger virus replication, its spread, and progression to disease are not known.

There are at least two equally attractive hypotheses that could account for the transition from the long clinically asymptomatic latency state to the onset of AIDS. First, it is possible that as shown for murine and feline AIDS, the virus that causes the immunodeficiency syndrome is not the fully infectious HIV but rather an aberrant replication-defective mutant (3-5). The long latency period may be required in order to generate this defective mutant that can only replicate in the presence of the helper virus (HIV). An example of such a newly formed recombinant virus that is the pathogenic agent is provided by retrovirus-induced T-cell lymphoma in AKR mice, where disease is also preceded by a long latent period. During this latent period replicating ecotropic virus recombines with the endogenous xenotropic virus thereby giving rise to a new recombinant virus, *mink-cell-focus-forming virus* (MCFV) that is the agent of the T-cell lymphomas in AKR mice (6). The low levels of HIV replication during the disease-free latent period in HIV-infected individuals may account for the long interval (about 8 years) that is required for the generation of such putative pathogenic virus. The second possibility that may explain the long interval for the transition to AIDS is that HIV persists in cells in an unexpressed form and that the onset of disease is triggered by activation of HIV replication by external stimuli. Implicit in this model is the assumption that increased virus replication leads to disease onset and progression.

At this point, both of these models are only hypothetical. Although aberrant HIV forms have been identified, their role in disease progression has not been established. Furthermore the HIV detected in AIDS patients may contain a new variant of HIV which replicates faster and is more pathogenic than the original HIV inoculum or than the virus expressed in infected disease-free, asymptomatic patients. The elucidation of the mechanisms that regulate the spread and progression of HIV infection may have important clinical applications and will provide a rational approach to therapy.

In the infected cells, HIV replication is regulated by a number of cellular and viral proteins. At least three virus encoded genes have been shown to play a role in this process. The product of the *tat* gene is a strong positive regulator. Enhancement of HIV replication by the *tat* protein can be seen both at the RNA and protein levels (7-10). The product of the *rev* gene regulates the transport of viral mRNA encoding the structural proteins from the nucleus to the cytoplasm (7,11,12), while the product of the *nef* gene may have negative regulatory effects (7,13). During the course of a single step infection, mRNA for the trans-activating genes (*tat* and *rev*) can be detected in cells during the early stage of infection before detection of mRNA coding for the structural genes and genomic RNA. This suggests that the early (*tat*, *rev*) and late (structural) genes are involved in the complex process of synthesis of viral gene products.

It is still not known whether the direct binding of the *tat* and *rev* gene products with the *cis*-regulatory region in the viral genome is necessary or whether these proteins interact with cellular factors to perform their regulatory functions. It has been demonstrated that a number of cellular transcriptional factors contribute, as well, to the transcriptional up-regulation of the HIV genome. Binding sites for several transcriptional factors are present in the HIV LTR. Thus, the U3 region of the LTR contains three Sp1 binding sites located -45 to -78 basepairs (bp) upstream, of the RNA cap site (Fig. 1).

Upstream from the Sp1 sites (-81 to -103 bp) two direct repeats can be found. These repeats are homologous to viral core enhancers found in the SV40 as well as to the NF-$\kappa$B binding site which was originally found in the enhancer region of immunoglobulin kappa gene (7,14,15). Binding sites for the transcriptional factor AP-1 have also been found in the negative regulatory element of HIV-LTR. Thus, it is likely that the transcriptional regulation involves a complex cooperation between the products of viral genes and cellular trans-activating factors. It is possible that viral trans-activating proteins increase or modify cellular factors.

The viral LTR can also be stimulated by a number of extracellular stimuli. It has been suggested that these stimuli can induce the expression of the latent

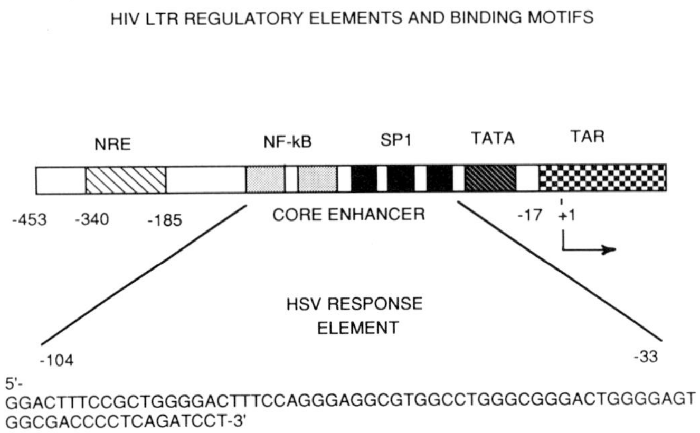


**Fig. 1.** HIV-LTR regulatory elements and binding motifs. NRE, negative regulatory element; NF-κB, NF-κB binding site; SP1, SP1 binding site; TATA, TATA box; TAR, *tat* responsive region.

form of HIV and thus, enhance virus replication, triggering the onset of disease. These extracellular signals include mitogens (7,15-17), cytokines (18-20), and differentiation factors (14,15,21,22). Interestingly, transcription of the HIV-LTR can also be activated by infection with several heterologous viruses, including viruses of the herpes group which often accompany HIV infection in AIDS patients. The molecular mechanism by which the virus stimulates the HIV-LTR is not clear; it was shown with several viruses of the herpes group that the activation of HIV- LTR occurs at the transcriptional level (23-28), and it is mediated by the products of the viral immediate early (IE) genes (24,26,28-30). Since a diverse group of IE viral proteins have a stimulatory effect on HIV-LTR, these data suggest that the activation proceeds through a stimulation of cellular transcriptional factors. In this review we will summarize the studies on the effect of heterologous viral infection on HIV-LTR activity, will discuss the possible mechanism by which the heterologous virus infection can activate the HIV-LTR, and consider the possible role of opportunistic infections in the induction of AIDS.

## TRANS-ACTIVATION OF HIV-LTR BY DNA VIRUSES

The trans-activation of expression from HIV-LTRcat hybrid plasmid in cells infected with viruses of the herpes group or co-transfected with fragments of herpes viral DNA containing the IE genes was demonstrated by several groups. Gendelman *et al.* (25,29) have shown that expression of CAT activity coded by the HIV-LTRcat plasmid in transfected human colon carcinoma (SW480) cells could be stimulated by co-transfection with plasmids containing the IE genes of both herpes simplex virus type 1 (HSV-1) and varicella-zoster virus (VZV) or with plasmids containing part of BKV, JLV, and LPV sequences which encode for the T-antigens. Similarly, Mosca *et al.* (23, 24) have shown that in permanent cell lines (Vero and Ltk$^-$) which contain the integrated HIV-LTRcat plasmid, but do not express the CAT activity constitutively, infection with HSV-1 results in the induction of CAT activity. The trans-activation by HSV-1 was shown to occur at the transcriptional level and to lead to an increase of relative levels of correctly initiated CAT mRNA in the cells.

After these two initial observations, a number of other DNA viruses were shown to be able to trans-activate the expression of the HIV-LTRcat plasmid. Thus, infection with human and simian cytomegalovirus (CMV) was shown to be able to induce CAT activity in Vero cells containing the integrated HIV-LTRcat plasmid (24,26), and the co-transfection of HIV-LTRcat plasmid with the DNA fragment of human CMV encoding the IE gene region (0.66 to 0.75 map unit) into HeLa cells trans-activated the HIV promoter. The trans-activation was also shown with the new herpes virus isolate HHV6 (27), which is antigenically dissimilar from other human herpesviruses. Co-culture of the human T-cell line Hut-78 transfected with HIV-LTRcat plasmid together with Hut-78 or HSB-2 cells infected with HHV-6 led to a dramatic increase (40- to 60-fold) in CAT activity (27). Thus, the ability to trans-activate the expression of HIV-LTR seems to be a property of a large variety of herpes viruses.

<u>Evidence for a role of trans-activating viral gene products in stimulation of HIV-LTR expression</u>.

In a first approach to identify the HSV-1 viral gene products which are essential for the activation of HIV-LTR, Mosca *et al.* (23) infected the Vero cell line containing integrated, unexpressed HIV-LTRcat plasmid with temperature-sensitive mutants of HSV-1 and showed that the expression of the HSV-1 genome is essential for HIV-LTR activation. Replication of HSV-1 is known to be coordinately regulated, and three classes of genes (alpha, beta, and gamma or IE, delayed early and late respectively) are expressed at different stages of infection (31). Trans-activation of the HIV-LTR occurred within a few hours after HSV-1 infection. During this time, the IE genes (ICP0, ICP4, ICP22, ICP27, and ICP47), which are known to have trans-activating activity on the expression of the delayed early (DE) and late genes of HSV, are synthesized. These authors, therefore, examined whether the function of HSV-1 IE genes is essential for HIV-LTR activation. To determine whether the ICP4 gene, that was shown to play an essential role in the activation of HSV-1 DE genes, is involved in HIV-LTR activation, Mosca and his colleagues used a temperature-sensitive HSV-1 mutant *tsk* that produces an ICP4 protein which is unable to trans-activate (32). Infection with the *tsk* mutant, both at the permissive and non-permissive temperatures, activated the expression of the HIV-LTRcat plasmid, while the same mutant at non-permissive temperatures failed to activate the HSV-1 DE genes. Similarly, infection with the *tsE6* mutant, which is defective in IE ICP27 protein (33), enhanced CAT activity from HIV-LTRcat both at permissive and non-permissive temperatures (29). These data indicate that neither ICP4 nor ICP27 proteins are involved in the activation of the HIV-LTR in Vero cells (24,29).

Direct identification of the IE genes which are essential for the activation of HIV-LTR, was done by co-transfection of the individual IE genes with the HIV-LTRcat plasmid into various cell lines followed by CAT activity assays at 48 hr after transfection (transient assay). The results showed that the ICP0 and ICP4 IE genes were able to activate the expression of HIV-LTR. Whether either ICP0 or ICP4 alone contributed the trans-activating activity, depended on the cell type used in the transfection experiments and on the presence of the viral trans-activator *tat*. Thus, in Vero cells containing integrated HIV-LTRcat, ICP0 or a combination of ICP0 and ICP4, but not ICP4 alone, could activate the HIV promoter region (29). On the other hand, in a transient expression assay both

ICP0 and ICP4 enhanced CAT expression only in the presence of the *tat* gene product but had no stimulating effect in its absence. In SW480 cells, however, ICP4 alone stimulated CAT expression, while ICP0 had no effect. !n Jurkatt cells, expressing *tat* protein, the ICP0 or combination of ICP0 and ICP4, but not ICP4 alone, were stimulatory (30). Thus, in every cell line tested, trans-activation could be achieved by the combination of ICP0 and ICP4 gene products, but the activity of ICP0 and ICP4 was cell type dependent. Additionally, the presence of the *tat* gene product, which enhances trans-activation coordinately with the IE genes, seems to not only quantitatively enhance the stimulation, but also change the activity of the IE gene(s).

Activation of the HIV-LTR expression by the product of IE genes was not limited to HSV infection. It was shown by Davis *et al.* (26) that the region of HCMV, which contains the IE genes, can trans-activate the HIV promoter in HeLa cells. When both the IE region of HCMV and the plasmid containing the *tat* region were co-transfected together with HIV-LTRcat plasmid, the stimulation was additive. Similarly, the region of Epstein-Barr virus (EBV) that encodes an IE nuclear protein with trans-activating properties, was able to stimulate CAT expression upon co-transfection with HIV-LTRcat into HeLa cells (28). Surprisingly, however, this stimulation could be demonstrated only when the CAT gene and not other genes were used as a reporter gene (Pagano, personal commun.), and this activation was found to occur largely at the post-transcriptional level (28).

The products of two other viral trans-activating genes were found to be able to enhance activity of the HIV promoter. One of them was a protein-encoded by the E1A gene of adenovirus which can either up- or down-regulate activity of a number of cellular and viral genes (34). Co-transfection of the E1A gene together with HIV-LTRcat into Jurkatt cells or Jurkatt cells expressing the HIV *tat* gene (Jurkatt-*tatIII* cells) significantly enhanced CAT activity (34). Another interesting observation was the trans-activation of the HIV promoter region by the hepatitis virus (HBV) *X* protein. The *X* protein of HBV was recently shown to be able to enhance the activity of both homologous and heterologous enhancers (35,36). It was shown, again in transient transfection studies, that co-transfection of the HBV segment containing the gene for the *X* protein with HIV-LTRcat plasmid into Jurkatt cells caused a significant increase in CAT activity (35,36).

The observation that viral gene products of unrelated viruses can modify the expression of HIV-LTR implies that it is unlikely that these proteins stimulate the activity of the HIV promoter by direct binding to its sequences, but rather suggests that viral trans-activators interact or modify cellular regulatory proteins that bind to sequences within the HIV promoter.

*Cis*-acting region within HIV-LTR responds to viral trans-activating factors.

The identical responsive region within HIV-LTR was identified for a majority of the trans-activating viruses (or viral trans-activating genes) studied so far, when activation was assayed in Vero or HeLa cells and in the absence of the *tat* gene product. Deletion analysis of the HIV-LTR indicated that the HSV-1 responsive element was localized to -104 and -32 bp from the HIV cap site (24). Insertion of this 73 nucleotide-long fragment 5' to the minimal murine alpha-interferon (IFN $\alpha$) promoter, which by itself was transcriptionally inactive, rendered this promoter region inducible by HSV infection or by the product of the HSV-1 ICP0 gene.

The HIV promoter region contains a number of *cis*-regulatory elements that can either positively or negatively regulate the activity of the promoter region. The 73 nucleotide-long region inducible by HSV-1 contains a sequence that was previously identified, by deletion mapping, to contain the HIV enhancer (Fig. 1). It contains two copies of an 11 bp sequence NF-$\kappa$B which was originally identified in the $\kappa$-immunoglobulin gene enhancer region (37). In B-cells, the binding of the transcriptional factor NF-$\kappa$B to this region was shown to correlate with the transcriptional activation of the $\kappa$ gene (37). This region also contains three Sp1 binding sites which are present in many cellular and viral promoter regions, and binds the transcriptional factor, Sp1 (38). It was found that alterations in either of these sites abolished or decreased the HSV-1-induced trans-activation. Thus, a mutation in the $\kappa$B binding site which inhibits its interaction with the NF-$\kappa$B motifs, also inhibited trans-activation by HSV-1 (Mosca, Bednarik, Raj, and Pitha, unpublished). Also, the deletion that removed all three Sp1 binding sites was not stimulated by HSV-1 infection (29). However, when the analysis was done in Jurkatt cells expressing the *tat* gene, there was not a single defined *cis*-elements which could confer the HSV-1 or ICP0 inducibility (34). In this system, the activation by ICP0 did not seem to even require the presence of the TATA box. The reason for this observed difference is not obvious. It may be due to the use of different cell lines for assay and the presence or absence of the *tat* protein in the assay system.

The requirement for the presence of the enhancer region in HIV-LTR trans-activation was also found for the hepatitis *X* protein in Jurkatt cells (36), where the deletion of the HIV enhancer region abolished the trans-activation.

In contrast to the findings for HSV-1, Mosca *et al*. (23) failed to detect the requirement for a specific *cis*-regulatory sequence for CMV trans-activation in the same cells (Vero). Also, a specific regulatory sequence was not found when CMV trans-activation was done in human fibroblast cells. Furthermore, a significant difference between the relative levels of CAT mRNA and CAT activity was observed in the CMV-infected cells. These results indicate that unlike HSV-1 which activates the HIV-LTR at the transcriptional level, the effect of CMV infection may be predominantly post-transcriptional. Similarly, it was suggested that the EBV IE gene product stimulates HIV-LTR expression at the post-transcriptional level (28).

Activation of HIV-LTR by HSV-1 infection is associated with the binding of a cellular factor to the NF-κB sequences.

The molecular mechanism of the HIV-LTR activation by the heterologous virus infection or by the products of the IE genes has not been established. It was shown, however, that the HSV-1-induced trans-activation occurs at the transcriptional level (23-25). It is not known whether the products of trans-activating IE genes can directly interact with HIV-LTR. However, the observations that stimulation shows cell type specificity and a variety of unrelated viral trans-activating proteins can mediate this effect, point to a complex interaction between viral and cellular trans-activators. It was recently shown by Gimble *et al*. (38) that the activation of HIV-LTR by HSV-1 infection in HeLa cells correlated with the induction of a NF-κB-like factor that was binding to the HIV enhancer region. Interestingly, protection against exonuclease digestion extended over both the NF-κB and Sp1 sites present in this region. The two NF-κB and the SP1 sites are also present in the 73 bp long region which was shown to be able to confer HSV inducibility to a heterologous promoter (24). The importance of NF-κB in HSV-1-induced stimulation of HIV-LTR was previously suggested by experiments, which show that the HIV enhancer region which contained the mutated NF-κB site, was not trans-activated by HSV-1 infection (Mosca, Bednarik, Raj, and Pitha, unpublished). Surprisingly, however, the mutation in the NF-κB site did not abort inducibility in *tat* expressing Jurkatt cells.

The induction of the NF-κB-like activity was also shown to be associated with activation of HIV-LTR in lymphocytes stimulated with the mitogen phytohemagglutinin (PHA) or treated with a differentiation compound such as phorbol ester (PMA). Similar treatment of HeLa cells with PMA, which activates the HIV-LTR, also led to the stimulation of NF-κB binding activity (40). Finally, it has been recently shown that the activation of HIV-LTR by tumor necrosis factor alpha (TNFα) proceeds via the induction of the NF-κB-like binding factor as well (72,41). Thus, the activation of HIV-LTR by various stimuli seems to be associated with the appearance of a protein binding to NF-κB sequences. It is, however, unlikely that the NF-κB binding activity present in different cell types such as B- and T-cells or HeLa cells represents binding of an identical protein. It was shown by Baldwin and Sharp (42) that at least two distinct factors, NF-κB and H2TFI, can interact with a single regulatory region suggesting that the observed binding may represent the interaction of NF-κB sequences with several distinct DNA binding proteins. Thus the activation of the HIV-LTR by different stimuli could be the result of a complex interplay among the different binding factors and the factors induced by external stimuli (e.g. cytokines) or viral infection (e.g. HSV-1). Alternatively these external stimuli could modify the activity of the trans-acting transcriptional factors.

## ACTIVATION/ENHANCEMENTOF HIV REPLICATIONIN INFECTEDCELLS

Although the activation of HIV-LTR expression by infection with heterologous DNA viruses such as the herpesviruses or their IE gene products has been well established, the crucial question remains whether co-infection with heterologous viruses can activate HIV replication in cells in which the HIV provirus is not expressed, and whether it can enhance virus yields in infected cells. Since in the infected cells virus replication is regulated not only by cellular factors, but also by a number of virus encoded factors, the question remains whether the trans-activation seen with the isolated HIV-LTR can also be demonstrated within the complex regulation involved in the expression of the entire HIV genome.

The results of at least three groups, including ours, indicate that HIV infection can be augmented by co-infection with HSV-1 virus or by co-transfection with ICP0 and ICP4. Ostrove *et al.* (29) has co-transfected the infectious clone of the HIV provirus in HeLa cells either in the absence or presence of ICP0 and ICP4 and followed the production of HIV progeny during 7 days after infection. The results showed that the virus yield (measured by reverse transcriptase activity) was much higher in the cells which were transfected both with HIV provirus and ICP0 and ICP4 genes than with HIV provirus alone. These data indicated that the enhancement of HIV-LTR activity positively contributed to the rate of HIV replication. Direct evidence that co-infection of HSV-1 and HIV within a single cell stimulated HIV replication was recently provided by Albrecht *et al.* (30). These authors compared the yield of HIV produced in a $CD4^+$ CEM cell line which was chronically infected with HSV-1 to the yield of HIV in uninfected CEM cells, and found that HSV-1 infection accelerated HIV replication in these cells, but the overall virus yield was comparable both in HSV-infected and -uninfected cultures. These authors also addressed the question of which IE gene of HSV-1 is essential for the stimulation of HIV replication in these cells. By using HSV-1 deletion mutants they were able to show that the ICP4, but not ICP0, gene product was involved in the stimulation. There is a difference between this observation and the results obtained in co-transfection studies, which showed that the essential gene is ICP0 or a combination of ICP0 and ICP4, but not ICP4 alone. This difference may be due not only to a difference in the cell lines used in these two sets of experiments, but it may also indicate that, within the complexity of the HIV genome, additional transcriptional and post-transcriptional factors encoded by the viral genome may play a role in the overall stimulation of HIV replication.

While infection with HSV-1 was shown to activate the expression of the HIV provirus, it has not yet been shown that infection with CMV, EBV, or HSV viruses activates or enhances replication of the HIV provirus. Multiple factors, such as the effect of the particular viral infection on protein synthesis or stability of viral RNAs, may exert an additional effect on HIV replication which cannot be demonstrated with simple systems such as HIV-LTRcat plasmids in which the virus encoded factors are absent. The lack of a direct correlation between the activity of HIV-LTR in the transfection assay and the ability of the virus to replicate was recently shown by Rabson *et al.* (in preparation). These authors introduced mutations into the infectious HIV proviral DNA that abolished the NF-$\kappa$B mediated transcriptional activity of the HIV-LTR and showed that the virus with the mutated LTR did not lose its capacity to replicate. These data indicate that viral trans-activating factors such as *tat* are able to negate the defect or inefficiency of the viral enhancer region.

Finally, our recent data support the hypothesis that opportunistic infection with HSV-1 may be one of the factors which trigger the onset of the transition from the latent to the replicative HIV state and progression towards AIDS. By using human colon carcinoma cells which contained the latent HIV provirus, we were able to induce virus replication by HSV-1 infection or by transfection with the ICP0 gene (Fig. 2). HIV produced by these cells was infectious and could replicate in mitogen-stimulated lymphocytes. We believe that the ability of HSV-1 to induce HIV expression in these cells may be analogous to trans-activation of expression of the latent HIV provirus in asymptomatic individuals. The important factor in this mechanism is that the initial trans-activation of virus transcription is amplified by the HIV early gene products (*tat* and *rev*). Once the product of *tat* and *rev* gene are synthesized, the virus can replicate in spite of its possible defect in viral LTR which may have rendered it initially inactive (see DNA methylation).

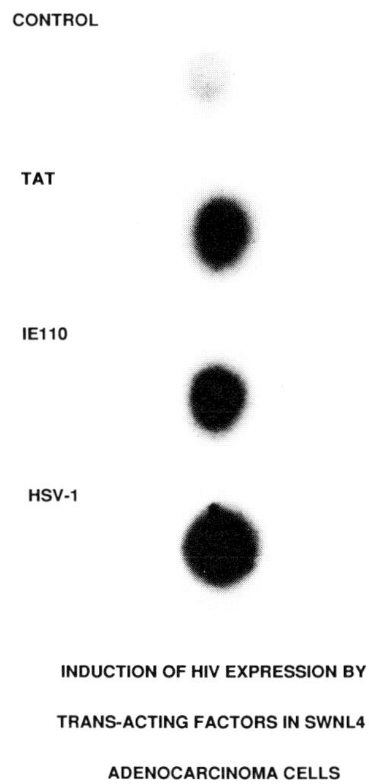


**Fig. 2**. Induction of HIV replication by trans-activating factors in SWNL4 adenocarcinoma cells containing integrated HIV provirus. The cells were transfected with HIV *tat* gene and HSV-1 gene ICP0, or were infected with HSV-1 (moi = 10). The culture medium was collected 24 hrs post-induction and the presence of HIV virions was determined by reverse transcriptase assay. Controls represent untreated SWNL4 cells.

## DNA METHYLATION-INDUCED LATENCY AND VIRUS TRANSACTIVATION.

The concept that DNA methylation is a controlling mechanism for virus latency is based on data acquired from studies of murine leukemia virus and adenovirus systems (43-46). It is first very important to understand the distinction between clinical and biochemical latency. Clinical latency consists of many variables, both viral and host, which suppress virus expression and pathological disease. Biochemical latency refers to the mechanism(s) by which integrated proviral DNA maintains transcriptional silence when harbored by the host cell. The definition of the word latent is derived from the latin term "*latens*" which means "to lie hidden" or "concealed." Today the term is used very loosely and these demarcations should be kept in mind. Our discussion refers to the mechanism(s) of biochemical latency of HIV. Much of the earlier work on DNA methylation involved regulation of gene expression in cellular differentiation and epigenetic inheritance. A very complete review of these studies is provided by Holliday *et al*. (47).

Work completed in our laboratory has focused upon the role of DNA methylation as a mechanism to induce quiescence of the viral genome. Control of cellular and viral gene expression has been shown to be modulated by DNA methylation (48-50). Extensive evidence has implicated the enzymatic conversion of cytosine to 5-methylcytosine in the dinucleotide sequence CpG as a mechanism for transcriptional inhibition of genes (43-51). Viruses are relevant to this concept since they contain sequences which are rich in CpG islands. Many such gene sequences as well as virus promoters depend upon induction of transcription by a protein which acts in *trans*. Over two years ago, our laboratory discovered that when the HIV-LTR was integrated into the host chromosomal DNA in permanent cell lines, its *cis*-level transcriptional activity became silent (Fig. 3; 52). When these cells were exposed to the HIV trans-acting factor *tat*, HSV, or various DNA-insulting agents, LTR transcriptional activity was increased to high levels (Fig. 3; 52,53). Upon further examination of the integrated HIV-LTR, several CpG and at least one non-CpG site(s) were mapped which were determined to be hypermethylated in the unexpressed, latent state. Only upon stimulation of the host cell was *de novo* transcription observed (54).

Previous studies have shown that promoter sequences, which are hypermethylated, demonstrate a greatly reduced capacity to bind some of the trans-acting nuclear proteins (54). However, the same sequences were able to form complexes with protein and show a nuclear protein binding capacity of 100% when the DNA was hypomethylated by 5-azacytidine (5-azaC) treatment (54). This observation is true for many cellular nuclear proteins but conflicts with the action of viral trans-activating factors. The elegant studies of

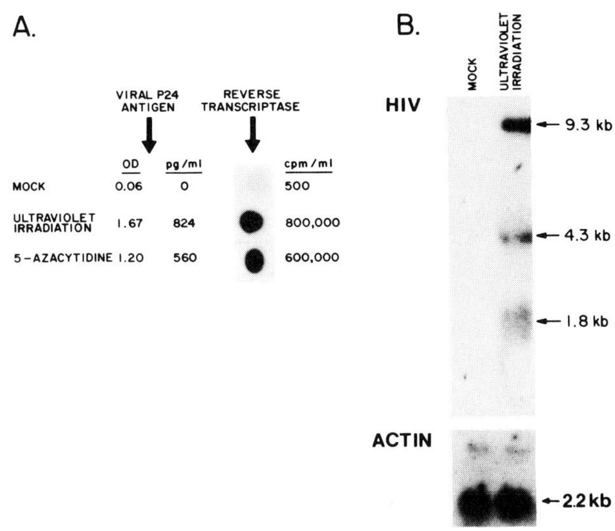


**Fig. 3.** Induction of HIV expression by UV light and 5-azacytidine treatment. SWNL4 cells containing the integrated HIV provirus were either exposed to UV irradiation or treated with 5-azacytidine. **A.** HIV production was measured 24 hrs. after treatment by assaying the levels of viral p24 antigen and reverse transcriptase activity in the medium. **B.** Analysis of viral mRNA present in the treated and untreated (mock) cells. The levels of actin mRNA show that the amount of RNA present on the filters are comparable in mock-treated and induced cells.

Weissharr *et al.* (55) have demonstrated that methylation and transcriptional inactivation of adenovirus regulatory elements can be overcome in *trans* by E1A protein without apparent demethylation of these sequences. In a similar manner, we have found that when HIV-LTR sequences are enzymatically methylated *in vitro*, and then transfected into T-cells, *cis*-level expression is abolished (53). The methylation-sensitive sites in the HIV promoter are within the boundaries of the enhancer/TAR region and are therefore influenced by trans-activating factors which interact with these regions of DNA. Our recent results indicate that the trans-activating factor *tat* overcomes this transcriptional blockade without affecting the apparent methylation status of the LTR (53). In contrast, the stimulation of HIV replication by HSV-1 infection may be due to the hypomethylation of latent HIV-LTR early after infection by HSV-1 (56). This hypomethylation activity of HSV-1 has been observed in other systems as well (56).

One possible mechanism of *tat* action may be the subversion or induction of cellular factors which can reactivate the hypermethylated HIV-LTR. Evidence for this mechanism was derived from our recent studies involving the binding of cellular proteins to *in vitro* methylated HIV-LTR DNA (53). The regions of HIV-LTR sequences sensitive to inactivation by methylation were mapped by deletion analysis to include upstream sequences containing both core enhancer/NF-κB binding sites and three Sp1 binding sites. Within this region we have identified two HpaII/MspI CpG sites located at -218 and -146 nucleotides upstream from the cap site which, when enzymatically methylated *in vitro*, renders the LTR transcriptionally inactive. In order to further examine the mechanism by which *tat* obviates the methylation-induced transcriptional block, we prepared nuclear extracts from SW480 adenocarcinoma cells or SW480 cells expressing *tat* protein. Incubation of nuclear extracts with labeled probe DNA containing HpaII/MspI sites at -218/-146 resulted in retardation of the probe by a nuclear protein (NPI) when analyzed by electrophoretic mobility shift assay. When either HpaII or MspI methylase was employed, the binding affinity of NPI was reduced several-fold. However, when the same probes were incubated with nuclear proteins from SW480 cells expressing *tat*, the binding affinity of NPI remained virtually the same, and a second factor, NPII, appeared. It is not known whether NPII plays any role in allowing the binding of NPI, but it is tempting to speculate such a scenario. Neither nuclear protein is the NF-κB protein as determined by mobility shift studies employing a mutant NF-κB HIV-LTR in which both GGG sequences in the two binding motifs are mutated to CTC (plasmids were obtained courtesy of Drs. D. Baltimore and G. Nabel). These observations are consistent with the interpretation that *tat* obviates the methylation-induced transcriptional block (Bednarik, Cook, and Pitha, in preparation).

The most intriguing aspect of these studies is the appearance of the second nuclear protein (NPII). Can *tat* orchestrate a cooperative mechanism which would permit the binding of transcriptional factors to the proviral DNA regardless of the methylation state? If this is true, then it would not be unreasonable to speculate a role for *tat* in triggering expression of key cellular genes which would not normally be transcribed.

The release of transcription from a methylation-induced latent state does not always correspond with loss of methylation in these regulatory sequences. Although HSV infection changes the methylation state of chromosomal DNA (56), it is unclear whether hypomethylation occurs within the HIV-LTR subsequent to HSV-1 infection (53). It appears that *tat* may initiate trans-activation without requiring removal of methyl groups in a manner analogous to the adenovirus system (55). In the absence of trans-activation, *cis*-level expression would remain inactive as long as the LTR was methylated, thereby preserving latency.

One may envision the hypermethylation of HIV-LTR CpG sequences after integration into host DNA. In such a model, the expression of virus would be minimal to non-existent until cellular activation occurred, or until a heterologous trans-activating factor was encountered. We believe that the results described in these studies support DNA methylation as a viable control mechanism for latency. Further studies, currently in progress, will elucidate the methylation profile of integrated, latent/active HIV by genomic sequencing techniques. This information will contribute to our understanding of the cellular control of retrovirus expression and to the mechanism of trans-activator function.

## FINAL COMMENTS

Epidemiological data indicate that infection with viruses of the herpes group (HSV and CMV) is very common in HIV infected individuals (57). Prolonged fever and chronic lymphadenophathy in AIDS patients, have been related to CMV and EBV infections. The AIDS-associated malignancies such as Kaposi sarcoma and primary central nervous system (CNS) lymphoma have been associated with CMV and EBV infections, respectively. Overall, the high frequency of infection with the herpesviruses and the diseases associated with these infections indicate that the herpesviruses are the most significant viral pathogens in AIDS patients (58).

There are three major consequences of the infection with herpesviruses in AIDS patients. First, these viruses are the major cause of AIDS-associated disease and death. Second, infection with these viruses may provide a trigger in the activation of the latent to the replicative HIV form (59). And third, the viral infection may accelerate the progression of HIV infection by elevating the proportion of activated circulating T-lymphocytes which support productive infection with HIV much more efficiently than resting T-cells. Thus, while the clinical significance of infection with the herpesviruses in AIDS patients is clear, further studies are needed in order to clarify their relationship to the disease process. However, if it is established that trans-activation of HIV replication by the herpesviruses plays an important role in transition to AIDS, a possible therapy for HIV could be targeted towards the inhibition of herpesvirus growth or prevention of herpesvirus infection, especially by blocking the expression of their IE genes. This is part of the rationale for the ongoing study of the role of acyclovir in AIDS therapy.

## ACKNOWLEDGEMENT

We want to thank Dr. A. Rabson for his helpful suggestions during the preparation of this manuscript.

## REFERENCES

1. Rosenberg, Z.F. and Fauci, A.S. Clinc. Immunol. and Immunopath. 50:5149-5156, 1989.
2. Blattner, W.A., Briggar, R.J., Weiss, S.W., Melbye, M. and Goedert, J.J. Ann. Intern. Med. 103: 665-669, 1985.
3. Overbaugh, J., Donahue, P.R., Quackenbush, S.L., Hoover, E.A. and Mullins, J.I. Science 239: 906-910, 1988.
4. Aziz, D.C., Hanna, Z. and Jolicoeur, P. Nature (London) 338: 505-508, 1989.
5. Chattopadhyay, S.K., Morse, H.C. III, Makino, M., Ruscetti, S.K. and Hartley, J.W. Proc. Natl. Acad. Sci. USA 86: 3862-3866, 1989.
6. Hartley, J.W., Wolford, N.K., Old, L.J. and Rowe, W.P. Proc. Natl. Acad. Sci. USA 74: 789-792, 1977.
7. Varmus, H. Genes and Development 2: 1055-1062, 1988.
8. Sodroski, J.G., Rosen, C., Wong-Staal, F., Salahuddin, Z.S., Popovic, M., Arya, S., Gallo, R. and Haseltine, W.A. Science 227: 171-173, 1985.
9. Cullin, B.R. Cell 46: 973-982, 1986.
10. Rosen, C.A., Sodroski, J.G., Goh, W.C., Dayton, A.I., Lippke, J. and

Haseltine, W.A. Nature (London) 319: 555-559, 1986.
11. Felber, B.K., Hadzopoulou-Cladaras, M., Cladaras, C., Copeland, T. and Pavlakis, G. Proc. Natl. Acad. Sci. USA 86: 1495-1499, 1989.
12. Hauber, J., Bouvier, M., Malim, M.H. and Cullen, B.R. J. Virol. 62: 4801-4804, 1988.
13. Neiderman, T.M.J., Thielan, B.J. and Ratner, L. Proc. Natl. Acad. Sci. USA 86: 1128-1132, 1989.
14. Kaufman, J.D., Valandra, G., Rodriquez, Bushar, G., Giri, C. and Norcross, M.A. Mol. Cell. Biol. 7:3759-3766, 1987.
15. Nabel, G. and Baltimore D. Nature (London) 326: 711-713, 1987.
16. Sen R. and Baltimore D. Cell 47: 921-928, 1986.
17. Siekevitz, M., Josephs, S.F., Dukovich, M., Peffer, N., Wong-Staal, F. and Greene, W.C. Science 238: 1575-1578, 1987.
18. Folks, T.M., Justement, J., Kinter, A., Dinarello, C.A. and Fauci, A.S. Science 238: 800-802, 1987.
19. Folks, T.M., Clouse, K.A., Justement, J., Rabson, A., Duh, E., Kerhl, J.A. and Fauci, A.S. Proc. Natl. Acad. Sci. USA 86: 2365-2368, 1989.
20. Matsuyama, T., Hamamoto, Y., Soma, G.-I., Mizuno, D., Yamamota, N. and Kobayashi, N. J. Virol. 63: 2504-2509, 1989.
21. Dinter, H., Chiu, R., Imagana, M., Karin, M. and Jones, K.A. EMBO J. 6: 4067-4071, 1987.
22. Harada, S., Koyanag, Y., Nakashima, H., Kobayashi, N. and Yamamoto, N. Virology 154: 249-258, 1986.
23. Mosca, J.D., Bednarik, D.P., Raj, N.B.K., Rosen, C.A., Sodroski, J.G., Haseltine, W.A. and Pitha, P.M. Nature (London) 325: 67-70, 1987.
24. Mosca, J.D., Bednarik, D.P., Rosen, C.A., Raj, N.B.K., Hayward, G.S., Haseltine, W.A. and Pitha, P.M. Proc. Natl. Acad. Sci. USA 84: 7408-7412, 1987.
25. Gendelman, H.E., Phelps, W., Feigenbaum, L., Ostrove, J.M., Adachi A., Howley, P.M., Khoury, G., Ginsberg, H.S. and Martin, M.A. Proc. Natl. Acad. Sci. USA 83: 9759-9763.
26. Davis, M.G., Kenney, S.C., Kamine, J., Pagano, J.S. and Huang, E.-S. Proc. Natl. Acad. Sci. USA 84: 8642-8646, 1987.
27. Horvat, R.T., Wood, C. and Balachandran, N. J. Virol. 63: 970-973, 1989.
28. Kenney, S., Kamine, J., Markovitz, D., Fenrick, R. and Pagano, J. Proc. Natl. Acad. Sci. USA 85: 1652-1656, 1988.
29. Ostrove, J.M., Leonard, J., Weck, K.E., Rabson, A.B. and Gendelman, H.E. J. Virol. 61: 3726-3732, 1987.
30. Albrecht, M.A. DeLuca, N.A., Byrn, R.A., Schaffer, P.A. and Hammer, S.M. J. Virol. 63: 1861-1868, 1989.
31. Wathen, W.M. and Stinski, M.F. J. Virol. 41: 462-477, 1982.
32. Preston, C.M. J. Virol. 32: 357-369, 1979.
33. Sacks, W.R., Greene, C.C., Aschman, D.P. and Schaffer, P.A. J. Virol. 55: 796-805, 1985.
34. Nabel, G.J., Rice, S.A., Knipe, D.M. and Baltimore, D. Science 239: 1299-1302, 1988.
35. Seto, E., Benedict, T.S., Peterline, B.M. and Ou, J.-H. Proc. Natl. Acad. Sci. USA 85: 8286-8290, 1988.

36. Twu, Jr.-S., Rosen, C.A., Haseltine, W.A. and Robinson, W.S. J. Virol. 63: 2857-2860, 1989.
37. Sen, R. and Baltimore, D. Cell 46: 705-716, 1986.
38. Harrich, D., Garcia, J., Wu, F., Mitsuyasu, R., Gonzalez, J. and Gaynor, R. J. Virol. 63: 2585-2591, 1989.
39. Gimble, J.M., Duh, E., Ostrove, J.M., Gendelman, H.E., Max, E.E. and Rabson, A.B. J. Virol. 62: 4104-4112, 1988.
40. Wu, F.K., Garcia, J.A., Harrich, D. and Gaynor, R.B. EMBO J. 7: 2117-2129, 1988.
41. Osborn, L., Kunkel, S. and Nabel, G.J. Proc. Natl. Acad. Sci. USA 86: 2336-2340, 1989.
42. Baldwin, A.S. and Sharp, P.A. Proc. Natl. Acad. Sci. 85: 723-727, 1988.
43. Bird, A.P. Nature (London) 321: 209-213, 1986.
44. Kruczek, J. and Doerfler, W. Proc. Natl. Acad. Sci. USA 80: 7586-7590, 1983.
45. Harbers, K., Schieke, N., Stuhlmann, H., Jahner, D. and Jaenisch, R. Proc. Natl. Acad. Sci. USA 78: 7609-7613, 1981.
46. Graessmann, M., Graessmann, A., Wagner, H., Werner, E. and Simon, D. Proc. Natl. Acad. Sci. USA 80: 6470-6474, 1983.
47. Holliday, R. Scientific American (June) pp.60-73, 1989.
48. Doerfler, W. Annu. Rev. Biochem. 52: 93-124, 1983.
49. Doerfler, W. Curr. Topics Microbiol. Immunol. 108: 79-98, 1984.
50. Keshet, I., Leiman-Hurwitz, J., Cedar, H. Cell 44: 535-543, 1986.
51. Lindsay, S. and Bird, A.P. Nature (London) 327: 336-338, 1988.
52. Bednarik, D.P., Mosca, J.D. and Raj, N.B.K. J. Virol. 61: 1253-1257, 1987.
53. Bednarik, D.P., Cook, J.A. and Pitha, P.M. UCLA Symposia publication, in press, 1989.
54. Michalowsky, L.A. and Jones, P.A. Mol. Cell. Biol. 7: 3076-3083, 1987.
55. Weissharr, B., Langner, K.D., Juttermann, R., Muller, U., Zock, C., Klimkait, T. and Doerfler, W. J. Mol. Biol. 202: 255-270, 1988.
56. Macnab, J.C.M., Adams, R.L.P., Rinaldi, A., Orr, A. and Clark, L. Mol. Cell. Biol. 8: 1443-1448, 1988.
57. Quinn, T.C., Piot, P., McCormick, J.B., Feinsod, F.M., Taelman, H., Kapita, B., Stevens, W. and Fauci, A.S. JAMA 257: 2617-2621, 1987.
58. Quinnan, G.V., Masur, H., Rook, A.H., Armstrong, G., Frederick, W.R., Epstein, J., Manischewitz, J.F., Macher, A.M., Jackson, L., Ames, J., Smith, H.A., Parker, M., Pearson, G.R., Parrillo, J., Mitchell, C. and Straus, S.E. JAMA 252: 72-77, 1984.
59. Holmberg, S.D., Stewart, J.A., Gerber, R., Byers, R.H., Lee, F.K., O'Malley, P.M., and Nahmias, A.J. JAMA 259: 1048-1050, 1988.

# 13

## SUBUNIT VIRAL VACCINES: PROPHYLACTIC AND THERAPEUTIC USE

L.R.STANBERRY

**Department of Pediatrics, University of Cincinnati College of Medicine and Division of Infectious Diseases, Children's Hospital Research Foundation, Cincinnati, Ohio**

### ABSTRACT

For viruses capable of establishing persistent infection such as the herpes simplex viruses (HSV) and human immunodeficiency viruses (HIV), subunit viral vaccines are likely to prove safer than live-attenuated or whole-inactivated viral vaccines. Numerous structural and nonstructural viral proteins have been shown to be immunogenic in both animals and humans. HSV vaccines containing such immunogenic components do not prevent primary infection in animals. However, they prevent the development of clinical disease, reduce latent infection and decrease the likelihood of subsequent herpetic recurrences. Limited evaluation of subunit HIV vaccines has demonstrated their immunogenicity and shown that they prevent experimental infection of chimpanzees. While protection from disease may be adequate for herpetic infections, it is likely that prevention of infection is an important goal for HIV vaccines. Improved understanding of both the protective as well as the detrimental immune responses produced by viral infection should permit rational selection of subunit immunogens. This, together with the development of safer and more potent synthetic adjuvants, should permit the production of new effective subunit viral vaccines. In addition to providing protection for the uninfected host, these vaccines may be useful as immunotherapeutic agents in the treatment of patients with latent/persistent infection.

### INTRODUCTION

The host, as a consequence of initial viral infection, engenders a variety of immune responses which limits the acute infection and provides long term protection against symptomatic reinfection. The principle of immunization is to induce these long term protective immune responses by administration of non- pathogenic immunogens. A variety of immunizing agents, such as those listed in Table 1, have been used to protect humans and animals against viral disease. Traditionally, most viral vaccines have contained either live-attenuated or inactivated whole virus. The principal advantage of whole viral preparations has been their potent immunogenicity. However, in the case of oncogenic viruses, a major disadvantage of such preparations is that they contain viral nucleic acids, raising the concern that the vaccines may also be potentially oncogenic. A further problem for viruses like HSV and HIV is the concern that live-attenuated whole virus vaccines might also establish persistent infection with potential deleterious effects. While one approach to circumventing these difficulties is to genetically engineer viral mutants which are non-oncogenic and cannot establish persistent infection, a more practical approach is the preparation of subunit vaccines free of viral nucleic acids.

TABLE 1. IMMUNOGENS USED IN VIRAL VACCINES.

| IMMUNOGEN TYPE | VIRAL DISEASE |
| --- | --- |
| Whole virus | |
| Antigenically-related non pathogenic virus | Smallpox |
| Killed (inactivated) whole virus | Polio |
| Live-attenuated virus | Measles |
| Genetically engineered avirulent virus | Pseudorabies |
| Subunit | |
| Solubilized viral proteins (detergent extracted whole virus) | Influenza |
| Purified viral proteins (chemical/mechanical methods) | Hepatitis B |
| Synthetic viral proteins (genetically engineered) | Hepatitis B |

## HSV IMMUNOGENS

The earliest HSV subunit vaccines were crude mixtures of viral proteins prepared from virus infected cells by detergent and/or organic solvent extraction. These preparations were generally rich in the immunogenic envelope glycoproteins but also contained undefined amounts of capsid protein, non-structural viral proteins, cell polypeptides and glycoproteins as well as cellular and viral DNA. Numerous strategies were developed to refine these preparations including the use of inhibitors of nucleic acid synthesis or DNase treatment to reduce viral DNA contamination, ultracentrifugation to separate virus from infected cells and viral nucleocapsids, exhaustive dialysis to remove low molecular weight contaminants, and lectin-sepharose chromatography to further enrich for the glycoprotein components. An interesting variation on the use of detergent and solvent extraction to obtain envelope proteins was developed by Pearson and Scott (1). They showed that chemical treatment of herpes virus saimiri infected cells induced vesicles of plasma membranes to be shed into the medium. Thornton *et al.* (2), capitalized on this observation to prepare HSV-2 vaccine from the plasma membranes of infected MRC-5 cells. Using various combinations of these methods several candidate vaccines were prepared and evaluated in animals and humans. While all of these vaccine preparations contained immunogenic viral polypeptides and glycoproteins, they suffered from three major shortcomings. First, they were of undefined composition, second, it was difficult to insure that they were free of contaminating viral DNA, and third, they were laborious (and hence expensive) to prepare.

A significant improvement in experimental vaccine preparation came with the development of affinity chromatography using monoclonal antibodies to specific viral glycoproteins. This method permitted further purification of the crude extracts reducing the likelihood of DNA contamination and providing defined immunogens. Affinity chromatography purified HSV-1 and HSV-2 glycoproteins, gB, gC, gD, and gE, were examined for protective efficacy. While it was shown that a defined vaccine could be prepared from HSV infected cells by the combined use of ultracentrifugation, detergent/solvent extraction, DNase treatment and affinity chromatography, there remained a small risk of viral DNA contamination and the process was both labor intensive and very inefficient [e.g. Roberts *et al.* (3), reported a recovery of approximately 300 µg of gB from an extract of $10^8$ HSV-1 infected Chang cells]. Genetic engineering provided a major breakthrough in producing defined immunogenic viral polypeptides. The initial strategy developed by Watson and colleagues was to insert the protein coding region of the HSV-1 gD gene into a bacterial expressor plasmid which enabled *Escherichia coli* to synthesize an immunoreactive non-glycosylated gD-like polypeptide (4). This technology made it theoretically possible to produce large amounts of a defined viral immunogen while avoiding most of the problems inherent in preparing a vaccine from virus infected cells. Lasky and co-workers improved the strategy by cloning a truncated gD gene lacking the sequence data coding for the transmembrane anchor domain into a mammalian expression vector (5). This construct not only permitted Chinese hamster ovary cells to synthesize a glycosylated form of gD, but because the protein lacked the anchor domain it was secreted into the medium. This facilitated recovery and purification of the protein and also allowed the cells to be used to continuously produce the immunogen. Pachl, Stuve and colleagues used this same approach to prepare recombinant HSV-1 and HSV-2 gB (6,7). These recombinant glycoproteins are as immunogenic and protective as glycoprotein vaccines derived from virus infected cells. The use of specific immunogens produced by recombinant DNA technology may be a preferable alternative to the use of semi-purified cell lysate vaccines with their batch to batch variability in glycoprotein representation, potential for viral DNA contamination and higher production costs.

Another novel approach to the construction of subunit HSV vaccines has involved the expression of the gD gene by a non-HSV viral vector (8). Vaccinia virus recombinants expressing HSV-1 and/or HSV-2 gD proteins have been constructed and shown to induce HSV-specific humoral and cellular mediated immunity in animals and to protect mice against lethal HSV challenge and guinea pigs against primary and recurrent cutaneous disease (9-12). While this is an intriguing concept, there are questions regarding the potential human safety and efficacy of such vaccines. For example, wild-type vaccinia can occasionally cause a fatal disseminated infection (13) and while these recombinants appear less virulent than the wild-type virus it is unknown whether the genetically altered recombinants can also produce disseminated infection (14,15). Furthermore, it is uncertain whether individuals previously immunized with vaccinia would mount an antibody response against gD when re-immunized with the vaccinia/HSV gD construct (16). A modification of this strategy that may avoid these potential difficulties has been to insert the HSV glycoprotein genes into other non-HSV vectors. For example, McDermott *et al.* (17), reported that vaccination of mice with an adenovirus vector containing HSV-1 gB protected animals from lethal challenge. Other investigators have cloned the gD gene into a baculovirus vector (18). Studies exploring these constructs may prove very interesting.

Another ingenious strategy to maximize the immunogenicity of gD was developed by Valenzuela and co-workers (19). Capitalizing on their extensive knowledge of the molecular biology of hepatitis B surface antigen (HBsAg) they inserted a segment of the HSV-1 gD gene into the region coding for the pre-S segment of HBsAg. Yeast cells transformed with a plasmid carrying this construct produced polymeric particles containing both HSV-1 gD and HBsAg epitopes. These particles were shown to be immunogenic. However, their protective efficacy as a vaccine has not been reported.

The idea of chemically synthesizing polypeptide vaccines was a logical extension of the research designed to genetically engineer native HSV glycoproteins. The concept was to first map the important antigenic sites on viral glycoproteins using monoclonal antibodies, then, based on DNA sequence data, identify those antigenic sites resulting from contiguous amino acid chains and synthesize the appropriate polypeptides. Several laboratories have reported mapping antigenic domains of HSV gB, gC, and gD (20-26). Using data on the the structure of gD antigenic sites, Cohen, Eisenberg and colleagues synthesized peptides corresponding to a type-common antigenic determinant of gD contained within the N-terminal amino acid residues of

HSV-1 gD. When bound to a carrier protein (keyhold limpet hemocyanin) these peptides were shown to be both immunogenic and efficacious in protecting mice from lethal challenge (27). These results are encouraging and certainly warrant further study.

## EVALUATION OF SUBUNIT HSV VACCINES IN MODELS OF PRIMARY INFECTION

Two types of animal models have generally been used in the evaluation of experimental HSV vaccines: lethal challenge models, and models that exhibit cutaneous herpetic disease. Each type of model has certain advantages and disadvantages when exploring the efficacy of an experimental vaccine. The archetypical lethal challenge model is the fatal infection resulting from intraperitoneal HSV injection of mice (5,21,28-30). This model is inexpensive, easily established in most laboratories, has an objective and easily assessable endpoint (death) and evidence of latent HSV infection can be sought in the survivors. It has numerous shortcomings, including: - it does not use a natural route of inoculation and hence the potential role of local immune responses cannot be assessed; - there is no simple way to determine the effect of immunization on the virologic course of infection; - because the animals do not develop cutaneous lesions it is impossible to determine the extent to which immunization prevents morbidity due to skin disease; - the surviving animals exhibit no evidence of recurrent infection and therefore the model is not useful in assessing the effect of immunization on recurrent disease; - because there is only one measure of efficacy (survival), the model is not suited for studies designed to compare the relative efficacy of immunization regimens; and - the small size of the animal does not permit the repetitive large volume blood sampling necessary for assays of cell-mediated immune (CMI) functions.

Other lethal challenge models that have been used or may prove useful in vaccine evaluation include intravenous HSV inoculation of rabbits and mice (28,31-33) intracerebral inoculation of mice (34,35), intranasal inoculation of rabbits and mice (36-38), footpad inoculation of mice (39,40), intravenous inoculation of rabbits and mice (41,42), intravenous inoculation of tree shrews (41), and the intravaginal inoculation of owl monkeys (35).

There are several very good models of cutaneous herpetic disease, one of which is the experimental genital HSV infection of female guinea pigs. This model shares many similarities with genital herpes in humans including a natural route of inoculation (intravaginal), a self-limited, quantifiable genital skin disease (vulvovaginitis) with associated neurologic and urologic complications. The virologic course of primary infection may be determined by plaque titration of vaginal secretions collected on swabs (42). After recovery from primary infection, the animals experience both spontaneous and inducible recurrent infections (42-44) and latent infection can be demonstrated in lumbosacral dorsal root ganglia (42,45). In addition, guinea pigs are of sufficient size as to permit repetitive collection of large volumes of blood for immunologic assays (46). The principal disadvantages of this model vis a vis the lethal challenge mouse model relate to increased costs and the necessity of becoming familiar with a subjective scoring system for quantifying genital skin disease.

Another well characterized model of cutaneous disease is the HSV infection of the mouse ear pinna, a model that has been extensively used in the study of viral pathogenesis and the evaluation of antiviral drugs and vaccines (47-49). Other models of cutaneous HSV infection include oral/labial (50,51) cutaneous (52), and genital (53,54) infection in mice and cutaneous inoculation of rabbits (55) and guinea pigs (56). Guinea pigs inoculated via the footpad have also been useful in vaccine evaluation (12).

There also exist several models of ocular HSV infection using rabbits (57,58), mice (59-62) and guinea pigs (63). The rabbit and mouse models have previously been used to evaluate HSV vaccines and recently the guinea pig model of ocular herpetic disease has also been successfully employed to investigate the efficacy of a HSV-1 envelope vaccine (H.C.Bubel, personal communication).

Animal studies of HSV subunit vaccines have universally demonstrated the immunogenicity of these preparations. In general, experimental vaccines have afforded laboratory animals protection against primary HSV infection. Compared to unimmunized controls, immunized animals have been shown to have less viral replication, reduced morbidity (eg. cutaneous disease) and reduced mortality (see Table 2). There are at least two exceptions to this generalization. Meignier *et al.* (35) reported that a mixture of HSV-2 glycoproteins protected intravaginally infected guinea pigs against disease while a similar vaccine failed to provide any protection to intravaginally inoculated owl monkeys. They suggested the owl monkey model may be a more sensitive system for assessing the efficacy of vaccines because of this animal's exquisite susceptibility to HSV. Another interesting vaccine failure was reported by Schrier and colleagues, who showed that a purified HSV-1 gC vaccine protected mice against HSV-1 but not HSV-2 intraperitoneal challenge, illustrating the importance of homotypic protection (28). Likewise, other investigators have shown that type specific glycoproteins or glycoprotein mixtures provide the greatest protection against homotypic viral challenge, while providing less protection against challenge with the heterologous virus (67-69).

Subunit glycoprotein vaccines also provided substantial protection against recurrent herpetic infections. Guinea pigs immunized with recombinant HSV-1 gD or a mixture of envelope derived purified HSV-2 glycoproteins prior to intravaginal challenge with type 2 virus showed little or no evidence of primary infection and experienced from 45 to 90 percent fewer recurrences than unimmunized controls (67,68). The observation that immunized animals developed recurrences can be taken as *a priori* evidence that these vaccines do not prevent the establishment of latent infection. Indeed, several reports have suggested that HSV subunit vaccines do not prevent latency (39,52,68,71), while other reports indicate a reduced recovery of latent virus from immunized animals (50,68,75). At best, protection from latent infection appears incomplete, possibly because subunit as well as live virus vaccines, can prevent the clinical manifestations of herpetic disease without actually preventing infection (75,76).

In considering the effectiveness of vaccines we need to distinguish between prevention of infection and prevention of disease. For example, vaccine studies using the guinea pig model of genital herpes have shown that animals immunized with specific glycoproteins or glycoprotein mixtures may have little

TABLE 2. ANIMAL MODELS AND THE EVALUATION OF HSV ENVELOPE VACCINES

| IMMUNOGEN[a] | MODEL[b] | INDICES[c] | | | | |
|---|---|---|---|---|---|---|
| | | Virology | Morbidity | Mortality | Latency | REF |
| Env-1 | HSV-1/rabbit/id | ND[d] | ND | Reduced | ND | 55 |
| Env-1 | HSV-1/rabbit/iv | ND | NA[e] | Reduced | ND | 31 |
| Env-1 | HSV-1/mouse/oral | ND | Reduced | Reduced | Reduced | 50 |
| Env-1 | HSV-2/mouse/ivag | Reduced | Reduced | Reduced | ND | 70 |
| Env-1 | HSV-1 or 2/mouse/id | Reduced | Reduced[f] | NA | ND | 28,49 |
| Env-1 | HSV-1/mouse/id | ND | Reduced | Reduced | No effect | 52 |
| Env-1 | HSV-1/mouse/cutaneous | ND | Reduced | Reduced | No effect | 71 |
| Env-1 | HSV-1/mouse/ivag | ND | ND | Reduced | No effect | 71 |
| Env-2 | HSV-2/rabbit/iv | ND | NA | Reduced | ND | 32 |
| Env-2 | HSV-2/mouse/iv | ND | NA | Reduced | ND | 32 |
| Env-2 | HSV-2/guinea pig/ivag | Reduced | Reduced | NA | ND | 2 |
| gProt-1 | HSV-1/mouse/ip | NA | NA | Reduced | ND | 29 |
| gProt-1 | HSV-2/guinea pig/ivag | Reduced | Reduced | NA | ND | 67 |
| gProt-1 | HSV-1/mouse/ic | NA | NA | Reduced | ND | 35 |
| gProt-2 | HSV-2/guinea pig/ivag | Reduced | Reduced | NA | ND | 35,67 |
| gProt-1+2 | HSV-2/owl monkey/ivag | ND | ND | No effect | ND | 35 |
| gB-1 | HSV-2/guinea pig/ivag | Reduced | Reduced | NA | ND | 67 |
| gB-1or2 | HSV-1/mouse/footpad | ND | NA | Reduced | No effect | 39 |
| gB-1or2 | HSV-2/mouse/footpad | ND | NA | No effect | ND | 39 |
| gC-1 | HSV-1/mouse/ip | ND | NA | Reduced | ND | 28 |
| gC-1 | HSV-2/mouse/ip | ND | NA | No effect | ND | 28 |
| gD-1 | HSV-1 or 2/mouse/ip | ND | NA | Reduced | ND | 5 |
| gD-1or2 | HSV-1/mouse/footpad | ND | NA | Reduced | No effect | 39 |
| gD-1 | HSV-2/guinea pig/ivag | ND | Reduced | NA | ND | 72 |
| gD-1 | HSV-2/guinea pig/ivag | Reduced | Reduced | NA | Reduced | 67,68,73 |
| gD-2 | HSV-2/guinea pig/ivag | Reduced | Reduced | NA | No effect | 68 |
| gD/pep | HSV-2/mouse/ip | ND | NA | Reduced | ND | 27 |
| gD/pep | HSV-2/mouse/footpad | ND | Reduced | Reduced | ND | 27,74 |

[a] Env-1 = HSV-1 envelope vaccine; Env-2 = HSV-2 envelope vaccine; gProt-1 = HSV-1 glycoproteins, gProt-2 = HSV-2 glycoproteins; gB-1 = HSV-1 glycoprotein B; gC-1 = HSV-1 glycoprotein C; gD-1 = HSV-1 glycoprotein D; gD-2 = HSV-2 glycoprotein D; gD/pep = synthetic glycoprotein D -related peptides.

[b] Challenge virus/animal model/route of inoculation; id = intradermal; iv = intravenous; ivag = intravaginal.

[c] Indices: virology refers to measures of viral replication in tissues or secretions, morbidity refers to evaluation of non-lethal manifestations of HSV infection (eg. cutaneous lesions).

[d] ND = Not done.

[e] NA = Not applicable

[f] Reduction in severity of primary infection was also associated with a subsequent reduction in recurrent disease.

or no clinical disease but are nevertheless infected as evidenced by viral replication in the genital tract (67,68,75,76) or by seroconversion (77). Bernstein *et al.* (77) have suggested that a functional definition of vaccine failure may be the detection of antibody to HSV proteins not present in the vaccine, since immunized, intravaginally challenged guinea pigs that develop such antibody are more prone to develop severe recurrent disease. The failure of vaccine-induced immune responses to prevent infection may not be surprising since even the immunity resulting from primary genital HSV-2 infection does not prevent the asymptomatic reinfection of the genital tract following HSV-2 reinoculation (76). However, these results raise an important public health question. If HSV vaccines protect humans against disease but not infection, will immunized but asymptomatically infected individuals serve as an important reservoir of communicable virus? While it is laudable to seek a vaccine that prevents not only disease but infection as well, such a goal may be unrealistic. The Committee on Issues and Priorities for New Vaccine Development of the United States National Academy of Sciences' Institute of Medicine has established as realistic goals a 50 percent reduction in the number of symptomatic primary infections, a 75 percent reduction in the number of recurrences, and a reduction in severity of episodes averaging approximately 60 percent (78). Based upon these criteria, if the results obtained using the guinea pig genital infection model are predictive of the protection that can be obtained in humans, then several currently available experimental HSV subunit vaccines may be effective in controlling HSV infections. The limited clinical experience with these vaccines in humans is outlined in a later section.

In addition to evaluating immunogen composition, studies using animal models have been useful in examining the effects of dose, route of administration, immunization schedule and adjuvant on the efficacy of subunit vaccines. Perhaps not surprisingly, numerous investigators have demonstrated enhanced efficacy when the vaccines were combined with adjuvants. Thompson *et al.* (51) reported that several adjuvants including aluminum hydroxide (alum), complete Freund's adjuvant (CFA), and poly-L-lysine complexed with carboxymethylcellulose (PICLC) enhanced the efficacy of a HSV-1 envelope vaccine in protecting mice against oral/labial infection. Similarly, it has been reported that CFA substantially enhanced the efficacy of glycoprotein vaccines in protecting guinea pigs against genital infection (67). More recently it has been observed that CFA is superior to alum in enhancing the protection against primary infection provided by glycoprotein vaccines (75). The extent of enhancement appears related to the immunopotentiating potency of the adjuvant. For example, Sanchez-Pescador and coworkers examined the immunogenicity and efficacy of a recombinant HSV-1 gD vaccine combined with either alum, CFA, a lipophilic derivative of muramyl tripeptide (MTP-PE) or a muramyl dipeptide covalently linked to the immunogen (MDP) (79). They reported that the mean antibody titers for guinea pigs immunized with the gD vaccine plus MTP, alum, or MDP were 57 percent, 16 percent, and 13 percent of the CFA titers, respectively. Protection paralleled the serologic responses with animals immunized with the gD vaccine plus alum or MDP exhibiting more severe genital HSV-2 infection than those immunized with gD plus MTP-PE or CFA. These results emphasize the importance of adjuvant in providing the most potent vaccine possible. Of course the reactogenicity of CFA has precluded its use in humans and while alum is approved for human use, it is clearly a poor adjuvant. The development of additional safe and potent adjuvants will be necessary to insure the effectiveness of HSV subunit vaccines in clinical trials. One interesting experimental adjuvant that warrants further study is MTP (N-acetyl-glucosamine-3yl-acetyl-L-alanyl-d-isoglutamine). Not only has this adjuvant been shown to enhance the efficacy of HSV glycoprotein vaccines in experimental animal infections (79) but it has recently been proven safe in a World Health Organization phase 1 clinical trial of a birth control vaccine (80). Such results are encouraging and raise the possibility of some very interesting vaccine combinations.

A novel approach to enhancing the efficacy of subunit glycoprotein vaccines was developed by Weinberg and Merigan (73). They demonstrated that recombinant interleukin 2 administered in conjunction with a mixture of HSV-2 envelope glycoproteins or recombinant non-glycosylated HSV-1 gD enhanced the efficacy of the immunogens in protecting guinea pigs against primary genital HSV-2 infection. They also observed that animals protected against primary infection exhibited no evidence of recurrent infections. Cytotoxic immune responses, which are important in the control of HSV disease, were enhanced by interleukin 2, hence, this cytokine may prove useful as an adjuvant for HSV subunit vaccines.

**IMMUNOGENICITYOF HSV SUBUNIT VACCINES IN HUMANS**

Both crude envelope vaccines as well as lectin purified, infected cell-derived glycoprotein vaccines have been shown to induce humoral and cellular immune responses in humans. Cappel and coworkers reported that an HSV-2 envelope vaccine elicited neutralizing, complement fixing and complement dependent cytotoxic antibodies in HSV seronegative vaccinees as well as producing anamnestic antibody rises in HSV seropositive recipients. Likewise, they showed that HSV seronegative vaccinees developed CMI, as evidenced by lymphocyte stimulation responses, while the HSV seropositive recipients exhibited enhancement of the CMI responses (81,82).

Using a HSV-1 envelope vaccine Skinner *et al.* (83) reported that 19 of 21 and 4 of 8 HSV seronegative volunteers receiving high and low doses of vaccine, respectively, developed neutralizing antibody titers, while 6 of 13 HSV seropositive individuals given the low dose of vaccine demonstrated elevation of their neutralizing anti-HSV antibody titers.

Early work examining the immunogenicity of a glycoprotein vaccine in humans was reported by Hileman in 1980. He showed that six volunteers given a vaccine composed of HSV-2 infected cell-derived glycoproteins adsorbed to alum developed low titer neutralizing antibody and lymphocyte stimulation responses to HSV- 1 and HSV-2 (84). Working with the same HSV-2 glycoprotein vaccine, Mertz *et al.* (85) examined the ontogeny of immune responses in 13 HSV seronegative and 10 HSV-1 seropositive vaccinees. The recipients were given 3 doses of the vaccine, with the second and third doses administered four and 22 weeks after the first dose. Twelve of the 13 seronegative vaccinees developed neutralizing antibody and antibody-dependent cell-mediated cytotoxicity (ADCC) against HSV. Although the neutralizing and ADCC titers waned over time, they were still detectable one year after the first immunization. Interestingly, only four of the 10 HSV-1 seropositive vaccinees developed a four-fold rise in their HSV-2 neutralizing titers. The investigators noted that this was similar to the percentage of HSV-1 seropositive patients who develop a four-fold rise in anti-HSV-2 neutralizing antibody after first episode genital HSV-2 infection. All 13 seronegative vaccine recipients were shown to develop *in vitro* CMI responses which, unlike the waning humoral responses, remained stable over the seven months following the third immunization. These results parallel the observations made in immunized laboratory animals and suggest that HSV subunit vaccines might be efficacious in protecting humans against primary HSV infections.

## VACCINE TRIALS IN HUMANS:PROTECTIONAGAINST PRIMARY HSV INFECTION

There are only a limited number of studies on the ability of subunit vaccines to protect humans against primary HSV infection. Skinner and co-workers reported on 60 subjects immunized with a HSV-1 envelope vaccine without adjuvant (86). The vaccine recipients were the regular sexual consorts of patients with recurrent genital herpes, and therefore, were at high risk of developing symptomatic primary genital HSV infection. On two occasions one month apart, the volunteers received either low dose ($2 \times 10^7$ infected cell equivalents) or high dose ($1 \times 10^8$ infected cell equivalents) vaccine. Neutralizing HSV antibody responses were noted in 59 percent and 90 percent of the low dose and high dose recipients, respectively. After a mean follow-up period of 18 months (range 4 - 26 months) 59 of the 60 vaccinees remained free of any symptoms indicative of genital HSV infection. One vaccine recipient reported an episode suggestive of a mild case of genital herpes one week following her first immunization. However, at follow-up two weeks later the vaccinee had no clinical evidence of genital herpes even though HSV was recovered from her genital tract. Because this was an open trial no data from unimmunized controls were collected. For comparison purposes the investigators offered historic control data. They noted that of 20 unimmunized consorts eight had contracted symptomatic genital herpes within one year of consortship. This attack rate is considerably higher than 10 cases per 40 consorts previously reported by these investigators (83).

In 1987, Skinner and his colleagues reported on 50 subjects immunized with the HSV-2 envelope vaccine (44 of whom received vaccine with alum adjuvant). As in their previous study, the vaccinees did not have a history of genital herpes but their sexual partners did. In follow-up periods ranging from 4 to 48 months only one of the 50 recipients developed symptoms suggestive of genital HSV infection. No pre- or post- immunization immune studies were reported. While there are numerous potential problems associated with non-placebo controlled, open trials, the results of these studies suggest that subunit HSV vaccines may be efficacious in protecting patients against symptomatic primary HSV infections.

A prospective, placebo-controlled double-blind human trial evaluating subunit vaccine efficacy for protection from primary HSV infection was conducted by Corey and co-workers (88). One hundred and sixty one HSV-2 seronegative volunteers whose sexual partners had a history of recurrent genital herpes were randomized to receive a vaccine containing HSV-2 glycoproteins with alum (50 µg at weeks 0,4, and 22) or placebo. At entry, 60 percent of the vaccinees were HSV-1 seropositive. Sera were collected for analysis to examine for HSV proteins not present in the vaccine (89). This permitted the investigators to detect vaccinees who became asymptomatically infected during the trial. During the first year, 7 of 83 vaccine recipients and 5 of 78 placebo recipients developed HSV-2 infection. Symptomatic genital herpes occurred in 80 percent of cases while asymptomatic seroconversion to HSV-2 proteins not present in the vaccinees was observed in 20 percent of cases. Half of the symptomatic cases among the vaccine recipients were brief atypical episodes with culture-negative lesions. The investigators determined an attack rate of 0.16 cases of genital HSV-2 infection per patient per year in HSV-1 seronegative individuals compared to 0.07 cases per patient per year in HSV-1 seropositive persons ($p<0.05$). Their results suggested that prior HSV-1 infection provided partial protection against acquisition of HSV-2, and while the subunit vaccine did not prevent HSV-2 infection it did ameliorate the severity of genital disease. One would predict that the failure of this vaccine to protect against HSV-2 infection might be related to its immunogenicity. Evaluation of the immunologic data from this study is ongoing and as yet has not been reported. This very well conceived and executed study illustrates the importance of a placebo controlled design, the significance of controlling for pre-existing HSV antibody, and the necessity of examining, by serologic means, for evidence of asymptomatic HSV-2 infection.

## HSV SUBUNIT VACCINES AS IMMUNOTHERAPEUTIC AGENTS

It has been periodically suggested that stimulation of the immune system of the latently infected host might result in better control of recurrent herpetic infections. The basic premise has been that host response to natural infection is incomplete or inadequate to control recurrent disease, although what specific defect(s) may be involved is only now being carefully investigated (90-93). Several strategies were developed in an attempt to augment the immune responses of the latently infected host. Early efforts included the use of smallpox vaccine to non-specifically stimulate host immunity (94,95) as well as autoinoculation with live non-attenuated HSV (96-98). In retrospect, the decision to administer virulent live viruses to patients with presumed immune defects seems dangerously naive. Neither of these approaches were successful in controlling herpetic recurrences. A more reasoned approach involved the use of formalin-inactivated whole HSV preparations. As early as 1938, Frank reported on the use of formalin-inactivated suspensions of HSV infected rabbit brain (99). This preparation was neither immunogenic nor efficacious. With the advent of cell culture systems, it became possible to prepare safer and immunogenic inactivated whole virus vaccines. Inactivated HSV vaccines are currently licensed in several European countries and have indications for the treatment of persons with recurrent herpetic disease (100). However, the evidence supporting the efficacy of these vaccines in the management of recurrent HSV infections is based upon retrospective, uncontrolled studies (101). In one prospective, controlled trial using an inactivated whole HSV vaccine, Kern and Schiff found that 70 percent of the vaccine recipients reported improvement (appreciably fewer recurrences) and

76 percent of the control group also reported improvement (102). This study illustrates both the dubious efficacy of these vaccines in controlling recurrent infections and the significant placebo effect that must be considered when evaluating any therapy for recurrent HSV disease.

More recent studies have examined the efficacy of subunit HSV vaccines in controlling recurrent herpetic infections. In his initial study, Skinner reported that 17 of 24 patients with a history of recurrent genital herpes improved following immunization with an adjuvant-free HSV-1 envelope vaccine (83). These vaccine recipients reported less frequent, less severe, or less extensive herpetic disease. While the study was encouraging it was of uncertain significance because of its uncontrolled, open trial design. A similar study was conducted by Cappel *et al.* (82) using a vaccine consisting of HSV-2 envelope proteins but lacking adjuvant. They reported that compared to 33 unimmunized control patients, the 59 vaccine recipients experienced fewer recurrences, significantly greater intervals between recurrences ($p<0.01$), and when recurrences did occur the vaccinees reported a significantly shorter duration of pain and time to healing ($p<0.01$). While this study included a control group, the unimmunized patients did not receive a placebo and hence a possible placebo effect in the immunized group could not be excluded. Interpretation of the data was further compromised by combining patients with genital HSV-2 recurrences and patients with oral HSV-1 recurrences in the control group.

Initial animal studies by European investigators did not substantiate the findings of the human trials. Scriba reported that the pattern of spontaneous recurrent HSV-2 infections observed in latently infected guinea pigs was not altered by the administration of a HSV-2 envelope vaccine with incomplete Freund's adjuvant (103). Further, Slichtova *et al.* (104) showed that a HSV-1 envelope vaccine with complete Freund's adjuvant given to latently infected mice did not diminish inducible cutaneous HSV-1 recurrences. The animal studies failed to provide any scientific support for the concept of immunotherapy, but once again illustrated the importance of a randomized, placebo- controlled experimental design when evaluating therapy for recurrent herpetic infections.

In 1988, Kutinova and co-workers reported on a small-scale, double-blind, placebo-controlled trial of a HSV-1 infected cell- derived glycoprotein vaccine for treatment of patients with frequent recurrent HSV-1 or HSV-2 infections (105). The 42 volunteers were immunized at least three times at three week intervals with either an alum adsorbed glycoprotein mixture or a placebo. The patients were evaluated for frequency of recurrences, lesion size, duration of the vesicular phase, or time to healing. Over the 14 month follow-up period the majority of patients exhibited improvement, however, the percentage of patients reporting improvement was about equally distributed between the two groups, 57.1 percent of the vaccine recipients versus 45.0 percent of the placebo controls. As with Kern and Schiff's 1964 study (102), a remarkable placebo effect was observed and no evidence of vaccine efficacy could be demonstrated. Unfortunately, this vaccine preparation was only minimally immunogenic, producing little or no rise in HSV antibody titers, hence, the only conclusion that can be drawn from this study (aside from the importance of the placebo effect) is that poorly immunogenic HSV vaccines are not useful in the treatment of recurrent herpetic disease. Presently, a multi-center, randomized, double-blind, placebo controlled trial of the Skinner inactivated HSV vaccine for treatment of recurrent genital herpes is nearing completion (C. Benson, personal communication). Because of the careful study design and large number of patients enrolled, this trial is expected to provide interesting (and meaningful) data regarding vaccine immunotherapy.

To further examine specific immunotherapy for recurrent HSV infections Stanberry *et al.* (106), administered vaccines containing HSV glycoproteins with complete Freund's adjuvant to latently infected guinea pigs. In two experiments, after recovery from primary genital HSV-2 infection, animals were randomized to receive either a glycoprotein vaccine or to serve as controls. Both a mixture of HSV-2 glycoproteins derived from infected cell cultures and a mixture of HSV-1 glycoproteins B and D, prepared by genetic engineering in mammalian cells, significantly reduced the frequency and severity of subsequent herpetic recurrences ($p<0.05$). Further, vaccine administration was associated with a 4- to 244- fold rise in HSV antibody titers. This was the first controlled experimental evidence that administering HSV glycoprotein vaccines to a latently infected host could alter the natural history of recurrent herpetic disease. Because latently infected guinea pigs, like humans, may shed virus from their genital tract in the presence or absence of clinically apparent lesions (42,107,108) there was concern that glycoprotein vaccines might reduce symptomatic recurrences simply by converting them to subclinical recurrences. To address this question Myers *et al.* (109), conducted a second immunotherapy experiment which demonstrated that administration of the HSV-1 gB + gD vaccine to animals experiencing recurrent HSV-2 infections reduced the frequency of both clinical recurrences and cervicovaginal viral shedding.

Subsequently, Kern and his colleagues reported that recombinant HSV-1 or HSV-2 gD vaccines containing an alum adjuvant protected guinea pigs against primary infection but failed to alter the frequency or duration of recurrent HSV-2 infections when administered to latently infected animals (68). The vaccine contained 50 μg of recombinant protein administered subcutaneously on days 29,43 and 57 after intravaginal HSV-2 inoculation. Unfortunately, no immunologic data were presented. It is likely that the conflicting results of these studies might be a consequence of vaccine composition, immunization schedule, adjuvant or route of administration. To address these possibilities Stanberry and co-workers examined the influence of adjuvant as well as the timing and route of administration upon the immunotherapeutic efficacy of a HSV-1 gB + gD vaccine (110). In one experiment vaccine with complete Freund's adjuvant was administered by footpad injection during primary infection (day 8), early after recovery from primary infection (day 15), or after the onset of recurrent disease (day 21) with a second dose of vaccine given 21 days after the first immunization. We observed that all the vaccinated groups experienced significantly fewer recurrences ($p<0.05$), with early initiation of glycoprotein immunotherapy on day 8 of primary infection resulting in the greatest reduction in subsequent recurrences. In a second experiment they demonstrated that the inclusion of a potent adjuvant such as complete

Freund's or muramyl tripeptide was essential for immunotherapeutic efficacy. In the final experiment, we observed that footpad administration of vaccine was more effective in reducing recurrences than was vaccine given intramuscularly. Interestingly, all immunized animals responded to the vaccine with significant elevations in HSV antibody titers, however, the greatest titer rises were noted in animals receiving vaccine with complete Freund's adjuvant administered via the footpad. In addition to inducing humoral responses, preliminary studies suggested that this vaccine combination induces enhanced lymphoproliferative, cytolytic, and lymphokine responses in HSV-2 infected guinea pigs (111) and antibody dependent cell cytotoxicity which was protective in the lethal neonatal mouse model (112).

Confirmation of the effectiveness of glycoprotein immunotherapy in the treatment of experimental recurrent herpetic disease was recently reported by Merigan's group (113). Using the guinea pig model of recurrent genital herpes, they demonstrated that a liposome preparation containing recombinant glycoprotein D with a mixture of muramyl tripeptide and phosphatidylethanolamine produced a 75 percent reduction in the frequency and severity of recurrent infections. This liposome-immunogen-adjuvant combination also enhanced antigen-specific lymphocyte stimulation.

Collectively, these studies suggest that specific immune augmentation can result in better control of recurrent HSV infections. The ineffectiveness of subunit vaccines alone or in combination with alum in reducing herpetic recurrences in controlled trials suggests that in order to be effective the immunogen must be administered with potent adjuvants. Given the substantial placebo effect observed in treatment studies of recurrent HSV disease in humans, it is essential that any human trial of viral glycoprotein immunotherapy involve a randomized, double-blind, placebo-controlled design. While the potential benefits of such therapy are great, the possible adverse reactions must also be carefully explored. Undesirable effects that warrant critical assessment include the possibility that viral immunogens may trigger a Stevens-Johnson reaction (erythema multiforme) or that the use of potent adjuvants may increase the risk of developing lymphoma. With the advent of genetic engineering and the co-development of new synthetic adjuvants it is very likely that virus-specific immunotherapy of persistent viral infections can be successfully developed.

## OTHER HERPESVIRUS SUBUNIT VACCINES

Our knowledge of the immunogenic glycoproteins of varicella-zoster virus (VZV), cytomegalovirus (CMV), and Epstein-Barr virus (EBV) is sufficiently advanced as to permit the development of subunit vaccines (114-118). Indeed, at least one gene fragment of VZV used for producing a polypeptide has been patented for use as a possible vaccine against chickenpox and VZV disease (119). The principal reason for the limited development of vaccines against other herpesvirus infections relates to the lack of suitable animal models with which to evaluate the candidate vaccines. Unlike HSV which has a broad host range and therefore, several excellent animal models, the other human herpesviruses have a very restricted host range and hence, few or no models that use the human virus to produce infection in animals. Little progress has been made in developing experimental animal infections with CMV and EBV. However, several laboratories have recently demonstrated that guinea pigs and marmosets can be experimentally infected with VZV (120-124). Future application of these models should facilitate the successful development of VZV subunit vaccines.

## RELATED SUBJECTS

Several excellent reviews pertaining to herpesvirus vaccines have recently been published. The reviews by Meignier (125), Hall and Katrak (126) and Dix (127) provide additional information regarding live and killed whole virus vaccines. Natural resistance mechanisms in HSV infections has been discussed by Lopez (128), the immunological reactivity of HSV proteins has been overviewed by Norrild and Teglbjaerg (129) and Rouse and colleagues have recently reviewed the role of the cytotoxic T lymphocyte in vaccine-induced protection against viral disease (130).

## HIV IMMUNOGENS

Unlike the large HSV genome, HIV-1 contains only seven known genomic coding regions located between two long terminal repeat sequences. All of the gene products coded for by these regions may be considered as potential immunogens (Table 3).

TABLE 3. HIV-1 GENE PRODUCTS AS POTENTIAL SUBUNIT VACCINES

| HIV GENE | PRODUCT | FUNCTION | VACCINE CANDIDATE |
|---|---|---|---|
| env | gp160 | Precursor of envelope glycoprotein | Yes |
| | gp120 | Envelope glycoprotein - CD4 binding site | Yes |
| | gp41 | Transmembrane protein | Yes |
| gag | p55 | Precursor of the structural proteins | Yes |
| | p24 | Major core protein | Yes |
| | p17 | Nucleocapsid protein | Yes |
| | p15 | ? | ? |
| pol | p51/p66 | Enzymes including reverse transcriptase | ? |
| | p31 | Endonuclease | ? |
| sor | p23 | Regulatory protein | ? |
| tat | p14 | Regulatory protein | ? |
| nef | p27 | Regulatory protein | ? |
| art | p18 | Regulatory protein | ? |

Unfortunately, what constitutes protective immunity to HIV is undefined, making it difficult to predict the best immunogens and the best strategies for developing a vaccine. While an inactivated whole virus vaccine could theoretically present all HIV-1 immunogens to the host, there are several potential problems associated with such a strategy. First, killed whole virion HIV vaccines contain retroviral genome which might integrate into the host genome producing deleterious effects. Second, the conformational integrity of the viral proteins can be altered by the process of inactivation. Cellular and humoral immune responses engendered by such altered proteins may be unable to protect against viral infection or may actually enhance disease produced by viral challenge. For example, young cats immunized with formalin inactivated feline leukemia virus (an onco-retrovirus) or goats immunized with formalin inactivated caprine arthritis-encephalitis virus (a lenti-retrovirus) developed more severe disease after viral challenge than did the unimmunized controls (131,132). In the case of HIV-1, it has been suggested that antibodies to gp120, the envelope glycoprotein, may cause enhancement of HIV infection and potentially result in auto-immunosuppression through a mechanism of molecular mimicry (133,134). Salk and colleagues have recently prepared a gp120-depleted, killed virion HIV vaccine which they propose to test in seronegative persons (135). Such human trials will undoubtedly yield new information regarding the immunopathogenesis of HIV disease.

In order to produce infection, HIV specifically binds to the CD4 molecule of T4 positive cells via the major viral envelope glycoprotein, gp120 (136), hence an immune response against gp120 might prevent HIV infection. Using a traditional approach, Arthur and co-workers have prepared an immunogenic gp120 vaccine from viral infected cells (137). However, such preparations may carry a small risk of contaminating viral nucleic acid. The use of recombinant DNA technology has made it possible to produce purified HIV envelope glycoproteins while abrogating the concern regarding contaminating viral RNA. An immunogenic non-glycosylated gp120-like molecule has been produced in bacteria by Putney *et al.* (138) and in yeast by Steimer and her coworkers at the Chiron Corporation (139). Interestingly, the non-glycosylated yeast product, while immunogenic, does not bind to the CD4 molecule. Currently, the importance of the carbohydrate moiety in either protective immunity or in immunogen-induced enhancement of disease is uncertain. Future studies comparing glycosylated and non-glycosylated immunogens will be able to address this important question. Presently, a glycosylated gp120 has been successfully produced by Snyder and coworkers using a recombinant vaccinia virus coinfection system (140), while the full length precursor protein, gp160, has been prepared by scientists at MicroGeneSyS Inc. using a baculovirus insect vector expression system (135). Human trials with purified gp120 or gp160 preparations are currently underway, however, two potential problems may limit the usefulness of such vaccines. The first, discussed earlier, is the theoretical risk that immunization with glycosylated envelope proteins might result in enhancement of disease after exposure to HIV. The second potential problem concerns the enormous genetic variation of the envelope (*env*) gene. Immune responses to the hypervariable region(s) of gp120 are highly specific for the immunizing antigen and may not recognize HIV strains with different epitopes. Conserved domains within the *env* gene yield group-specific epitopes, but it is currently unknown whether immune responses to these group-specific epitopes will be sufficient to provide protection against multiple different HIV strains. This issue may be particularly important if HIV-1 gp120/gp160 vaccines are evaluated in Africa where HIV-2 is prevalent. Although HIV-1 and HIV-2 demonstrate considerable homology, the envelope glycoproteins are distinct, suggesting that a HIV-1 *env* polypeptide vaccine may not protect against HIV-2 infection (141).

In addition to intact envelope glycoproteins, other HIV immunogens are being explored as possible vaccine candidates. These include synthetic peptides from immunogenic domains of the envelope glycoproteins (142) and core proteins (p17) (143); intact recombinant core proteins (*gag* polypeptides p55 and p24) (139); virus-like particles composed of yeast retrotransposon:HIV *env* or *gag* protein hybrids (144); and the expression of envelope glycoproteins via live virus vectors (eg. vaccinia) (145,146). While these preparations appear to be immunogenic, their protective efficacy has not been established. The use of anti-idiotype antibodies is also being investigated. Anti-idiotype antibodies to the CD4 molecule have been shown to neutralize HIV isolates *in vitro* (147). However, the safety of such a strategy has not been established and the potential for undesirable autoimmune responses must be carefully evaluated.

## ANIMAL MODELS OF HIV INFECTION

A major shortcoming in the development and evaluation of HIV vaccines has been the lack of suitable animal models. The best model to date is the chimpanzee. When inoculated with HIV these animals become viremic and engender immune responses, however, they do not exhibit clinical illness nor do they progress to an immunodeficiency state. Further, the number of animals available for HIV related experimentation is very limited and the cost of chimpanzee studies is staggering ( exceeding $30,000 per animal) (135). Only limited HIV vaccine studies have been conducted using chimpanzees. Gibbs and coworkers observed seroconversion by Western blot analysis in a chimpanzee following administration of a gp120-depleted whole virion vaccine (135), while Arthur and colleagues reported that chimpanzees immunized with a gp120/alum vaccine developed waning neutralizing antibody (137). However, no data regarding protective efficacy was provided for either vaccine. Hu *et al.* (145) immunized chimpanzees with a recombinant vaccinia virus expressing the HIV envelope protein. These animals developed both HIV specific antibody and T-cell responses but on viral challenge still became viremic, indicating that immunization did not prevent HIV infection. Recently, Eichberg reported testing seven HIV vaccines in chimpanzees: a synthetic peptide; a vaccinia-*env* recombinant; a vaccinia-*gag* recombinant; a mixture of the vaccinia-*env* and vaccinia-*gag* recombinants; a recombinant gp120 expressed in mammalian cells; a recombinant gp160 peptide; and anti-CD4 idiotypic antibodies (148). He reported that all vaccines induced humoral responses but that CMI responses were only observed with the vaccinia constructs and with the gp120 vaccine. Animals immunized with the synthetic peptide, the vaccinia-*env* recombinant, or the gp120 vaccine were subsequently challenged with HIV-1 and, disappointingly, all demonstrated evidence of viremia. Clearly, we need a better understanding of which immune responses are important in the prevention of infection. Because of the limits of the chimpanzee model, additional animal model infections are needed. Recently, two groups reported having successfully infected rabbits with HIV-1 (149, 150), while another group reported infecting baboons with HIV-2 (151). The further development of these and future models should facilitate vaccine development.

While not useful in evaluating vaccines for human use, several animal model infections employing other retroviruses are proving important in exploring the immunopathogenesis of retroviral infection. One exciting model is the simian immunodeficiency virus (HIV) infection of macaques (152). This lenti-retrovirus has considerable homology with HIV-1 and HIV-2 and produces an AIDS- like syndrome in cynomolgus and rhesus monkeys (153,154). Ongoing studies using this model to investigate inactivated HIV vaccines may provide important leads in designing HIV vaccines (155). Recently a lenti-retrovirus has been isolated from cats with an AIDS-like syndrome (156). Laboratory cats can be experimentally infected, developing lymphadenopathy, immunosuppression, and possible opportunistic infections. The greater availability of animals for this model infection, along with reduced costs may encourage further development of this system. Ungulate lentiviruses including visna virus infection of sheep, caprine arthritis-encephalitis virus infection of goats, and equine infectious anemia virus infection of horses have been useful in studying the pathogenesis of lenti-retrovirus infection (157). While these viruses produce a chronic infection, they do not produce an AIDS-like syndrome. Clearly, the size of these animals and their housing requirements limit their universal applicability.

## HUMAN TESTING OF HIV SUBUNIT VACCINES

Limited evaluations of candidate HIV vaccines have been performed in humans. Both Zagury *et al.*, and Cooney and co-workers have reported that a vaccinia virus vaccine expressing HIV gp160 induced both humoral and CMI responses in volunteers (158, 159). Likewise, a study sponsored by the National Institutes of Health evaluating the MiroGeneSys HIV gp160 baculovirus produced vaccine has shown that volunteers develop antibodies against gp160 without apparent adverse effects (160-162). A synthetic peptide from p17 produced by Viral Technologies, Inc. has been shown to be immunogenic in animals and is soon to be tested in humans in the United Kingdom (163). No studies to date have addressed the efficacy of any of these experimental vaccines.

At least two studies exploring the usefulness of HIV vaccines in the treatment of HIV infected patients are currently underway. Jonas Salk and co-workers have immunized HIV seropositive asymptomatic volunteers with an inactivated gp120-depleted vaccine (135). Likewise, investigators at the Walter Reed Army Medical Center have begun evaluation of the MicroGeneSys HIV gp160 vaccine in HIV seropositive persons (R. Redfield, personal communication). It is too early in these studies to determine either beneficial or untoward effects.

## SUMMARY

Our expanded knowledge of the HSV and HIV genomes coupled with recent advances in recombinant DNA technology have made possible the production of defined viral immunogens. These immunogens will provide the basis for a new generation of viral vaccines. To facilitate the rational (rather than the empiric) development of safe and effective subunit viral vaccines however, we need to further define the protective immune responses produced by natural infection, we need to identify safe and potent new adjuvants and we need to continue to develop predictive animal models for vaccine evaluation.

## ACKNOWLEDGEMENTS

This work was supported in part by grant number AI 22667 from the National Institutes of Health. I wish to thank Barbarie Hill and her staff of librarians who managed to locate even the most obscure references. I would also like to thank Sandra Baur Orlando for excellent editorial assistance and Drs. Sandra L. Fowler, Beverly L. Connelly and David I. Bernstein for their thoughtful review of this manuscript.

## REFERENCES

1. Pearson, G.R. and Scott, R.E. Proc. Natl. Acad. Sci. USA. 74:2546-2550, 1977.
2. Thornton, B., Baskerville, A., Bailey, N.A., Melling, J.M. and Hambleton, P. Vaccine 2:141-148, 1984.
3. Roberts, P.L., Duncan, B.E., Raybould T.J.G. and Watson, D.H. J. Gen. Virol. 66:1073-1085, 1985.
4. Watson, R.J., Weis, J.H., Salstrom,J.S. and Enquist, L.W. Science 218:381-384, 1982.
5. Lasky. L.A., Dowbenko, D., Simonsen, C.C. and Berman, P.W. Biotechnology 2:527-532, 1984.
6. Pachl, C., Burke, R.L., Struve, L.L., Sanchez-Pescador, L., Van Nest, G., Masiarz, F. and Dina, D. J.Virol. 61:315-325, 1987.
7. Struve, L.L., Brown-Shimer, S., Pachl, C., Najarian, R. Dina, D. and Burke, R.L. J. Virol. 61:326-335, 1987.
8. Paoletti, E., Perkus, M.E. and Piccini, A. Antiviral Res. 1(suppl.1):301-307, 1985.
9. Paoletti, E., Lipinskas, B.R., Samsonoff, C., Mercer S. and Panicali, D. Proc. Natl. Acad. Sci. USA. 81:193-197, 1984.

10. Cremer, K.J., Mackett,M., Wolhenberg, C., Notkins, A.L. and Moss, B. Science 228:737-740, 1985.
11. Martin, S., Moss, B., Berman P.W., Laskey, L.A. and Rouse, B.T. J. Virol. 61:726-734, 1987.
12. Wachsman, M., Aurelian, L., Smith, C.C., Perkus, M.E. and Paoletti, E. J. Infect. Dis. 159:625-634, 1989.
13. Lane, J.M. and Millar, J.D. Am. J. Epidemiol. 93:238-240, 1971.
14. Moss, B. In: **High Technology Route to Vaccines** (Eds. G.R.Dreesman, L.G.Bronson, and R.C.Kennedy), Am. Soc. Microbiol., Washington, D.C., 1985, pp.69-74.
15. Buller, R.M.L., Chakrabarti, S., Cooper, J.A., Twardzik, D.R. and Moss, B. J. Virol. 62:866-874, 1989.
16. Beale, J. Nature (London). 302:476-477, 1983.
17. McDermott, M.R., Graham,F.L., Hanke, T. and Johnson, D.C. Virology 169:244-247, 1989.
18. Krisna, S., Blacklaws, B.A., Overton, H.A., Bishop, D.H.L. and Nash, A.A. J. Gen. Virol. 70:1805-1914, 1989.
19. Valenzuela, P., Coit, D., Medina-Selby, M.A., Kuo, C.H., VanNest, G., Burke, R.L., Bull, P., Urdea, M.S. and Graves, P. Biotechnology 3:323-326, 1985.
20. Pellet, P.P., Kousoulas, K.G., Pereira, L. and Roizman, B. J. Virol. 53:243-253, 1985.
21. Kikuchi, G.E., Coligan, J.E., Holland, T.C., Levine, M., Glorioso, J.C. and Nairn, R. J. Virol. 52:806-815,1984.
22. Marlin, S.D., Holland, T.C., Levine, M. and Glorioso, J.C. J. Virol. 53:128-136, 1985.
23. Eisenberg, R.J., Long, D., Pereira, L., Hampar, B., Zweig, M. and Cohen, G.H. J. Virol. 41:478-488, 1982.
24. Cohen, G.H., Dietzschold, B., Ponce de Leon, M., Long, D., Golub, E.A., Varrichio, A., Pereria, L. and Eisenberg, R.J. J. Virol. 49:102-108, 1984.
25. Dietzschold, B., Eisenberg, R.J., Ponce de Leon, M., Golub, E., Hudecz, F., Varricho, A. and Cohen, G.H. J. Virol. 52:431-435, 1984.
26. Eisenberg, R.J., Long, D., Ponce de Leon, M., Matthews, J.M., Spear, P.G., Gibson, M.G., Lasky, L.A., Berman, P., Golub, E. and Cohen, G.H. J. Virol. 53:634-644, 1985.
27. Eisenberg, R.J., Cerini, C.P., Heilman, C.J., Joseph, A.D., Dietzschold, B., Golub, E., Long, D., Ponce de Leon, M. and Cohen, G.H. J. Virol. 56:1014-1017, 1985.
28. Schrier, R.D., Pizer, L.I. and Moorhead, J.W. J.Immunol. 130:1413-1418, 1983.
29. Chan, W.L. Immunology 49:343-352, 1983.
30. Long, D., Madara, T.J., Ponce de Leon, M., Cohen, G.H., Montgomery, P.C. and Eisenberg, R.J. Infect. Immun. 43:761-764, 1984.
31. Cappel, R. Arch. Virol. 52:29-35, 1976.
32. Cappel, R., de Cuyper, F., and Rickaert, F. Arch. Virol. 65:15-23, 1980.
33. Koff, W.C., Showalter, S.D., Hampar, B., and Fidler, I.J. Science 228:495-497, 1985.
34. Renis, H.A., Eidson, E.E., Mathews, J. and Gray, J.E. Infect. Immun.

14:571-578, 1976.
35. Meignier, B., Jourdier, T.M., Norrild, B., Pereira, L., and Roizman, B. J. Infect. Dis. 155:921-930, 1987.
36. Stroop, W.G. and Schaefer, D.C. J. Infect. Dis. 153:721-731, 1986.
37. Richards, J.T., Kern, E.R., Overall Jr., J.C., and Glasgow, L.A. J. Infect. Dis. 144:464-471, 1981.
38. Rooney, J.F., Wohlenberg, C., Cremer, K.J., and Notkins, A.L. J. Infect. Dis. 159:974-976, 1989.
39. Dix, R.D. and Mills, J. J. Med. Virol. 17:9-18, 1985.
40. Balachandran, N., Bacchetti, S., and Rawls, W.E. Infect. Immun. 37:1132-1137, 1982.
41. Darai, G., Schwaier, A., Komitowski, D., and Munk, K. J. Infect. Dis. 137:221-226, 1978.
42. Stanberry, L.R., Kern, E.R., Richards J.T., Abbott, T.M., and Overall Jr., J.C. J. Infect. Dis. 146:397-404, 1982.
43. Stanberry, L.R., Kern, E.R., Richards, J.T., and Overall Jr., J.C. Intervirology 24:226-231, 1985.
44. Stanberry, L.R. J. Med. Virol. 28:125-128, 1989.
45. Stanberry, L.R., Kit, S., and Myers, M.G. J. Virol. 55:322-328, 1985.
46. Harrison, C.J., Jenski, L., Voychehovski, T., and Bernstein, D.I. Antiviral Res. 10:209-224, 1988.
47. Hill, T.J., Field, H.J., and Blyth, W.A. J. Gen. Virol. 28:341-353, 1975.
48. Field, H.J. In: **Antiviral Agents: The Development and Assessment of Antiviral Chemotherapy Volume I** (Ed. H.J.Field), CRC Press, Boca Raton, Florida, 1988, pp. 67-84.
49. Slichtova, V., Kutinova, L., and Vonka, V. Arch. Virol. 66:207-214, 1980.
50. Kitces, E.N., Morahan, P.S., Tew, J.G., and Murray, B.K. Infect. Immun. 16:955-960, 1977.
51. Thomson, T.A., Hilfenhaus, J., Moser, H., and Morahan, P.S. Infect. Immun. 41:556-562, 1983.
52. Klein, R.J., Buimovici-Klein, E., Moser, H., Moucha, R., and Hilfenhaus, J. Arch. Virol. 68:73-80, 1981.
53. Baker, M.B., Larson, C.L., Ushijima, R.N., and Anderson, F.D. Infect. Immun. 10:1220-1234, 1974.
54. Walz, M.A., Price, R.J., Hayashi, K., Katz, B.J. and Notkins, A.L. J. Infect. Dis. 135:744-752, 1977.
55. Zaia, J.A., Palmer, E.L. and Feorino, P.M. J. Infect. Dis. 132:660-666, 1975.
56. Spruance, S.L. and McKeough, M.B. Antiviral Res. 9:295-313, 1988.
57. Nesburn, A.B., Cook, M.L., and Stevens, J.G. Arch. Ophthalmol. 88:412-417, 1972.
58. Green, M.T., Dunkel, E.C., and Courtney, R. Invest. Ophthalmol. Vis. Sci. 25:1436-1449, 1984.
59. Rector, J.T., Lausch, R.N., and Oakes, J.E. Infect. Immun. 38:168-174, 1982.
60. Tullo, A.B., Shimfeld, C., Blyth, W.A., Hill, T.J., and Easty, D.L. Arch. Ophthalmol. 101:961-964,1983.
61. Atherton, S.A. and Streilein, J.W. Current Eye Res. 6:133-139, 1987.

62. Ksander, B.R. and Hendricks, R.L. Invest. Ophthalmol. Vis. Sci. 28:1986-1993, 1987.
63. Wander, A.H., Bubel, H.C., and McDowell, S.G. Arch. Virol. 95:197-209, 1987.
64. Willey, D.R., Cantin, E.M., Hill, L.R., Moss, B., Notkins, A.L., and Openshaw, H. J. Infect. Dis. 158:1382-1386, 1988.
65. Hendricks, R.L., Tao, M.S.P., and Glorioso, J.C. J. Immunol. 142:263-269, 1989.
66. Meignier, B., Longnecker, R., and Roizman. B. J. Infect. Dis. 158:602-614, 1988.
67. Stanberry, L.R., Bernstein, D.I., Burke, R.L., Pachl, C., and Myers, M.G. J. Infect. Dis. 155:914-920, 1987.
68. Berman, P.W., Vogt, P.E., Gregory, T., Lasky, L.A., and Kern, E.R. J.Infect. Dis. 157:897-902, 1988.
69. Stanberry, L.R., Ward, M.A., Burke, R.L., and Myers, M.G. Pediatr. Res. 25:191A, 1989.
70. Skinner, G.R.B., Williams, D.R., Buchan, A., Whitney, J., Harding, M., and Bodfish, K. Med. Microbiol. Immunol. 166:119-132, 1978.
71. Hilfenhaus, J., Moser, H., Herrmann, A., and Mauler, R. In: **Developments in Biological Standardization, Volume 52 : Herpes Virus of Man and Animal: Standardization of Procedures** (Eds. M. Bonneau and H. Hennessen), S. Karger, Basel, 1982, pp. 321-331.
72. Berman, P., Gregory, T., Crase, D., and Lasky, L. Science 227:1490-1492, 1985.
73. Weinberg, A. and Merigan, T.C. J. Immunol. 140:294-298, 1988.
74. Watari, J., Dietzschold, B., Szokan, G. and Heber-Katz, E. J. Exp. Med. 157:459-470, 1987.
75. Stanberry, L.R., Myers, M.G., Stephanopoulos, D.E., and Burke, R.L. J. Gen. Virol. 70:3177-3185, 1989.
76. Stanberry, L.R., Bernstein, D.I., Kit, S., and Myers, M.G. J. Infect. Dis. 153:1055-1061, 1986.
77. Bernstein, D.I., Stanberry, L.R., Kappes, J.C., Burke, R.L. and Myers, M.G. J. Infect. Dis. 157:1178-1186, 1988.
78. Committee on Issues and Priorities for New Vaccine Development, In: **New Vaccine Development: Establishing Priorities,** National Academy Press, Washington, D.C., 1985, pp 280-312.
79. Sanchez-Pescador, L., Burke, R.L., Ott, G. and Van Nest, G. J. Immunol. 141:1720-1727, 1988.
80. Jones, W.R., Judd, S.J., Ing, R.M.Y., Powell, J., Bradley, J., Denholm, E.H., Mueller, U.W., Griffin, P.D. and Stevens, V.C. Lancet 1:1295-1298, 1988.
81. Cappel, R., Sprecher, S., Rickaert, F. and de Cuyper, F. Arch. Virol. 73:61-67, 1982.
82. Cappel, R., Sprecher, S., de Cuyper, F. and De Braekeleer, J. J. Med. Virol. 16:137-145, 1985.
83. Skinner, G.R.B., Woodman, C., Hartley, C., Buchan, A., Fuller, A., Wiblin, C., Wilkins, G. and Melling, J. In: **Developments in Biological Standardization, Volume 52 : Herpes Virus of Man and Animal: Standardization of Procedures** (Eds. M. Bonneau and H. Hennessen),

S. Karger, Basel, 1982, pp 333-344.
84. Hilleman, M. R., Larson, V.N., Lehman, E.D., Salerno, R.A., Conard, P.G. and McLean, A.A. In: **The Human Herpesviruses: An Interdisciplinary Perspective** (Eds. A. J. Nahmias, W.R. Dowdle and R.F. Schinazi), Elsevier, New York, 1981, pp 503-506.
85. Mertz, G.J., Peterman, G., Ashley, R., Jourden, J.L., Salter, D., Morrison, L., McLean, A. and Corey, L. J. Infect. Dis. 150:242-249, 1984.
86. Woodman, C.B.J., Buchan, A., Fuller, A., Hartley, C., Skinner, G.R.B., Stocker, D., Sugrue, D., Clay, J.C., Wilkins, G., Wilbin, C. and Melling, J. Bri. J. Vener. Dis. 59:311-313, 1983.
87. Skinner, G.R.B., Fink, C.G., Cowan, M., Buchan, A., Fuller, A., Hartley, C.E., Durham, J., Wiblin, C. and Melling, J. Med. Mircobiol. Immunol. 176:161-168, 1987.
88. Corey, L., Mertz, G., Ashley, R. and Benedetti, J. Abstracts of the 27th Interscience Conference on Antimicrobial Agents and Chemotherapy, New York City, October, 1987, Abst. #1288.
89. Ashley, R., Mertz, G.J. and Corey, L. J. Virol. 61:264-268, 1987.
90. Sheridan, J.F. and Aurelian, L. Diagn. Immunol. 1:245-256, 1983.
91. Donnenberg, A.D., Bell, R.B. and Aurelian, L. Cell. Immunol. 56:526-539, 1980.
92. Donnenberg, A.D., Chaikof, E. and Aurelian, L. Infect. Immun. 30:99-109, 1980.
93. Iwasaka, T., Sheridan, J.F. and Aurelian, L. Infect. Immun. 42:955-964, 1983.
94. Schiff, B.L. and Kern, A.B. Postgrad. Med. 15:32-36, 1954.
95. Kern, A.B. and Schiff, B.L. J. Invest. Dermatol. 33:99-109, 1959.
96. Lazar, M.P. Amer. Med. Assoc. Arch. Dermatol. 71:70-74, 1955.
97. Goldman, L. Arch. Dermatol. 84:1025-1030, 1961.
98. Blank, H. and Haines, H.G. J. Invest. Dermatol. 61:223-225, 1973.
99. Frank, S.B. J. Invest. Dermatol. 1:267-282, 1938.
100. Wise, T.G., Pavan, P.R. and Ennis, F.A. J. Infect. Dis. 136:706-711, 1977.
101. Dundarov, P., Andonov, P., Bakalov, B., Nechev, K. and Tomov, C. In: **Developments in Biological Standardization, Volume 52 : Herpes Virus of Man and Animal: Standardization of Procedures** (Eds. M. Bonneau and H. Hennessen), S. Karger, Basel, 1982, pp.351-358.
102. Kern, A.B. and Schiff, B.L. Arch. Virol. 89:844-845, 1964.
103. Scriba, M. Med. Microbiol. Immunol. 171:33-42, 1982.
104. Slichtova, V., Kutinova, L. and Vonka, V. Arch. Virol. 71:75-78, 1982.
105. Kutinova, L., Benda, R., Kalos, Z., Dbaly, V., Votruba, T., Kvicalova, E., Petroska, P., Doutlik, S., Kaminkova, J., Domorazkova, E., Bergmonov, V., Krcmar, A., Suchankova, A., Nemeckova, S. and Vonka, V. Vaccines 6:223-228, 1988.
106. Stanberry, L.R., Burke, R.L. and Myers, M.G. J. Infect. Dis. 157:156-163, 1988.
107. Ekwo, E.E., Wong, Y.W. and Myers, M.G. Amer. J. Obstet. Gynecol. 134:102-103, 1979.
108. Adam, E., Dreesman, G.E., Kaufman, R.H. and Melnick, J.L. Amer. J. Obstet. Gynecol. 137: 827-830, 1980.

109. Myers, M.G., Bernstein, D.I., Harrison, C.J. and Stanberry, L.R. Antiviral Res. 10:83-88, 1988.
110. Stanberry, L.R., Harrison, C.J., Bernstein, D.I., Burke, R.L., Shukla, R., Ott, G. and Myers, M.G. Antiviral Res. 11:203-214, 1989.
111. Harrison, C.J., Bernstein, D,I., Jenski, L., Stanberry, L.R. and Myers, M.G. Abstracts of the 28th Interscience Conference on Antimicrobial Agents and Chemotherapy, Los Angeles, California, October, 1988, pp 236.
112. Bernstein, D.I., Loo, L.S. and Kohl, S. Antiviral Res. 10:279-284, 1988.
113. Ho, R.J.Y., Burke, R.L. and Merigan, T.C. J. Virol. 63:2951-2958, 1989.
114. Pereira, L. In: **The Herpesviruses, Volume 3** (Ed. B. Roizman), Plenum Press, New York, 1985, pp 383-404.
115. Brunell, P.A., Novelli, V.M., Keller, P.M. and Ellis, R.W. J. Infect. Dis. 156:430-435, 1987.
116. Jones, F. and Grouse, C. J. Virol. 62:2701-2711, 1988.
117. Dolyniuk, M., Pritchett, R. and Kieff, E. J. Virol. 17:935-949, 1976.
118. Qualtiere, L.F., Chase, R. and Pearson, G.R. J. Immunol. 129:814-818, 1982.
119. Patent Report, Vaccine 5:251, 1987.
120. Myers, M.G., Duer, H.I. and Hausler, C.K. J. Infect. Dis. 142:414-420, 1980.
121. Matsunaga, Y., Yamanishi, K. and Takahashi, M. Infect. Immun. 37:407-412, 1982.
122. Myers, M.G., Stanberry, L.R. and Edmond, B.J. J. Infect. Dis. 151:106-113, 1985.
123. Walz-Cicconi, M.A., Rose, R.M., Dammin, G.J. and Weller, T.H. Arch. Virol. 88:265-277, 1987.
124. Provost, P.J., Keller, P.M., Banker, F.S., Keech, B.J., Klein, H.J., Lowe, R.S., Morton, D.H., Phelps, A.H., McAleer, W.J. and Ellis, R.W. J. Virol. 61:2951-2955, 1987.
125. Meignier, B. In: **The Herpesviruses, Volume 4** (Eds. B. Roizman and C. Lopez), Plenum Press, New York, 1985, pp 265-296.
126. Hall, M.J. and Katrak, K. Vaccine 4: 138-150, 1986.
127. Dix, R.D. Prog. Med. Virol. 34:89-128, 1987.
128. Lopez, C. In: **The Herpesviruses, Volume 4,** (Eds. B. Roizman and C. Lopez), Plenum Press, New York, 1985, pp 37-68.
129. Norrild, B. and Teglbjaerg, C.S. In: **Herpesvirus,** (Ed. F. Rapp), Alan R. Liss, Inc, New York, 1984, pp 23-36.
130. Rouse, B. T., Norley, S. and Martin, S. Rev. Infect. Dis. 10:16-33, 1988.
131. Schaller, J.P., Hoover, E.A. and Olsen, R.G. J. Natl. Cancer Inst. 59:1441-1450, 1977.
132. McGuire, T.C., Adams, D.S., Johnson, G.C., Klevjer-Anderson, P., Barbee, D.D. and Gorham, J.R. Amer. J. Vet. Sci. 47:537-540, 1986.
133. Martinez-A, C., de la Hera, A., Alonso, I. M., Marcos, M.A.R., Marquez, C., Toribio, M.L. and Coutinho, A. Lancet 1:454-457, 1988.
134. Robinson Jr., W.E., Montefiori, D.C. and Mitchell, W.M. Lancet 1:830-831, 1988.
135. Fauci, A. S., Gallo, R.C., Koenig, S., Salk, J. and Purcell, R.H. Ann.

Intern. Med. 110:373-385, 1989.

136. Klatzmann, D., Champagne, E., Chamaret, S., Gruest, J., Guetard, D., Hercend, T., Gluckman, J. and Montagnier, L. Nature (London) 312:767-768, 1984.

137. Arthur, L.O., Pyle, S.W., Bess Jr.,J.W., Gilden, R.V., Nara, P.L., Robey, W.G., Fischinger, P.J. and Kelliher, J.C. In: **Vaccines** (Cold Spring Harbor), (Ed. H.M. Ginsberg), Cold Spring Harbor Laboratory, Cold Spring Harbor, New York, 1988, pp 277-282.

138. Putney, S.D., Matthews, T.M., Robey, W.G., Lynn, D.L., Robert-Guroff, M., Mueller, W.T., Langlois, A.J., Ghrayeb, J., Petteway Jr., S.R., Weinhold, K.,J., Fischinger, P.J., Wong-Staal,F., Gallo, R.C. and Bolognesi, D.P. Science 234:1392-1395, 1986.

139. Steimer, K.S., Van Nest, G.A., Haigwood, N.L., Tillson, E.M., George-Nascimento, C., Barr, P.J. and Dina, D. In: **Vaccines** (Cold Spring Harbor), (Ed. H.M Ginsberg), Cold Spring Harbor Laboratory, Cold Spring Harbor, New York, 1988, pp 347-355.

140. Snyder Jr., C.E., Earl, P.L., Todd, W. and Barstad, P.A. In: **Technological Advances in Vaccine Development** (Ed. L. Lasky), Alan R. Liss, Inc, New York, 1988, pp 311-320.

141. Gnann Jr., J.W., McCormick, J.B., Mitchell, S., Nelson, J.A. and Oldstone, M.B.A. Science 237:1346-1348, 1987.

142. Rusche, J.R., Javaherian, K., McDanal, C., Petro, J., Lynn, D.L., Gramaila, R., Langlois, A., Gallo, R.C. and Arthur, L.O. Proc. Natl. Acad. Sci. USA. 85:3198-3202, 1988.

143. Naylor, P.H., Naylor, C.W., Badamchian, M., Wada, S., Goldstein, A.L., Wang, S.S., Sun, D.K., Thornton, A.H. and Sarin, P.S. Proc. Natl. Acad. Sci. USA, 84:2951-2955, 1987.

144. Adams, S.E., Senior, J.M., Kingsman, S.M. and Kingsman, A.J. In: **Technological Advances in Vaccine Development** (Ed. L. Lasky), Alan R. Liss, Inc, New York, 1988, pp 117-126.

145. Hu, S.L., Fultz, P.N., McClure, H.M., Eichberg, J.W., Thomas, E.K., Zarling, J., Singhal, M.C., Kosowski, S.G., Swenson, R.B., Anderson, D.C. and Todaro, G. Nature 328:721-723, 1987.

146. Zarling, J.M., Eichberg, J.W., Moran, P.A., McClure, J., Sridhar, P. and Hu, S.L. J. Immunol. 139:988-990, 1987.

147. Dalgleish, A.G., Thomson, B.J. and Chanh, T.C. Lancet 2:1047-1050, 1987.

148. Eichberg, J.W. Abstracts of the 5th International Conference on AIDS, Montreal, Quebec, June, 1989, pp 513.

149. Filici, G., Cereda, P.M. and Varnier, O.E. Nature 335:366-369, 1988.

150. Kulaga, H., Folks, T., Rutledge, R., Truckenmiller, M.E., Gugel, E. and Kindt, T.J. J. Exp. Med. 169:321-326, 1989.

151. Letvin, N.L., Daniel, M.D., Sehgal, P.K., Yetz, J.M., Solomon, K.R., Kannagi, M., Schmidt, D.K., Silva, D.P., Montagnier, L. and Desrosiers, R.C. J. Infect. Dis. 156:406-407, 1987.

152. Daniel, M.D., Letvin, N.L., King, N.W., Kannagi, M., Sehgal, P.K., Hunt, R.D., Kanki, P.J., Essex, M. and Desrosiers, R.C. Science 228:1201-1204, 1985.

153. Daniel, M.D., Letvin, M.D., Sehgal, P.K., Desrosiers, R.C., Hunt, R.D.

Waldron, L.M., MacKey, J.J., Schmidt, D.K., Chalifoux, L.V. and King, N.W. Science 230:71-73, 1985.

154. Franchini, G., Gurgo, C., Guo, H.G., Gallo, R.C., Collati, E., Fargnoli, K.A., Hall, L.F., Wong-Staal, F. and Reitz, Jr., M.S. Nature 328:539-542, 1987.

155. Sutjipto, S., Pedersen, N., Gardner, M.B., Hanson, C.V., Miller, C., Gettie, A., Jennings, M., Higgins, J. and Marx, P.A. Abstracts of the 5th International Conference on AIDS, Montreal, Quebec, June, 1989, pp 541.

156. Pedersen, N.C., Ho, E.M., Brown, M.L. and Yamamoto, J.K. Science 235:790-793, 1987.

157. Haase, A.T. Nature 322:130-136, 1986.

158. Cooney, E.L., Zarling, J., Hu, S.L., Watson, A., Corey, L. and Greenberg, P. Abstracts of the 5th International Conference on AIDS, Montreal, Quebec, June, 1989, pp 538.

159. Zagury, D., Picard, O., Bernard, J., Salaun, J.J., Lurhuma, Z. and Gallo, R.C. Abstracts of the 5th International Conference on AIDS, Montreal, Quebec, June, 1989, pp 549.

160. Dolin, R., Graham, B., Greenberg, S., Tackett, C., Belshe, R., Clements, M., Fernie, B., Stablein, D., Smith, G. and Koff, W. Abstracts of the 5th International Conference on AIDS, Montreal, Quebec, June, 1989, pp 538.

161. Garrison, L., Midthun, K., Clayman, B., Smith, G., Visicidi, R, and the NIAID AIDS Vaccine Evaluation Group. Abstracts of the 5th International Conference on AIDS, Montreal, Quebec, June, 1989, pp 543.

162. Keefer, M.C., Bonnez, W., Roberts Jr., N.J., Dolin, R. and Reichman, R.C. Abstracts of the 5th International Conference on AIDS, Montreal, Quebec, June, 1989, pp 543.

163. Goldstein, A.L., Naylor, P.H., Kirkley, J.E., Naylor, C.W., Gibbs, C.J. and Sarin, P.S. Abstracts of the 5th International Conference on AIDS, Montreal, Quebec, June, 1989, pp. 544.

# 14

## ANTISENSE NUCLEIC ACID ANALOGUES AS POTENTIAL ANTIVIRAL AGENTS

PAUL S. MILLER

**Department of Biochemistry, The Johns Hopkins University, School of Hygiene and Public Health, Baltimore, Maryland, 21205**

### ABSTRACT

Antisense oligodeoxyribonucleotides and nuclease resistant, phosphate-modified oligonucleotide analogues can be used to inhibit virus gene expression both *in vitro* and in cell culture. Oligomers targeted to bind to functional regions of viral mRNAs such as the initiation codon region or splice junction regions of pre-mRNA, can interfere with mRNA expression in a sequence specific manner. Antisense oligonucleotides and oligonucleotide analogues have been used to inhibit virus protein synthesis and replication in cells infected with vesicular stomatitis virus, herpes simplex virus and human immunodeficiency virus. Because the inhibitory effects of the oligomers depend upon the oligomer sequence, it appears that antisense oligomers could be used as research tools to study and manipulate virus gene expression in cell culture and could be developed into highly specific and effective antiviral agents.

### INTRODUCTION

In recent years a great deal of interest has focused on the synthesis of oligonucleotides and oligonucleotide analogues which are complementary to nucleotide sequences in mRNA. Such antisense oligonucleotides are designed to hybridize with specific mRNA species and as a result, specifically suppress the expression or function of the targeted mRNA. When used in cell culture, antisense oligonucleotides have the potential of being able to specifically control gene expression at the mRNA level. Reagents of this type would be useful not only as research tools but could also be developed for practical uses such as therapeutic agents (1-4).

Viruses generally introduce "foreign" nucleic acid sequences into cells. Because these sequences are usually unique to the virus, they should provide ideal targets for antisense oligonucleotides. Efforts have been underway in a number of laboratories to use antisense oligonucleotides to study and control virus gene expression in cell culture. Because antisense oligonucleotides can in theory home in on a single viral mRNA without binding to or perturbing cellular mRNAs, they hold the promise of being highly specific antiviral agents.

This chapter will describe the properties of some of the oligonucleotides and oligonucleotide analogues which have been developed as antisense reagents. Attention will focus on those oligomers which have been specifically targeted against virus mRNAs and the experiments which have been carried out both *in vitro* and in virus infected cells in culture.

## ANTISENSE OLIGONUCLEOTIDE ANALOGUES

Antisense oligonucleotides have been synthesized with a variety of sugar-phosphate backbones. In each case, the oligomers consist of 2'-deoxyribonucleoside units linked *via* (3'- 5') internucleotide bonds. In general, the oligomers can be divided into two types, those which contain a negatively charged sugar-phosphate backbone, and those which contain nonionic internucleotide linkages.

ß-Oligodeoxyribonucleotides [1] contain the same sugar-backbone found in naturally occurring DNA. The phosphodiester linkage is negatively charged and is susceptible to hydrolysis by both exo- and endonucleases. The oligomers can form stable hybrids with both single stranded DNA and RNA. The strength and fidelity of hybridization is controlled to a large extent by the ionic strength of the medium.

ß-Oligodeoxyribonucleotides targeted against various regions of mRNA have been used both in cell free systems and in cell culture to block translation (5-25). It appears that these oligomers can inhibit translation by at least two different mechanisms. The oligomers can bind to the mRNA and physically block binding or translocation of the ribosome during translation. The effectiveness of this mechanism may depend upon the binding site of the

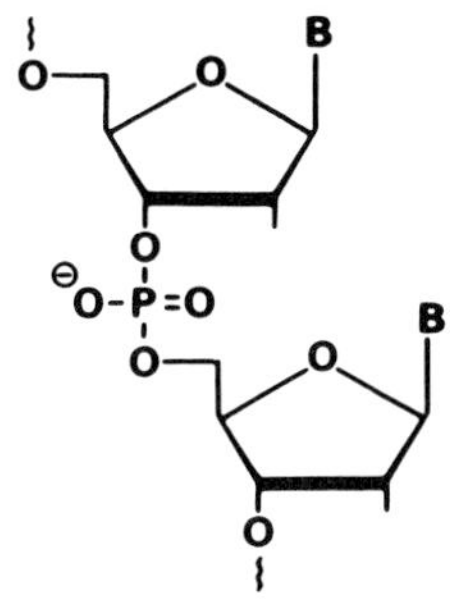

1

antisense oligomer, because ribosomes appear to have an unwinding activity which can disrupt secondary structure during the elongation step of protein synthesis (8). Thus, oligomers bound to the coding regions of the mRNA may be displaced during protein synthesis. In this case, oligomers bound to the initiation codon region or the 5-cap site of the mRNA might be able to successfully prevent ribosome binding, thus preventing the initiation of translation.

ß-Oligodeoxyribonucleotides bound to mRNA form hybrid duplexes which can serve as substrates for ribonuclease H, an activity which cleaves the RNA strand of RNA/DNA hybrids (10,16,20,23). This activity, which is found in a wide variety of both prokaryotic and eukaryotic cells, may in some cases be responsible for the antisense activity of oligodeoxyribonucleotides. Because the oligomer is not cleaved by the action of this enzyme, it can "catalyze" the cleavage of a large number of mRNA molecules. This could account for the ability of these oligomers to effectively inhibit translation at very low concentrations in cell free systems and to function effectively at low concentrations in some cell culture systems. In support of this mechanism, Haeuptle *et. al.* found that oligomers targeted against the coding regions of mRNA resulted in the formation of truncated proteins in *Xenopus* oocytes (12).

In an effort to inhibit translation or processing of the mRNA, ß-oligodeoxyribonucleotides have been targeted against functional regions of mRNA or against the splice junction regions of precursor mRNA. This targeting strategy relies on the known ability of the oligomers to form duplexes with single stranded RNA. Under certain conditions, pyrimidine-containing oligomers can bind to polypurine tracts in double stranded DNA with the formation of triple stranded complexes (26,27). This property offers the possibility of targeting genes directly at the DNA level rather than at the mRNA level. The ability of oligonucleotides to form triple stranded complexes with heteronucleotide sequences in DNA is currently being explored (28-31). It appears that such triple strand formation can be used to inhibit gene expression in cell culture (32).

Although the hydrogen bonding capabilities of ß-oligodeoxyribonucleotides make them quite suitable for use as antisense reagents, their sensitivity to nuclease hydrolysis could limit their utility as antisense reagents, especially in cell culture or in animals. In addition, the negatively charged sugar-phosphate backbone of these oligomers may reduce their ability to penetrate the phospholipid bilayer of cellular membranes. In an effort to increase the nuclease resistance and/or to enhance the uptake of the antisense oligomers, a variety of analogues have been synthesized which have specific properties designed to enhance their usefulness as antisense reagents.

α-Anomeric oligodeoxyribonucleotides [**2**] contain a negatively charged phosphodiester backbone which is very resistant to hydrolysis by both exo- and endonucleases (33-39). These oligomers form very stable duplexes with both single stranded DNA and RNA. It appears that RNA/oligomer duplexes are considerably more stable than DNA/oligomer duplexes. Despite these promising properties, the α-anomeric oligomers do not inhibit mRNA translation in cell-free translating systems and have not shown inhibitory properties in cell culture systems (40-42). It appears that α-oligomer/RNA duplexes are not substrates for ribonuclease H, which may account for their lack of inhibitory activity. It is unclear, however, why the oligomers cannot prevent translation by simply blocking ribosome binding during initiation of elongation.

Substitution of one of the nonesterified oxygens of the phosphodiester bond with a sulfur atom produces negatively charged phosphorothioate analogs [**3**] (2). Because the phosphorous atom of this internucleotide linkage is bound to four nonequivalent atoms, it is chiral. Thus, the linkage can exist in either

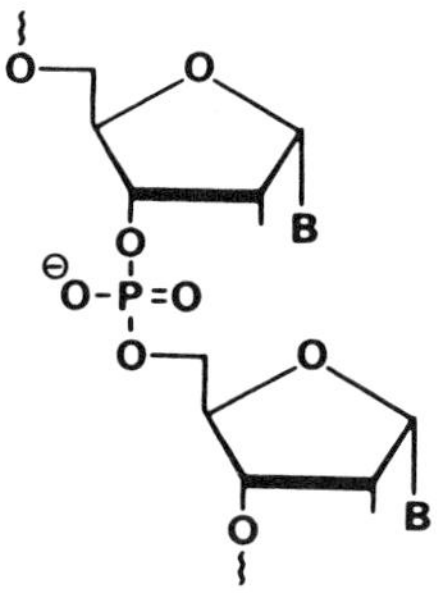

2

the $R_p$ [3a] or $S_p$ [3b] configuration. Unless the linkage is synthesized stereospecifically, which is usually not the case, phosphorothioate oligomers will consist of $2^n$ diastereoisomers, where *n* is the number of internucleotide linkages in the oligomer.

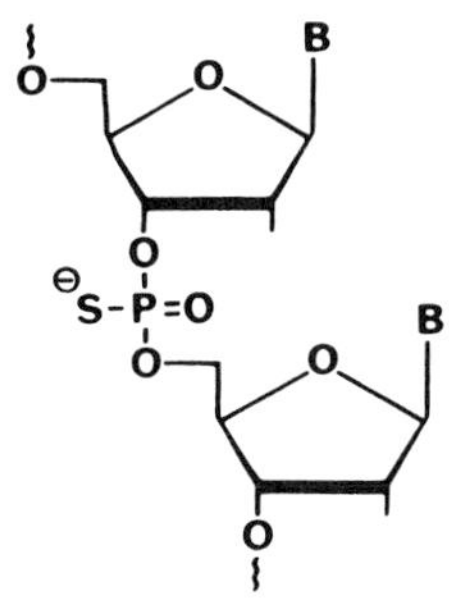

3a

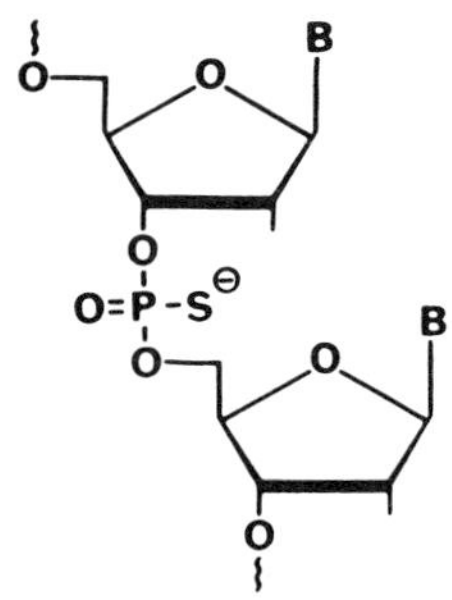

3b

The diastereomeric phosphorothioate linkages are less susceptible to nuclease hydrolysis than is the phosphodiester internucleotide bond. However, both the $R_p$ and $S_p$ isomers are slowly hydrolyzed by either exo- or endonucleases (43,44). Phosphorothioate oligomers form normal Watson-Crick base paired duplexes with complementary nucleic acid sequences. The stabilities of these duplexes appear to be dependent upon the configuration of the internucleotide bond (45). Duplexes formed between phosphorothioate oligomers and RNA serve as substrates for ribonuclease H.

The effect of phosphorothioate oligomers on human immunodeficiency virus (HIV) in cell culture has been studied rather extensively and will be described below. The greater polarizability of the sulfur atom in this linkage would suggest that phosphorothioate oligomers might be taken up more readily by cells than are the oligodeoxyribonucleotides. However, the association of these two types of oligomers with mammalian cells in culture has not been directly compared.

Oligonucleotide analogues which contain noncharged internucleotide linkages have properties which make them quite suitable for use as antisense reagents in cell culture (2,3). Two of the most extensively studied analogs are the oligodeoxynucleoside methylphosphonates [4] and the oligodeoxynucleoside alkylphosphoramidates [5]. The internucleotide bonds of oligonucleoside methylphosphonates and oligonucleoside alkylphosphoramidates result from replacement of one of the nonesterified oxygens with a methyl group or an alkylamino group, respectively. As in the case of the phosphorothioate oligomers, the methylphosphonate and

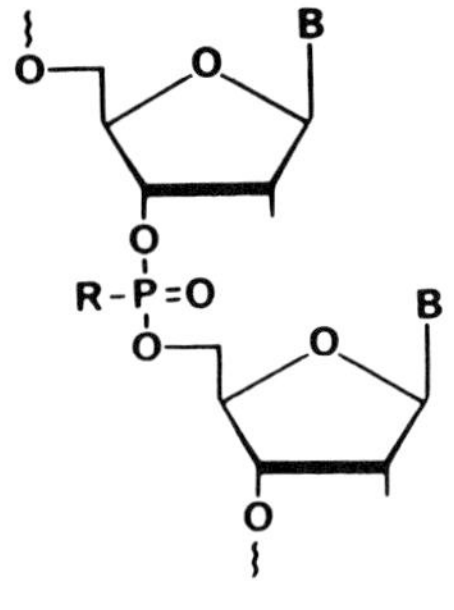


4 R = $CH_3$–

5 R = R'N(H)–

alkylphosphoramidate linkages occur in an $R_p$ or $S_p$ configuration. Unlike oligodeoxyribonucleotides or phosphorothioate oligomers, the sugar phosphate backbones of the methylphosphonate or alkylphosphoramidate oligomers are not negatively charged. The methylphosphonate backbone is completely resistant to both exo- and endonucleases (46-48). Because the backbone is electroneutral, the oligomers are much more lipophilic than negatively charged oligodeoxyribonucleotides. As a result of their lipophilic properties and their nuclease resistance, oligonucleoside methylphosphonates are readily taken up intact by mammalian cells in culture (49,50). Similar uptake can occur in bacterial cells for oligomers up to four nucleotides in length (51). The stability and uptake of oligonucleoside alkylphosphoramidates have not been reported.

Both oligonucleoside methylphosphonates and alkylphosphoramidates form stable, hydrogen bonded duplexes with complementary single stranded DNA (ssDNA) and with mRNA (52-58). The stability ($T_m$) of duplexes formed by oligonucleoside methylphosphonates and ssDNA is the same or slightly lower than comparable duplexes formed between oligodeoxyribonucleotides and ssDNA (53,54,57). It appears that the configuration of the backbone may play a role in determining the $T_m$'s of the duplexes, although this role has not been clearly defined, especially for oligomers containing all four nucleoside units (47,59). The stability of the phosphoramidate duplexes is dependent upon the size of the alkyl group and decreases as the size of the group increases. In general, it appears that the methylphosphoramidate duplexes have slightly lower $T_m$'s than do methylphosphonate duplexes of the same base sequence. Because the backbones of methylphosphonate and alkylphosphoramidate oligomers are electroneutral, the $T_m$'s of their duplexes with ssDNA are essentially independent of the ionic strength of the medium. This is in contrast to the behavior of oligodeoxyribonucleotide duplexes, whose $T_m$'s decrease as the ionic strength decreases. Methylphosphonate/RNA duplexes are not cleaved by RNase H (60), and this is probably also true for phosphoramidate/RNA duplexes.

## ANTIVIRAL EFFECTS OF ANTISENSE OLIGONUCLEOTIDE ANALOGUES

In an effort to prevent translation of mRNA, antisense oligonucleotides have been targeted against functional regions of mRNA, such as the 5'-cap region, the initiation codon region, or the nucleotide sequences in the coding region near the initiation codon. Inhibition of splicing of precursor mRNA is also possible by targeting donor or acceptor splice sites at the intron/exon junctions of the precursor mRNAs. In some cases other functional regions of viral RNAs, such as the primer binding sites or redundant sequences at the 5' and 3' ends of retrovirus RNA, have been targeted.

Antiviral activity *in vitro*

The first example of antisense oligodeoxyribonucleotide mediated inhibition of translation of a viral mRNA was described by Stephenson and Zamecnik (61). A deoxyribonucleotide 13-mer, d-AATGGTAAAATGG, which is complementary to the redundant sequences found at the 5'- and 3'-ends of Rous sarcoma virus (RSV) 70S RNA, was prepared. This oligomer was shown to bind to the 70S RNA as evidenced by its ability to serve as a primer for reverse transcriptase catalyzed copying of the 70S RNA *in vitro*. The oligomer inhibited translation of the 70S RNA 99% and 88% in a wheat germ extract when present at concentrations of 5 μM and 0.5 μM, respectively. Similar inhibition was obtained using a cDNA 100 nucleotides in length, which was complementary to the 5'-end of RSV 70S RNA (5). These results show that the 5'-end is required for translation of the 70S RNA.

The 13-mer also effectively inhibited translation of avian myeloblastosis virus (AMV) 70S RNA (93% and 57%) at these concentrations, but had a much smaller inhibitory effect on translation of rabbit globin mRNA (47% and 25%) or brome mosaic virus (BMV) RNA (63% and 30%). A noncomplementary 8-mer and 10-mer had little inhibitory effect at these concentrations. The specificity of inhibition by the 13-mer is consistent with its ability to bind selectively to RSV 70S RNA. Subsequent work in other laboratories has demonstrated that the wheat germ translation system contains substantial levels of ribonuclease H activity (10). This activity could account for the ability of the 13-mer to inhibit translation at very low concentrations.

A 15-mer, d-TTGGGATAACACTTA, complementary to nucleotides 17-31 near the 5'-end of vesicular stomatitis virus (VSV) M protein mRNA, inhibited translation of VSV mRNA in a rabbit reticulocyte lysate (15). The binding site for the oligomer appears to occur in a single stranded region of the mRNA as predicted by a computer-generated secondary structure for the VSV M mRNA. The oligodeoxyribonucleotide inhibited translation to the extent of 41% and 88% at 5 μM and 40 μM, respectively, and also inhibited translation of BMV 2% and 43% under the same conditions. Although inhibition appeared to be selective for VSV versus BMV translation, specific inhibition of M protein synthesis was not observed.

In contrast to these observations, Agris *et. al.* observed selective inhibition of translation of VSV mRNA by antisense oligonucleoside methylphosphonates in a rabbit reticulocyte lysate (62). In this system only the N, NS and M mRNAs are translated. Oligomers nine nucleotides in length which are complementary to the initiation codon regions of VSV N, NS and G protein mRNAs were synthesized. The N specific and NS specific oligomers at concentrations of 100 μM and 150 μM inhibited N (36% and 77%) and NS (25% and 35%), respectively in a selective manner, whereas the G specific oligomer had essentially no inhibitory effect. None of the oligomers inhibited translation of either rabbit α- or ß-globin mRNA.

Antisense oligodeoxyribonucleotides complementary to various regions of Sendai virus nucleocapsid protein and phosphoprotein mRNAs were found to have different inhibitory effects depending upon the location of the binding site of the oligomers (19). In these experiments, which were carried out in a rabbit reticulocyte lysate, oligomers which are complementary to the 5'-noncoding region of the mRNAs and to regions upstream of the initiation codon, were the most effective inhibitors of translation. Oligomers complementary to the initiation codon region or the coding region were less effective inhibitors. For example, a 20-mer complementary to nucleotides 1-20 of nucleocapsid protein mRNA inhibited translation 90% at 0.75 μM, whereas a 20-mer complementary to nucleotides 56-74 inhibited translation 29%. The initiation codon of the mRNA is located at positions 66-68. No inhibition was observed by a 20-mer complementary to nucleotides 67-85 located in the coding region. Similar inhibitory effects were observed for inhibition of translation of phosphoprotein mRNA.

These results suggest that oligomers complementary to the 5'-end of the mRNA may prevent binding of 40S ribosomes to the mRNA or scanning by the 40S ribosomes prior to assembly of the 80S ribosome initiation complex. They also suggest that once formed, the 80S ribosome complex may be able to displace oligonucleotides hybridized to sequences in the coding region, possibly through the action of an unwinding activity present in the complex

effects were observed on mouse L cell protein synthesis.

The mechanism of action of the poly(L-lysine) derivatized oligodeoxyribonucleotides is unknown. The increased efficiency of these antisense oligomers compared to those of oligodeoxyribonucleotides or methylphosphonate oligomers could result in part from increased uptake or compartmentalization of the derivatized oligomer. Alternatively, the poly-(L-lysine) may serve to protect the oligomer from degradation by nucleases or to increase its affinity for its target.

Oligonucleoside methylphosphonates have been found to be effective inhibitors of herpes simplex virus type 1 (HSV-1) replication in virus infected cells (67,68). An 8-mer, d-TpCCTCCTG, was prepared which is complementary to the acceptor splice site of immediate early precursor mRNAs 4 & 5. When added to the Vero cells before or at the time of infection with HSV-1, the oligomer reduced virus titer 50% and 90% at concentrations of 25 μM and 75 μM, respectively. The oligomer had a lower inhibitory effect on virus titer in HSV-2 infected cells. At 250 μM concentration, the 8-mer inhibited HSV-1 DNA synthesis 70% to 75%. Synthesis of HSV-1 proteins was inhibited 90%, including synthesis of immediate early species, a 130-kDa DNA-binding protein and glycoprotein gB. Under these conditions no inhibition of Vero cell growth, DNA synthesis or protein synthesis was detected as a result of treating noninfected cells with the 8-mer.

Highly specific inhibition of HSV-1 versus HSV-2 replication was observed with the 12-mer methylphosphonate, d-TpTCCTCCTGCGG (68). This oligomer is complementary to the 6 nucleotides on either side of the acceptor splice site of HSV-1 immediate early pre-mRNA 4 & 5. At concentrations of 15 μM and 100 μM, the 12-mer caused reductions of 50% and 98%, respectively, in virus titer in HSV-1 infected Vero cells, but caused no inhibition of virus replication in HSV-2 infected cells. As in the case of the 8-mer, the 12-mer specifically inhibited virus protein synthesis and growth, and did not affect uninfected cells. It appears that the 12-mer inhibits splicing of immediate early pre-mRNA 4 & 5 because increased levels of unspliced immediate early mRNA 4 were detected in virus infected cells treated with the oligomer.

The ability of the methylphosphonate oligomers to inhibit HSV-1 replication is extremely sensitive to the sequence of the oligomer. For example, when the two middle bases of d-TpCCTCCTG were exchanged, the resulting "mutated" oligomer, d-TpCCCTCTG, only inhibited replication approximately 10% at 75 µM concentration. Because this "mutated" oligomer would be expected to form a duplex of reduced stability due to the presence of two mismatched base pairs, it would be expected to have a reduced ability to interfere with splicing of immediate early mRNAs 4 & 5.

Although the 12-mer, d-TpTCCTCCTGCGG, was very effective as an inhibitor of HSV-1 replication, other 12-mers complementary to the splice junction region were less effective. For example, d-GpCGTTCCTCCTG, which is complementary to 9 nucleotides in the exon and 3 nucleotides in the intron, and d-TpTCCTGCGGGAAG, which is complementary to 2 nucleotides in the exon and 10 nucleotides in the intron of immediate early mRNA 4 & 5, only inhibited replication 30% and 9%, respectively, at 100 µM concentration. Neither oligomer inhibited HSV-2 replication. The reduced inhibitory effects of these oligomers may be due to secondary structural features of the pre-mRNA which affect the ability of the oligomers to bind to the splice junction site. The results do show that the choice of oligomer sequence can be critical to obtaining effective antisense inhibitors.

The effectiveness of d-TpTCCTCCTGCGG could be considerably increased by derivatizing the 5'-end of the oligomer with aminoethyltrimethylpsoralen (68). This functional group allows the oligomer to crosslink to its targeted nucleic acid when irradiated with long wavelength (365 nm) ultraviolet light. Thus, psoralen-derivatized d-TpTCCTCCTGCGG caused 90% to 95% inhibition of HSV-1 replication in infected cells treated with 5 µM oligomer and irradiated one to three hours after infection.

The effect of antisense oligonucleotides on human immunodeficiency virus (HIV) in cell culture was first reported by Matsukura *et. al.* (69). Four types of oligomers were studied: oligodeoxyribonucleotides, oligonucleotide phosphorothioates, oligonucleoside methylphosphonates and oligodeoxyribonucleotides containing $N^3$-methylthymidine residues. The oligomers were complementary to either nucleotides 5091-6005 or nucleotides 6018-6031 of the first exon of the *art/trs* gene in HIV. In addition to the antisense sequences, normal and phosphorothioate oligomers with the same sequence as nucleotides 5091-6005 were synthesized. A noncomplementary, random sequence phosphorothioate oligomer as well as the oligodeoxycytidine phosphorothioate, S-$dC_n$, were also studied.

The antiviral effects of these oligomers in immortalized T (ATH8) cells were studied by measuring cell viability as assayed by trypan blue exclusion after infection of the cells with HIV and treatment with the oligomers. Neither the antisense oligodeoxyribonucleotides or the antisense oligonucleoside methylphosphonates showed any significant antiviral effect over the concentration range 1 $\mu$M to 25 $\mu$M. In contrast, the antisense, the sense, the random and the S-$dC_n$ phosphorothioate oligomers all showed significant antiviral activity over the concentration range 5 $\mu$M to 25 $\mu$M. For example, the antisense oligomer complementary to nucleotides 5091-6005 protected 56% of the cells at 5 $\mu$M concentration and S-$dC_{14}$ gave complete protection of the cells. It appeared that the inhibitory effects of the phosphorothioate oligomers were dependent upon the ability of the oligomers to form hydrogen bonded complexes with their targets. The antisense phosphorothioate containing $N^3$-methylthymidine residues, an oligomer which is incapable of forming stable duplexes, showed no antiviral activity in this assay.

The most effective inhibitor of HIV was found to be S-$dC_{28}$. This oligomer gave essentially complete protection at a concentration of 1 $\mu$M. The oligomer was also shown to inhibit *de novo* viral DNA synthesis in virus infected cells but it had little or no inhibitory effect on HIV *gag* protein synthesis, even at oligomer concentrations of 25 $\mu$M.

The anti-HIV activity of oligonucleotide phosphorothioates was also examined by Agrawal *et. al.* (48). In these experiments, the ability of the oligomers to inhibit syncytia formation and synthesis of protein p24 were assayed. Antisense phosphorothioate oligomers 15 or 20 nucleotides long, which are complementary to either the 5'-capped end of the virus RNA (nucleotides 1-15), a donor splice site near the 5'-end of the virus (nucleotides

280-299), or an acceptor splice site in the *tat* gene (nucleotides 5349-5368) were studied. These oligomers all showed inhibitory effects over the concentration range 0.6$\mu$M to 15$\mu$M. A noncomplementary phosphorothioate 20-mer gave no inhibition at 0.6$\mu$M but inhibited both syncytia formation and p24 synthesis 89% and 100% at 3 $\mu$M and 15 $\mu$M, respectively. Homooligonucleotide phosphorothioates were also found to have significant inhibitory effects, even at low concentrations.

The results of these studies suggest that oligonucleotide phosphorothioates may inhibit HIV functions by two different mechanisms. At low concentrations, the heterooligomers may act in a sequence dependent manner and thus function as true antisense agents. Homooligomers and heterooligomers at high concentrations may function by serving as inhibitors of HIV reverse transcriptase. Recent studies by Majumdar *et. al.* have shown that S-$dC_{28}$ is a linear competitive inhibitor of HIV reverse transcriptase catalyzed DNA synthesis when poly(rA)-oligo(dT) is used as a template primer (70). This oligomer was also shown to inhibit murine leukemia virus reverse transcriptase as well as mammalian $\alpha$- and $\gamma$-polymerases.

In contrast to the results of Matsukura *et. al.*, Zamecnik *et. al.*, Goodchild *et. al.* and Agrawal *et. al.* found that oligodeoxyribonucleotides and oligonucleoside methylphosphonates were effective inhibitors of HIV activity as measured by their effects on reverse transcriptase activity, on syncytia formation, and on p24 synthesis (14,22,48). Oligodeoxyribonucleotides 20 nucleotides in length were targeted to various functional regions of HIV RNA (22). All the oligomers tested, with the exception of a noncomplementary control 20-mer, inhibited syncytia formation by HIV infected Molt-3 cells over the concentration range 0.6$\mu$M to 15$\mu$M. The most effective inhibitors were oligomers targeted to the cap (nucleotides 1-20), the 5'-untranslated region (nucleotides 54-73), a region near the primer binding site (nucleotides 162-181), two splice acceptor sites (nucleotides 5349-5368 and 7947-7966) and the poly(A) signal region (nucleotides 9183-9202). For example, the latter oligomer inhibited syncytia formation 71% and 85% and inhibited p24 synthesis 53% and 67% at 3$\mu$M and 15$\mu$M respectively. None of the oligomers appeared to be toxic to noninfected cells.

Not all of the oligomers inhibited both syncytia formation and p24 synthesis. For example a 20-mer complementary to the initiation codon region of the *tat* mRNA inhibited syncytia formation 61% at 3μM but had no inhibitory effect on p24 protein synthesis at this concentration. It is clear from these studies that oligomers targeted to different regions of HIV give different inhibitory effects and that scoring for an inhibitory effect by a particular oligomer depends upon the assay that is used. This may explain in part why the results of this study differ from those of Matsukura *et. al.* (69). The mechanism by which the various oligomers inhibit is not known and probably differs, depending upon the site being targeted by the oligomer.

Oligonucleoside methylphosphonates targeted to different regions of HIV RNA also gave inhibitory effects in HIV infected cells in culture. Zaia *et. al.* synthesized an antisense methylphosphonate, d-CpCAATTCT, complementary to the first splice junction of the *tat*-3 gene of HIV (71). In addition, a "sense" oligomer, d-ApGAATTGG, and an oligomer complementary to the splice acceptor site of HSV immediate early pre-mRNA 4 & 5, d-TpCCTCCTG, was also synthesized. The latter sequence has been shown to effectively inhibit HSV replication in cell culture as described above (67). The antisense oligomer significantly reduced syncytia formation (94%) and reverse transcriptase activity (92%) when added to H9 cells at a concentration of 100μM prior to infection with HIV. This oligomer also reduced the amount of HIV RNA 98% as measured by a polymerase chain reaction assay followed by Southern blot hybridization analysis. In contrast, the sense oligomer only inhibited syncytia formation 12% and reverse transcriptase activity 23%, but did reduce HIV RNA formation 88%. The HSV specific oligomer had no effect on syncytia formation, reverse transcriptase activity or HIV RNA synthesis.

Longer methylphosphonate oligomers proved to be more effective inhibitors of HIV in cell culture (54). For example, a 20-mer, d-ApCACCCAATTCTGAAAATGG, which is complementary to the first acceptor splice site of the *tat*-3 gene inhibited syncytia formation 98% and p24 synthesis 100% at a concentration of 3 μM. A 15-mer, d-CpTAACCAGAGAGACC, complementary to the 5'-cap region of HIV RNA inhibited syncytia formation 51% and p24 synthesis 69% at a concentration of 22 μM. The inhibitory activities of oligomers containing both methylphosphonate groups and phosphodiester groups appeared to depend upon the number of methylphosphonate groups in the oligomer. This effect may reflect the increased sensitivity of the phosphodiester linkages in these oligomers to nuclease hydrolysis.

Antisense phosphoramidate oligomers containing butylamidate, morpholidate or piperazidate linkages were shown to be effective inhibitors of syncytia formation and p24 synthesis in HIV infected cells (48). Oligomers 15 and 20 nucleotides in length gave inhibitory effects of 50% to 70% at a concentration of 0.6 μM. The oligomers were not toxic to the cells at this concentration, although phosphopiperazidates were toxic at 15 μM. As in the case of the phosphorothioate oligomers, homophosphoromorpholidates also gave inhibitory effects. For example, the morpholidate analogue of d-$C_{15}$ inhibited syncytia formation 44% at 0.6 μM whereas the morpholidate analogue of d-$T_{15}$ inhibited both syncytia formation and p24 synthesis 35% at this concentration. Thus, it appears that like the phosphorothioate oligomers, the phosphoramidate oligomers may function by a sequence independent mechanism in addition to sequence dependent inhibition.

## PROSPECTS AS ANTIVIRAL AGENTS

It is clear from the results described above that antisense oligodeoxyribonucleotides and their sugar-phosphate backbone modified analogues can function as sequence specific inhibitors of virus gene expression in cell culture. The mechanism of action of these antisense oligomers remains to be defined and probably differs for the different types of analogues. Once these mechanistic principles are clearly established, such antisense oligomers should be valuable research tools for studying the molecular biology of viruses.

Antisense oligomers also hold promise for development as clinically important antiviral agents. At this stage of our knowledge, it would appear that the nuclease resistant analogues are the most promising candidates for such applications. However, a number of important issues must be addressed before practical application becomes a reality.

The pharmacological effects of antisense oligomers in animals or humans is completely unknown. Preliminary toxicity experiments have been reported for oligonucleotide phosphorothioates and phosphoromorpholidates in mice (48). However, the stability of the oligomers, their distribution within the animal, and the immunogenicity of the oligomers remains to be determined. In addition, it is not clear how the oligomers will be targeted to sites of infection within the animal.

In order to answer these questions, it will be necessary to conduct large scale testing both in cell culture and in animal models. This requirement will necessitate the development of methods to synthesize gram or possibly kilogram quantities of these molecules. Present technology allows synthesis of milligram quantities on automated synthesizers, and synthesizers are currently being developed which are capable of preparing up to 0.5 gram of oligomer. Further advances in solid support syntheses or solution phase syntheses will be needed to produce the required amounts of compounds.

In spite of these unknowns, it appears that antisense oligonucleotides could provide a unique means to rationally design agents which can be used to control viral infections in a highly specific and effective manner.

## REFERENCES


1. Stebbing, N. Pharmac. Ther., 6:291-332, 1979.
2. Stein, C.A. and Cohen, J.A. Cancer Res., 48:2659-2668, 1988.
3. Miller, P.S. and Ts'o, P.O.P. Ann. Reports Med. Chem., 23:295-304, 1988.
4. van der Krol, A.R. and Stuitje, J.N.N.M. BioTechniques 6:958-975, 1988.
5. Perdue, M.L., Borchelt, D. and Resnick, R.J. J. Biol. Chem. 257:655-1-6555, 1982.
6. Blake, K.R., Murakami, A. and Miller, P.S. Biochemistry 24:6132-6138, 1985.
7. Cazenave, C., Loreau, N., Toulme, J.J. and Helene, C. Biochimie 68:1063-1069, 1986.
8. Shatkin, S.H. and Liebhaber, S.A. J. Biol. Chem. 21:16018-16025, 1986.
9. Toulme, J.J., Krisch, H.M., Loreau, N., Thuong, N.T. and Helene, C. Proc. Natl. Acad. Sci. (USA) 83:1227-1231, 1986.

10. Walder, J.A., Eder, P.S., Engman, D.M., Crentanano, S.T., Walder, R.Y.,Knutzon, D.S., Dorfman, D.M. and Donelson, J.E. Science 233:569-571, 1986.
11. Cornelissen, A.W.C.A., Verspieren, M.P., Toulme, J.J., Swinkels, B.W. and Borst, P. Nucleic Acids Res. 14:5605-5614, 1986.
12. Haeuptle, M.T., Frank, R. and Dobberstein, B. Nucleic Acids Res. 14:1427-1448, 1986.
13. Holt, J.T., Redner, R.L. and Nienhuis, W. Mol. Cell. Biol. 8:963-973, 1988.
14. Zamecnik, P.C., Goodchild, J., Taguchi, Y. and Sarin, P.S. Proc. Natl. Acad. Sci. (USA), 83:4143-4146, 1986.
15. Wickstrom, E. , Simonet, W.S., Medlock, K. and Ruiz-Robles, I. Biophys. J. 49:15-17, 1986.
16. Minshull, J. and Hunt, T. Nucleic Acids Res. 14:6433-6451, 1986.
17. Heikkila, R., Schwab, G., Wickstrom, E., Lake, S.L., Pluznik, D.H., Watt, R. and Neckers, L.M. Nature 328:445-449, 1987.
18. Maher III, L.J. and Dolnick, B.J. Arch. Biochem. Biophys. 253:214-220, 1987.
19. Gupta, K.C. J. Biol. Chem. 262:7492-7496, 1987.
20. Cazenave, C., Lorequ, N., Thuong, N.T., Toulme, J.J. and Helene, C. Nucleic Acids Res. 15:4717-4736, 1987.
21. Goodchild, J., Carroll III, E. and Greenberg, J.R. Arch. Biochem. Biophys. 263:401-409, 1988.
22. Goodchild, J., Agrawal, S., Civeira, M.P., Sarin, P.S., Sun, D. and Zamecnik, P.C. Proc. Natl. Acad. Sci. (USA) 85:5507-5511, 1988.
23. Shuttleworth, J. and Colman, A. EMBO J. 7:427-434, 1988.
24. Crum, C., Johnson, J.D., Nelson, A. and Roth, D. Nucleic Acids Res. 16:4569-4581, 1988.
25. R.Y. Walder and J.A. Walder. Proc. Natl. Acad. Sci. (USA) 85:5011-5015, 1988.
26. Morgan, A.R. and Wells, R.D. J. Mol. Biol. 37:63-80, 1968.
27. Wells, R.D., Collier, D.A., Hanvey, J.C., Shimizu, M. and Wohlrab, F., FASEB J. 2:2939-2949, 1988.
28. Strobel, S.A., Moser, H.E. and Dervan, P.B. J. Am. Chem. Soc. 110:7927-7929, 1988.
29. Fedorova, O.S., Knorre, D.G., Podust, L.M. and Zarytova, V.F. FEBS Letters 228:273-276, 1988.
30. Francois, J.C., Saison-Behmoaras, T. and Helene, C. Nucleic Acids Res. 16:11431-11440, 1988.
31. Povsic, T.J. and Dervan, P.B. J. Am. Chem. Soc. 111:3059-3061, 1989.
32. Cooney, M., Czernuszewicz, G., Postel, E.H., Flint, S.J. and Hogan, M.E. Science 241:456-459, 1988.
33. Morvan, F., Rayner, B., Imbach, J.L., Chan, D.K. and Lown, J.W. Nucleic Acids Res. 14:5019-5035, 1987.
34. Sun, J.S., Asseline, U., Rouzaud, D., Montenay-Garestier, T., Thuong, N.T., and Helene, C. Nucleic Acids Res. 15:6149-6158, 1987.
35. Praseuth, D., Chassignol, M., Takasugi, M., Le Doan, T., Thuong, N.T., and Helene, C. J. Mol. Biol. 196:939-942, 1987.
36. Morvan, F., Rayner, B., Imbach, J.-L., Lee, M., Hartley, J.A., Chang, D.K.

and Lown, J.W. Nucleic Acids Res. 15:7027-7044, 1987.

37. Thuong, T., Asseline, U., Roig, V., Takasugi, M. and Helene, C. Proc. Natl. Acad. Sci. (USA) 84:5129-5133, 1987.
38. Morvan, F., Rayner, B., Leonetti, J.P. and Imbach, J.L. Nucleic Acids Res. 16:833-847, 1988.
39. Durand, M., Maurizot, J.C., Thuong, N.T. and Helene, C. Nucleic Acids Res. 16:5039-5053, 1988.
40. Cazenave, C., Chevrier, M., Thuong, N.T. and Helene, C. Nucleic Acids Res. 15:10507-10521, 1987.
41. C. Gagnor, C., Bertrand, J.R., Thenet, S., Lemaitre, M., Morvan, F., Rayner, B., Malvy, C., LeBleu, B., Imbach, J.L. and Paoletti, C. Nucleic Acids Res. 15:10419-10436, 1987.
42. C. Helene, C. and Thuong, N.T. Biochem. Pharm. 37:1797-1798, 1988.
43. Stec, W.J., Zon, G., Egan, W. and Stec, B. J. Am. Chem. Soc. 108:6077-6079, 1984.
44. Connolly, B.A., Potter, B.V.L., Eckstein, F., Pingoud, A. and Grotjahn, L. Biochemistry 23:3443-3453, 1984.
45. LaPlanche, L.A., James, T.L., Powell, C., Wilson, W.D., Uznanski, B., Stec, W.J., Summers, M.F. and Zon, G. Nucleic Acids Res. 14:9081-9093, 1986.
46. Agarwal, K.L. and Riftina, F. Nucleic Acids Res. 6:3009-3024, 1979.
47. Miller, P.S., Yano, J., Yano, E., Carroll, C., Jayaraman, K. and Ts'o, P.O.P. Biochemistry 18:5134-5142, 1979.
48. Agrawal, S., Goodchild, J., Civeira, M.P., Thornton, A.H., Sarin, P.S. and Zamecnik, P.C. Proc. Natl. Acad. Sci. (USA) 85:7079-7083, 1988.
49. Miller, P.S., McParland, K.B., Jayaraman, K. and Ts'o, P.O.P. Biochemistry 20:1874-1880, 1981.
50. Marcus-Sekura, C.J., Woener, A.M., Shinozuka, K., Zon, G. and Quinnan,Jr.,G.V. Nucleic Acids Res. 15:5749-5763, 1987.
51. Jayaraman, K., McParland, K., Miller, P. and Ts'o, P.O.P. Proc. Natl. Acad. Sci. 78:1537-1542, 1981.
52. Murakami, A., Blake, K.R., and Miller, P.S. Biochemistry 24:4041-4046, 1985.
53. Froehler, B., Ng, P. and Matteucci, M. Nucleic Acids Res. 16:4831-4839, 1988.
54. Sarin, P.S., Agrawal, S., Civeira, M.P., Goodchild, J., Ikeuchi, T. and Zamecnik, P.C. Proc. Natl. Acad. Sci. (USA) 85:7448-7451, 1988.
55. Jager, A., Levy, M.J. and Hecht, S.M. Biochemistry 27:7237-7246, 1988.
56. Letsinger, R.L., Singman, C.N., Histand, G. and Salunkhe, M. J. Am. Chem. Soc. 110:4470-4471, 1988.
57. Lin, S.B., Blake, K.R., Miller, P.S. and Ts'o, P.O.P. Biochemistry 28:1054-1061, 1989.
58. Quartin, R.S. and Wetmur, J.G. Biochemistry 28:1040-1047, 1989.
59. Miller, P.S., Dreon, N., Pulford, S.M. and McParland, K.B. J. Biol. Chem. 255:9659-9665, 1980.
60. Walder, J.A. Gene Dev. 2:502-504, 1988.
61. Stephenson M.L. and Zamecnik, P.C. Proc. Natl. Acad. Sci. (USA) 75:285-288, 1978.

62. Agris, C.H., Blake, K.R., Miller, P.S., Reddy, M.P. and Ts'o, P.O.P. Biochemistry 25:6268-6275, 1986.
63. Lawson, T.G., Lee, K.A., Maimone, M.M., Abramson, R.D., Dever, T.E., Merrick, W.C. and Thach, R.E. Biochemistry 28:4729-4734, 1989.
64. Kawasaki, E.S. Nucleic Acids Res. 13:4991-5004, 1985.
65. Zamecnik, P.C. and Stephenson, M.L. Proc. Natl. Acad. Sci. (USA) 75:280-284, 1978.
66. Lemaitre, M., Bayard, B. and Lebleu, B. Proc. Natl. Acad. Sci. (USA) 84:648-652, 1987.
67. Smith, C.C., Aurelian, L., Reddy, M.P., Miller, P.S. and Ts'o, P.O.P. Proc. Natl. Acad. Sci. (USA) 83:2787-2791, 1986.
68. Kulka, M., Smith, C.C., Aurelian, L., Fishelevich, R., Meade, K., Miller, P. and Ts'o, P.O.P. Proc. Natl. Acad. Sci. (USA) 86, in press 1989.
69. Matsukura, M., Shinozuka, K., Zon, G., Mitsuya, H., Reitz, M., Cohen, J.S. and Broder, S. Proc. Natl. Acad. Sci. (USA) 84:7706-7710, 1987.
70. Majumdar, C., Stein, C.A., Cohen, J.S., Broder, S. and Wilson, S.H. Biochemistry 28:1340-1346, 1989.
71. Zaia, J.A., Rossi, J.J., Murakawa, G.J., Spallone, P.A., Stephens, D.A., Kaplan, B.E., Eritja, R., Wallace, R.B. and Cantin, E.M. J. Virol. 62:3914-3917, 1988.

# INDEX